# SPECIAL DISTILLATION PROCESSES

**Zhigang Lei**
**Biaohua Chen**
**Zhongwei Ding**

*Beijing University of Chemical Technology*
*Beijing 100029*
*China*

2005

ELSEVIER

Amsterdam – Boston – Heidelberg – London – New York – Oxford – Paris
San Diego – San Francisco – Singapore – Sydney – Tokyo

**ELSEVIER B.V.**    ELSEVIER Inc.    ELSEVIER Ltd.    ELSEVIER Ltd.
**Radarweg 29**     525 B Street, Suite 1900 The Boulevard, Langford Lane 84 Theobalds Road
**P.O. Box 211, 1000 AE Amsterdam** San Diego, CA 92101-4495 Kidlington, Oxford OX5 1GB London WC1X 8RR
**The Netherlands**    USA      UK       UK

First edition 2005

Library of Congress Cataloging in Publication Data
A catalog record is available from the Library of Congress.

British Library Cataloguing in Publication Data
A catalogue record is available from the British Library.

ISBN: 0-444-51648-4

♾ The paper used in this publication meets the requirements of ANSI/NISO Z39.48-1992 (Permanence of Paper).
Printed in The Netherlands.

# Preface

With its unique advantages in operation and control, distillation is a very powerful separation tool in the laboratory and industry. Although many promising separation methods are constantly proposed by engineers and scientists, most of them are not able to compete with distillation on a large product scale. In this book, a new term, "special distillation processes", is proposed by the authors and is the title of this book. This term signifies the distillation processes by which mixtures with close boiling points or those forming azeotropes can be separated into their pure constituents. Among all distillation processes, special distillation processes occupy an important position.

Special distillation processes can be divided into two types: one with separating agent (i.e. the third component or solvent added; separating agent and solvent have the same meaning in some places of this book) and the other without separating agent. The former involves azeotropic distillation (liquid solvent as the separating agent), extractive distillation (liquid and/or solid solvents as the separating agent), catalytic distillation (catalyst as the separating agent by reaction to promote the separation of reactants and products), adsorption distillation (solid particle as the separating agent) and membrane distillation (membrane as the separating agent); the latter involves pressure-swing distillation and molecular distillation. However, the former is implemented more often than the latter, and thus more attention is paid to the special distillation processes with separating agent. At the same time, the techniques with a close relationship to special distillation processes are also mentioned. But it should be noted that molecular distillation is different in that its originality does not originate from the purpose of separating mixtures with close boiling points or forming azeotropes, but for separating heat-sensitive mixtures in medicine and biology. Hence, the content on molecular distillation is placed in the section of other distillation techniques and clarified only briefly.

Undoubtedly, special distillation processes are a very broad topic. We have tried to be comprehensive in our courage, but it would be nearly impossibly to cite every reference. Until now, some subjects in special distillation processes are still hot research topics. From this viewpoint, special distillation processes are always updated.

This book is intended mainly for chemical engineers, especially those engaged in the field of special distillation processes. It should be of value to university seniors or first-year graduate students in chemical engineering who have finished a standard one-year course in chemical engineering principles (or unit operation) and a half-year course in chemical engineering thermodynamics. This book will serve as teaching material for graduate students pursuing a master's or doctor's degree at Beijing University of Chemical Technology. In order to strengthen the understanding, some examples are prepared.

I hope that I have been able to communicate to the readers some of the fascination I have experienced in working on and writing about special distillation processes. In writing this book I have become aware that for me, the field of special distillation is a pleasure, as well as an important part of my profession. I shall consider it a success if a similar awareness can be awakened in those students and colleagues for whom this book is intended.

This book is the culmination of my past labors. Acknowledgements must recognize all those who helped to make the way possible:

(1) Thanks to my Ph.D. supervisors, Professors Zhanting Duan and Rongqi Zhou, who directed my research in separation process and told me that I should continue to replenish the knowledge about chemical reaction engineering from Professor Chengyue Li; therefore,

(2) Thanks to my postdoctoral supervisor, Professor Chengyue Li, from whom I made up the deficiencies in chemical reaction engineering and constructed an integrated knowledge system in chemical engineering;

(3) Thanks to my supervisors in Japan, Professors Richard Lee Smith (the international editor of the *Journal of Supercritical Fluids*) and Hiroshi Inomata (the international editor of *Fluid Phase Equilibria*), who offered me a research staff position at the research center of SCF in Tohoku University and gave me sufficient time to write this book;

(4) Thanks to my German supervisor, Professor W. Arlt, who as my German host helped me to gain the prestigious Humboldt fellowship and will direct me in the fields of quantum chemistry and density function theory on ionic liquids;

(5) Thanks to co-authors, Professors Zhongwei Ding (writing chapter 6) and Biaohua Chen (writing chapter 8), who are expert in the corresponding branches of special distillation processes;

(6) Thanks to Ms. A. Zwart and Mr. D. Coleman, the editors at Elsevier BV who gave me some instructions on preparing this book;

(7) Thanks to the reviewers, whose efforts are sincerely appreciated in order to achieve a high quality;

(8) Thanks to my colleagues, Professors Shengfu Ji, Hui Liu, I.S. Md Zaidul, etc., who encouraged me to finish this book, and the graduate students in room 104A of the integration building who were working with me on numerous nights;

(9) Thanks to my wife, Ms. Yanxia Huang, who always encouraged me while writing this book, and who had to give up her job in Beijing to stay with me overseas for many years ;

(10) Thanks to the financial support from the National Nature Science Foundation of China under Grant No. (20406001).

(11) Thanks to the many who helped me under various circumstances. I am deeply grateful to all.

Finally, due to authors' limitation in academic research and the English language, I believe that some deficiencies will inevitably exist in the text. If any problem or suggestion arises, please contact me.

Dr. Zhigang Lei
The key laboratory of science and technology of controllable chemical reactions
Ministry of Education
Box 35
Beijing University of Chemical Technology
Beijing, 100029
P.R. China
Email: leizhg@mail.buct.edu.cn

# Contents

# Chapter 1. Thermodynamic fundamentals

This chapter presents the thermodynamic fundamentals closely related with special distillation processes, which may facilitate readers to understand the separation principle discussed in the subsequent chapters. Some novel contents, which are commonly not involved in many classic thermodynamic texts, but necessary for the understanding, such as salt effect, nonequilibrium thermodynamic analysis and multi-component mass transfer, are covered. In accordance with the step-by-step rule, the section of vapor-liquid phase equilibrium is present at the beginning, and thus there is a little overlapping with common texts.

## 1. VAPOR-LIQUID PHASE EQUILIBRIUM

Phase equilibrium can be sorted into vapor-liquid, liquid-liquid, vapor-liquid-liquid, etc. Vapor-liquid phase equilibrium as the most applied and calculable form of phase equilibrium plays a major role in special distillation processes, particularly with respect to energy requirements, process simulation and sizing equipment. Thus, in this chapter this content is more highlighted. But, for the sake of immiscibility, sometimes the occurrence of two-liquid phase in the distillation column is inevitable. So on the foregoing basis vapor-liquid-liquid equilibrium is also briefly mentioned in this chapter. The details about how to numerically calculate the problem on vapor-liquid-liquid equilibrium are described in chapter 3 (azeotropic distillation) where vapor-liquid-liquid equilibrium is involved.

### 1.1. The equilibrium ratio

Equilibrium is defined as a state that will be returned to its initial state after any short, small mechanical disturbance of external conditions. According to the knowledge of physical chemistry, within any closed system where phase equilibrium exists, the total Gibbs free energy for all phases is a minimum because at this time there is no heat and mass transfer in the system. This is the starting point for studying phase equilibrium. On the other hand, from classical thermodynamics, the total Gibbs free energy and the change of Gibbs free energy in a single phase, $i$-component system are:

$$G = G(T, P, n_1, n_2, ..., n_i, ....)$$ (1)

and

$$dG = -SdT + VdP + \sum_i \mu_i dn_i$$ (2)

respectively, where $T$ (temperature), $P$ (pressure), $n_1$, $n_2$, $\cdots$, $n_i$ (mole number of components 1, 2,...$i$,..., etc, respectively) are independent variants.

Thus, the change of total Gibbs free energy in a $p$-phase, $i$-component system is:

$$dG = \sum_p (-S^{(p)}dT + V^{(p)}dP + \sum_i \mu_i^{(p)}dn_i^{(p)}) \qquad (3)$$

At equilibrium, at constant temperature and pressure,

$$dG = \sum_p (\sum_i \mu_i^{(p)}dn_i^{(p)}) = 0 \qquad (4)$$

For each component $i$, the total number of moles is constant and equal to the initial before equilibrium:

$$\sum_p dn_i^{(p)} = 0 \qquad (5)$$

Combined Eq. (5) with Eq. (4), it is shown to be

$$\mu_i^{(1)} = \mu_i^{(2)} = \mu_i^{(3)} = \bullet\bullet\bullet = \mu_i^{(p)} \quad \text{(equality of chemical potential)} \qquad (6)$$

In addition, in terms of the phase equilibrium definition,

$$T^{(1)} = T^{(2)} = T^{(3)} = \bullet\bullet\bullet = T^{(p)} \quad \text{(thermal equilibrium)} \qquad (7)$$
$$P^{(1)} = P^{(2)} = P^{(3)} = \bullet\bullet\bullet = P^{(p)} \quad \text{(mechanical equilibrium)} \qquad (8)$$

Eqs. (6), (7) and (8) constitute the criteria of evaluating any phase equilibrium (including vapor-liquid phase equilibrium, of course).

In terms of the relationship of Gibbs free energy and fugacity in the multi-component mixture, it is given that

$$d\mu_i^{(p)} = RTd\ln \overline{f}_i^{(p)} \qquad (9)$$

where $R$ (8.314 J mol$^{-1}$ K$^{-1}$) is universal gas constant. For a pure component, the partial fugacity, $\overline{f}_i^{(p)}$, becomes the pure-component fugacity $f_i^{(p)}$.

Another form of Eq. (9) is expressed as

$$\overline{f}_i^{(p)} = C \exp(\frac{\mu_i^{(p)}}{RT}) \qquad (10)$$

where $C$ is a temperature-dependent constant. If Eq. (10) is substituted into Eq. (6), then we obtain

$$\overline{f}_i^{(1)} = \overline{f}_i^{(2)} = \overline{f}_i^{(3)} = \bullet\bullet\bullet = \overline{f}_i^{(p)} \qquad (11)$$

For vapor-liquid two phases, Eq. (11) is simplified as

$$\overline{f}_i^{V} = \overline{f}_i^{L} \qquad (12)$$

which is the starting point of derivation of many commonly used equations for vapor-liquid phase equilibrium.

In this and subsequent chapters, some examples are designed and can be used as exercises for the interested readers to strengthen the understanding. But it should be noted that the solutions to these examples may be neither exclusive nor complete.

**Example:** Use the Eqs. (4) and (5), and deduce Eq. (6).
**Solution:** Eqs. (4) and (5) can be respectively rewritten as

$$\sum_i (\mu_i^{(1)} dn_i^{(1)} + \mu_i^{(2)} dn_i^{(2)} + \bullet\bullet\bullet + \mu_i^{(p)} dn_i^{(p)}) = 0 \tag{13}$$

$$dn_i^{(1)} + dn_i^{(2)} + \bullet\bullet\bullet + dn_i^{(p)} = 0 \tag{14}$$

Since the total number of component $i$ in $p$ phases is constant, there are $p$-1 independent variants among $n_i^{(1)}, n_i^{(2)}, \cdots, n_i^{(p)}$ and it is assumed that $n_i^{(2)}, n_i^{(3)} \cdots, n_i^{(p)}$ are independent variants. Therefore, Eq. (14) can be rearranged as

$$dn_i^{(1)} = -dn_i^{(2)} - dn_i^{(3)} - \bullet\bullet\bullet - dn_i^{(p)} \tag{15}$$

If Eq. (15) is incorporated into Eq. (13), then

$$dG = \sum_{N=2}\sum_i (\mu_i^{(p)} - \mu_i^{(1)}) dn_i^{(p)} = 0 \tag{16}$$

Because the term $dn_i^{(p)}$ as an independent variable in Eq. (16) can be arbitrarily given, it is reasonably presumed that $dn_i^{(p)} > 0$, which gives rise to Eq. (6).

For a vapor-liquid two-phase multi-component system, from the relationships of partial fugacity and pressure, as well as partial fugacity and activity coefficient, the following equations can be written:

$$\overline{f}_i^V = Py_i \overline{\phi}_i^V, \quad \overline{f}_i^L = Px_i \overline{\phi}_i^L \tag{17}$$

$$\overline{f}_i^V = y_i \gamma_{iV} \overline{f}_{iV}^0, \quad \overline{f}_i^L = x_i \gamma_{iL} \overline{f}_{iL}^0 \tag{18}$$

where $\overline{\phi}_i$ and $\gamma_i$ are respectively fugacity and activity coefficients, and the superscript "0" stands for pure component.

At vapor-liquid two phase equilibrium, $\overline{f}_i^V = \overline{f}_i^L$, that is to say,

$$y_i \overline{\phi}_i^V = x_i \overline{\phi}_i^L \tag{19}$$

$$Py_i \overline{\phi}_i^V = x_i \gamma_{iL} f_{iL}^0 \tag{20}$$

Herein, the concept of equilibrium ratio $K_i$ (or phase equilibrium constant) is introduced and defined as the ratio of mole fractions of a component in vapor and liquid phases. By Eqs. (19) and (20), $K_i$ is

$$K_i = \frac{y_i}{x_i} = \frac{\overline{\phi}_i^L}{\overline{\phi}_i^V} \tag{21}$$

and

$$K_i = \frac{y_i}{x_i} = \frac{\gamma_{iL} f_{iL}^0}{\overline{\phi}_i^V P} \tag{22}$$

where

$$\overline{\phi}_i^V = \phi(T, P, y_1, y_2, \bullet \bullet \bullet, y_{n-1}) \tag{23}$$

$$\overline{\phi}_i^L = \phi(T, P, x_1, x_2, \bullet \bullet \bullet, x_{n-1}) \tag{24}$$

$$\gamma_{iL} = \gamma(T, P, x_1, x_2, \bullet \bullet \bullet, x_{n-1}) \tag{25}$$

$$f_{iL}^0 = f(T, P) \tag{26}$$

Now we analyze the degree of freedom. For a vapor-liquid two phases $N$-component system, the number of total variants is $2(N-1) + 2 = 2N$, i.e. $(T, P, x_1, x_2, \bullet \bullet \bullet, x_{N-1}, y_1, y_2, \bullet \bullet \bullet, y_{N-1})$. Because $\sum_i x_i = 1$ and $\sum_i y_i = 1$, the number of degree of freedom equals $N-2+2 = N$. Thus, the number of independent variants is $2N-N = N$. In practice, there are four types of problems on vapor-liquid equilibrium:

(1) Total pressure and mole fraction in the liquid phase, i.e. $P, x_1, x_2, \bullet \bullet \bullet, x_{N-1}$, are given, and

need to solve bubble temperature and composition, i.e. $T, y_1, y_2, \bullet \bullet \bullet, y_{N-1}$.

(2) Total pressure and mole fraction in the vapor phase, i.e. $P, y_1, y_2, \bullet \bullet \bullet, y_{N-1}$, are given, and

need to solve dew temperature and composition, i.e. $T, x_1, x_2, \bullet \bullet \bullet, x_{N-1}$.

(3) Temperature and mole fraction in the liquid phase, i.e. $T, x_1, x_2, \bullet \bullet \bullet, x_{N-1}$, are given, and

need to solve bubble pressure and composition, i.e. $P, y_1, y_2, \bullet \bullet \bullet, y_{N-1}$.

(4) Temperature and mole fraction in the vapor phase, i.e. $T, y_1, y_2, \bullet \bullet \bullet, y_{N-1}$, are given, and

need to solve dew pressure and composition, i.e. $P, x_1, x_2, \bullet \bullet \bullet, x_{N-1}$.

The solutions to those problems have already been programmed in many business software programs as public subroutines in the mathematic models with respect to special distillation processes. When solving them, Eqs. (21) or (22) should be involved. However, sometimes it is more convenient to use Eq. (22) for calculating $K_i$ because in the right side

of Eq. (21) $\overline{\phi}_i^L$ is relatively difficult to be derived. This is attributed to the difficulty of finding a reliable equation of sate to accurately describe the liquid phase behavior. In Eq. (22),

$\overline{\phi}_i^V$ is relative only with vapor phase, irrespective of liquid phase. On the contrary, $\gamma_{iL}$ is relative only with liquid phase, irrespective of vapor phase. The term $f_{iL}^0$ depends on the system's temperature and pressure. Yet Eq. (22) isn't too straightforward in this form and should be transformed to conveniently fit available data.

From classic thermodynamics, it is known that

$$d \ln f_i = \frac{V_i}{RT} dP \qquad (27)$$

Integrate the above equation from $P = P_i^s$ to $P =$ pressure of the system, and we obtain:

$$\ln \frac{f_{iL}^0}{f_{iL}^s} = \int_{P_i^s}^{P} \frac{V_{iL}}{RT} dP \qquad (28)$$

Rearrange Eq. (28), and it is written that

$$f_{iL}^0 = f_{iL}^s \left[ \exp\left( \int_{P_i^s}^{P} \frac{V_{iL}}{RT} dP \right) \right] = f_{iL}^s (PF)_i \qquad (29)$$

where $(PF)_i$ is Poynting factor, and

$$(PF)_i = \exp\left[ \int_{P_i^s}^{P} \frac{V_{iL}}{RT} dP \right] \qquad (30)$$

If Eq. (29) is incorporated into Eq. (20), then

$$P y_i \overline{\phi}_i^V = x_i \gamma_{iL} f_{iL}^s (PF)_i = x_i \gamma_{iL} P_i^s \bullet \frac{f_{iL}^s}{P_i^s} (PF)_i = x_i \gamma_{iL} P_i^s \phi_i^s (PF)_i \qquad (31)$$

Eq. (31) can be written in another form:

$$K_i = \frac{y_i}{x_i} = \frac{\gamma_{iL} P_i^s \phi_i^s (PF)_i}{P \overline{\phi}_i^V} \qquad (32)$$

Until now, no any assumption is made, and thus Eq. (32) is suitable for any pressure, including low pressure (< 200 kPa), middle pressure (< 200-2000 kPa) and high pressure (> 2000 kPa). However, in most cases, special distillation processes are implemented at low and middle pressures. Here only vapor-liquid equilibrium at low and middle pressures is discussed.

### 1.1.1. The equilibrium ratio $K_i$ at low pressure

In this case, it is supposed that the vapor phase is ideal gas, which means $\overline{\phi}_i^V = 1$ and

$\bar{\phi}_i^s = 1$. In addition, the ratio of molar volume of component $V_{iL}$ to $RT$, i.e. $V_{iL}/RT$, can be negligible. That is:

$$(PF)_i = \exp\left[\int_{P_i^s}^{P} \frac{V_{iL}}{RT} dP\right] = 1 \tag{33}$$

Thus,

$$Py_i = x_i \gamma_{iL} P_i^s \tag{34}$$

In terms of the definition of $K_i$, Eq. (34) becomes

$$K_i = \frac{y_i}{x_i} = \frac{\gamma_{iL} P_i^s}{P} \tag{35}$$

If it goes a further step to assume that the liquid phase is in ideal solution, which means that $\gamma_{iL} = 1$, then

$$Py_i = x_i P_i^s \tag{36}$$

or

$$K_i = \frac{y_i}{x_i} = \frac{P_i^s}{P} \tag{37}$$

It should be noted that the assumption for the vapor phase is reasonable to some extent because the P-V-T behavior of most real gases at low pressure conforms to the ideal gas law. But the assumption for the liquid phase isn't accepted in some cases, except for such systems as benzene / toluene, hexane / heptane, etc., which have similar molecular sizes and chemical properties.

### 1.1.2. The equilibrium ratio $K_i$ at middle pressure

If the assumption that $(PF)_i \approx 1$ at middle pressure is substituted into Eq. (31), then

$$Py_i \bar{\phi}_i^V = x_i \gamma_{iL} P_i^s \phi_i^s \tag{38}$$

That is,

$$K_i = \frac{y_i}{x_i} = \frac{\gamma_{iL} P_i^s \phi_i^s}{P \bar{\phi}_i^V} \tag{39}$$

At not too high pressure, the assumption of ideal gas for the vapor phase is still valid. Therefore, Eq. (39) becomes

$$Py_i = x_i \gamma_{iL} P_i^s \tag{40}$$

which is consistent with Eq. (35).

In order to calculate $K_i$ more accurately, it is better not to neglect $\overline{\phi}_i^V$ and $\phi_i^s$, especially for polar gases. From classic thermodynamics we know:

$$\ln \phi_i = \int_0^P [Z_i - 1] \frac{dP}{P} \tag{41}$$

for the pure component, and

$$\ln \overline{\phi}_i^V = \int_0^P [\overline{Z}_i - 1] \frac{dP}{P} \tag{42}$$

for the multi-component mixture. There are many famous equations of state (EOS), such as van der Waals equation, R-K (Redlich and Kwong) equation, S-R-K (Soave-Redlich-Kwong) equation, P-R (Peng-Robinson) equation and their revisions, which can be used to determine the compressibility factors, $Z_i$ and $\overline{Z}_i$. The interested readers can refer to any classic thermodynamic text [1-12]. One of equations of state is Virial equation of state, which has a sound theoretical foundation and can be derived from statistical mechanics. Herein, a two-term Virial equation of state is introduced as follows:

$$Z = 1 + \frac{BP}{RT} \tag{43}$$

where $B$ ($m^3 \, mol^{-1}$) is the second Virial coefficient.

One advantage of two-term Virial equation of state is that it is simple and accurate to some extent. After Eq. (43) is incorporated into Eqs. (41) and (42) and integrated, we will obtain:

$$\ln \phi_i^s = \frac{B_{ii} P_i^s}{RT} \tag{44}$$

and

$$\ln \overline{\phi}_i^V = \left( 2 \sum_{j=1}^n y_j B_{ij} - B_m \right) \frac{P}{RT} \tag{45}$$

where

$$B_m = \sum_i \sum_j y_i y_j B_{ij} \tag{46}$$

For pure-component non-polar molecules, the second Virial coefficient is commonly determined by the Tsonopoulos equation:

$$\frac{BP_c}{RT_c} = f^{(0)} + \omega f^{(1)} \tag{47}$$

where

$$f^{(0)} = 0.1445 - \frac{0.330}{T_r} - \frac{0.1385}{T_r^2} - \frac{0.0121}{T_r^3} - \frac{0.000607}{T_r^8} \tag{48}$$

$$f^{(1)} = 0.0637 + \frac{0.331}{T_r^2} - \frac{0.423}{T_r^3} - \frac{0.008}{T_r^8} \tag{49}$$

$T_r$ is the reduced temperature ($T_r = \frac{T}{T_c}$), and $\omega$ is acentric factor.

For pure-component polar molecules, the second coefficient is commonly determined by the modified Tsonopoulos equation:

$$\frac{BP_c}{RT_c} = f^{(0)} + \omega f^{(1)} + f^{(2)} \tag{50}$$

$$f^{(2)} = \frac{a}{T_r^6} - \frac{b}{T_r^8} \tag{51}$$

$$a = -2.140 \times 10^{-4} \mu_r - 4.308 \times 10^{-21} \mu_r^8 \tag{52}$$

where

$$\mu_r = \frac{10^5 \mu_p^2 P_c}{T_c^2} \tag{53}$$

In Eq. (53), $\mu_p$ is dipole moment (debyes), $P_c$ is critical pressure (atm), and $T_c$ is critical temperature (K).

The constant $b$ in Eq. (50) is zero for components exhibiting no hydrogen bonding, e.g. ketones, aldehydes, nitrides and ethers. And $b$ usually ranges from 0.04 to 0.06 for hydrogen-bond components, e.g. alcohols, water and organic acids.

For multi-component mixtures, the most important is to derive the interaction Virials coefficient $B_{ij}$ in order to calculate $B_m$. The simple way is by employing Eqs. (47) or (50).

But for this purpose, the combination rules must be devised to obtain $T_{cij}$, $P_{cij}$ and $\omega_{ij}$.

$$T_{cij} = (T_{ci}T_{cj})^{1/2}(1 - k_{ij}) \tag{54}$$

$$V_{cij} = \left( \frac{V_{ci}^{1/3} + V_{cj}^{1/3}}{2} \right)^3 \tag{55}$$

$$Z_{cij} = \frac{Z_{ci} + Z_{cj}}{2} \tag{56}$$

$$\omega_{ij} = \frac{\omega_i + \omega_j}{2} \tag{57}$$

$$P_{cij} = \frac{Z_{cij} RT_{cij}}{V_{cij}} \tag{58}$$

where, for a first-order approximation, $k_{ij}$ can be set to equal to zero.

On the other hand, although Eq. (39) can be used for calculating $K_i$, it seems somewhat sophisticated since sometimes $\overline{\phi}_i^V$ isn't easily derived. So one wishes to expand a simple rigorous formulation with enough accuracy. In this case, an important assumption is adopted that at middle pressure the vapor mixture can be regarded as ideal solution, but not as ideal gas. It indicates that $\gamma_{iV} = 1$ in Eq. (18), which is reasonable in most cases, especially for nonpolar gases. The reason is that the long-rang forces between the gas molecules prevail and they are much weaker than in the liquid mixture. The assumption of ideal solution is virtually different from the assumption of ideal gas in that in the latter the ideal gas law is tenable. According to the ideal gas law, not only $\gamma_{iV} = 1$, but also $\overline{\phi}_i^V = 1$. However, for ideal solution either for gas phase or for liquid phase, just $\gamma_{iV} = 1$ or $\gamma_{iL} = 1$. Thus, we can deduce another equation expressing $K_i$ on the basis of Eqs. (18) and (32):

$$K_i = \frac{y_i}{x_i} = \frac{\gamma_{iL} P_i^s \phi_i^s (PF)_i}{f_{iV}^0} \tag{59}$$

When Eq. (41) is applied to the vapor phase at the system pressure $P$, we have

$$\ln \phi_{iV}^0 = \ln \frac{f_{iV}^0}{P} = \frac{B_{ii} P_i^s}{RT} \tag{60}$$

Substituting Eqs. (33), (44) and (60) into Eq. (59), $K_i$ is replaced by

$$K_i = \frac{y_i}{x_i} = \frac{\gamma_{iL} P_i^s}{P} \exp\left[ \frac{(V_{iL} - B_{ii})(P - P_i^s)}{RT} \right] \tag{61}$$

This form is desirable since, comparing with Eq.(39), the physical quantities needed for calculation are relatively easily available.

Furthermore, when contrasting Eq. (35) and Eq. (61), it is found that only one new factor, $\exp\left[ \dfrac{(V_{iL} - B_{ii})(P - P_i^s)}{RT} \right]$, is added in Eq. (61). Now we will check the influence of this factor on $K_i$.

*Example:* Cumene as a basic chemical material is produced by alkylation reaction of benzene with propylene in a catalytic distillation column [13-17]. In order to avoid the polymerization of propylene, the mole ratio of benzene to propylene is often excessive, in the range of 1.0 to 4.0. Therefore, the concentration of benzene in the vapor and liquid phases is predominant. The operation conditions are: average temperature 110℃ (383.15K) and pressure 700kPa. Based on the physical constants given below and assuming that $\gamma_{iL} = 1$ in nonpolar mixture, estimate the values of the second Virial coefficient $B_{ii}$ and equilibrium ratio $K_i$ of benzene (in SI units).

Antoine vapor-pressure equation: $\ln P = A - \dfrac{B}{T+C}$ ($P$, mmHg; $T$, K).

For benzene, $A = 15.9008$, $B = 2788.51$, $C = -52.36$.

$T_c = 562.1\,\text{K}$, $P_c = 4894\,\text{kPa}$, $\omega = 0.212$, $V_{iL} = 8.8264 \times 10^{-5}\,\text{m}^3\,\text{mol}^{-1}$.

*Solution:* From the Antoine equation ($\ln P_i^s = A - \dfrac{B}{T+C}$ at $T = 383.15\,\text{K}$),

$$\ln P_i^s = 15.9008 - \frac{2788.51}{383.15 - 52.36}$$

$P_i^s = 1756.3\,\text{mmHg} = 234.2\,\text{kPa}$

$$T_r = \frac{T}{T_c} = \frac{383.15}{562.1} = 0.6816$$

From Eq. (48), after substituting the value of $T_r$,

$$f^{(0)} = 0.1445 - \frac{0.330}{0.6816} - \frac{0.1385}{0.6816^2} - \frac{0.0121}{0.6816^3} - \frac{0.000607}{0.6816^8} = -0.6890$$

Similarly, from Eq. (49), $f^{(1)} = 0.0637 + \dfrac{0.331}{0.6816^2} - \dfrac{0.423}{0.6816^3} - \dfrac{0.008}{0.6816^8} = -0.7314$

From Eq. (47),

$$B_{ii} = \frac{RT_c}{P_c}(f^{(0)} + \omega f^{(1)}) = \frac{8.314 \times 562.1}{4894000}(-0.6890 - 0.212 \times 0.7314) = 8.0599 \times 10^{-4}\,\text{m}^3\,\text{mol}^{-1}$$

From Eq. (35), $K_i = \dfrac{y_i}{x_i} = \dfrac{\gamma_{iL} P_i^s}{P} = \dfrac{1.0 \times 234.2}{700} = 0.3346$

From Eq. (61), $K_i = \dfrac{y_i}{x_i} = \dfrac{\gamma_{iL} P_i^s}{P} \exp\left[\dfrac{(V_{iL} - B_{ii})(P - P_i^s)}{RT}\right]$

$$= \frac{1.0 \times 234.2}{700} \exp\left[\frac{(8.8264 - 80.599)(700 - 234.2) \times 10^{-2}}{8.314 \times 383.15}\right]$$

$$= 0.3012$$

It can be seen that the difference of $K_i$ between Eqs. (35) and (61) is apparent and the

relative deviation is $\left| \dfrac{0.3346 - 0.3012}{0.3012} \right| = 11.09\%$, which indicates that the assumption of ideal

gas at middle pressure is a little inappropriate. The reason is that the factor,

$\exp\left[ \dfrac{(V_{iL} - B_{ii})(P - P_i^s)}{RT} \right]$, isn't so approximate to unity, in this case, 0.9004. So it is advisable

to apply Eq. (61) to calculate $K_i$ at middle pressure.

As mentioned above, the equilibrium ratio $K_i$ (or phase equilibrium constant) is an

important physical quantity in solving four types of problems about $P$-$T$-$x$-$y$ relation. Besides, for special distillation processes, especially with separating agents, we should pay more attention to another physical quantity, i.e. relative volatility (or separation factor). The magnitude of relative volatility can be used as the index of whether the separation process can occur or which one among all the possible separating agents is the best. For vapor-liquid two

phases, relative volatility $\alpha_{ij}$, as a dimensionless physical quantity, is defined as

$$\alpha_{ij} = \frac{K_i}{K_j} \tag{62}$$

which holds for either binary or multi-component systems. For the multi-component system, just one heavy key component and one light key component are investigated. In general, in a same distillation column, the heavy components are obtained as the bottom product, whilst the light component as the top product. The heavy and light key components are those which

are the most difficult to be separated among the multi-component mixture. In Eq. (62), $K_i$ is

often relative to the light component, $K_j$ relative to the heavy component. In other words,

$\alpha_{ij}$ is always equal or larger than unity. For special distillation processes, the meaning of

relative volatility is:

(1) If $\alpha_{ij}$ is equal or close to unity, then the separation process isn't worthwhile to be carried

out. Otherwise, the investment on equipment and operation may be unimaginable. It is

generally required that $\alpha_{ij} > 1.20$ at least.

(2) The larger $\alpha_{ij}$ is, the more easily the separation process will be carried out. So, for

special distillation processes with separating agents, we prefer to select the separating agent which has the largest $\alpha_{ij}$. Of course, under this condition, other factors such as price, toxics, chemical stability and so on, should be considered.

From the definition of $\alpha_{ij}$ and Eqs. (35) and (61), it is deduced that

$$\alpha_{ij} = \frac{\gamma_i P_i^0}{\gamma_j P_j^0} \tag{63}$$

at the low pressure,
and

$$\alpha_{ij} = \frac{\gamma_i P_i^0}{\gamma_j P_j^0} \exp\left[\frac{(V_{iL} - B_{ii})(P - P_i^s)}{RT}\right] / \exp\left[\frac{(V_{jL} - B_{ii})(P - P_j^s)}{RT}\right] \tag{64}$$

at the middle pressure.

*Example:* Based on the last example, estimate the $\alpha_{ij}$ values of propylene to benzene (in SI units). The physical constants of propylene are given below, and it is also assumed that $\gamma_{iL} = 1$ for both propylene and benzene in non-polar mixture.

Antoine vapor-pressure equation: $\ln P = A - \dfrac{B}{T+C}$ ($P$, mmHg; $T$, K)

For propylene, $A = 15.7027$, $B = 1807.53$, $C = -26.15$

$T_c = 365.0$ K, $P_c = 4620$ kPa, $\omega = 0.148$, $V_{iL} = 6.8760 \times 10^{-5}$ m$^3$ mol$^{-1}$

*Solution:* Since the boiling point of propylene (225.4K) is much smaller than that of benzene (353.3K), propylene is denoted by "1" as the light component, and benzene by "2" as the heavy component.

From the Antoine equation for propylene ($\ln P_1^s = A - \dfrac{B}{T+C}$ at $T = 383.15$K),

$$\ln P_1^s = 15.7027 - \frac{1807.53}{383.15 - 26.15}$$

$P_1^s = 41755.7$ mmHg $= 5567.0$ KPa

From Eq. (63), $\alpha_{12} = \dfrac{\gamma_1 P_1^0}{\gamma_2 P_2^0} = \dfrac{5567.0}{234.2} = 23.77$

In the same way as the last example,

$B_{11} = 2.0194 \times 10^{-4}$ m$^3$ mol$^{-1}$

The factor,

$$\exp\left[\frac{(V_{1L} - B_{11})(P - P_1^s)}{RT}\right] = \exp\left[\frac{(6.8760 - 20.194)(700 - 5567.0) \times 10^{-2}}{8.314 \times 383.15}\right] = 1.2257$$

From Eq. (64),

$$
\alpha_{12} = \frac{\gamma_1 P_1^0}{\gamma_2 P_2^0} \exp\left[\frac{(V_{1L} - B_{11})(P - P_1^s)}{RT}\right] \Big/ \exp\left[\frac{(V_{2L} - B_{22})(P - P_2^s)}{RT}\right]
$$

$$
= \frac{5567.0}{234.2} \cdot \frac{1.2257}{0.9004}
$$

$$
= 32.36
$$

As can be seen, similar as $K_i$, it is advisable to apply Eq. (64) to calculate $\alpha_{ij}$ at middle pressure because the difference of the results between Eqs. (63) and (64) is apparent and the relative deviation $\left|\dfrac{23.77 - 32.36}{32.36}\right|$ is 26.55%.

At some time, volatility $v_i$ is introduced as

$$
v_i = \frac{P_i}{x_i}, \quad v_j = \frac{P_j}{x_j} \tag{65}
$$

At lower and middle pressures, apply Dalton law for the vapor phase and we have

$$
\frac{v_i}{v_j} = \frac{P_i / x_i}{P_j / x_j} = \frac{y_i / x_i}{y_j / x_j} = \alpha_{ij} \tag{66}
$$

This equation uncovers the relationship between volatility and relative volatility.

Apart from volatility, there is another important physical quantity used in special distillation processes, i.e. selectivity, defined as

$$
S_{ij} = \frac{\gamma_i}{\gamma_j} \tag{67}
$$

Eq. (67) is different from Eq.(63) in that the term, $\dfrac{P_i^0}{P_j^0}$, is omitted. As we know, this term is

dependent of temperature. Evaluation for special distillation processes and the corresponding separating agents is frequently made at constant temperature. So sometimes selectivity can be regarded as the alternative to relative volatility. However, compared with relative volatility, selectivity as the evaluation index isn't complete in the vapor-liquid equilibrium where we can't judge whether relative volatility $\alpha_{ij}$ is enough larger than unity. As a matter of fact, it is more appropriate to consider selectivity in the liquid-liquid equilibrium because at this time, the term, $\dfrac{P_i^0}{P_j^0}$, needn't to be concerned. But liquid-phase activity coefficient is indispensable

and plays an important role in phase equilibrium calculation.

## 1.2. Liquid-phase activity coefficient in binary and multi-component mixtures

From the classic thermodynamics, we know that activity coefficient is introduced as the revision and judgment for non-ideality of the mixture. As the activity coefficient is equal to unity, it means that the interactions between dissimilar or same molecules are always identical and the mixture is in the ideal state; as the activity coefficient is away from unity, the mixture is in the non-ideal state. The concept of activity coefficient is often used for the liquid phase. The activity coefficient in the liquid phase must be determined so as to derive the equilibrium ratio $K_i$ and relative volatility $\alpha_{ij}$, and thus establish the mathematics model of special distillation processes. The liquid-phase activity coefficient models are set up based on excess Gibbs free energy. The relation of activity coefficient $\gamma_i$ and excess Gibbs free energy $G^E$ is given below:

$$\left[ \frac{\partial(nG^E)}{\partial n_i} \right]_{T,P,n_j} = RT \ln \gamma_i \tag{68}$$

$$n = \sum_i n_i \tag{69}$$

The liquid-phase activity coefficient models are divided into two categories:
(1) The models are suitable for the non-polar systems, for instance, hydrocarbon mixture, isomers and homologues. Those include regular solution model, Flory-Huggins no-heat model.
(2) The models are suitable for the non-polar and/or polar systems. Those models are commonly used in predicting the liquid-phase activity coefficient, and include Margules equation, van Laar equation, Wilson equation, NRTL (nonrandom two liquid) equation, UNIQUAC (universal quasi-chemical) equation, UNIFAC (UNIQUAC Functional-group activity coefficients) equation and so on.

Among those, Wilson, NRTL, UNIQUAC and UNIFAC equations are the most widely used for binary and multi-component systems because of their flexibility, simplicity and ability to fit many polar and nonpolar systems. Besides, one outstanding advantage of those equations is that they can be readily extended to predict the activity coefficients of multi-component mixture from the corresponding binary-pair parameters. In fact, in special distillation processes, multi-component mixture is often involved.

The formulations of Wilson, NRTL and UNIQUAC equations are listed in Tables 1 and 2 for binary and multi-component mixtures, respectively. In some famous chemical engineering simulation software programs, such as ASPEN PLUS, PROII and so on, the formulations of those equations have been embraced and even the binary-pair parameters are able to be rewritten to meet various requirements. But is should be aware of the unit consistency. For instance, in Table 1 for Wilson equation, if the unit of $\lambda_{12} - \lambda_{11}$ is cal mol$^{-1}$, then $R = 1.987$ cal mol$^{-1}$ K$^{-1}$. Otherwise, if the unit of $\lambda_{12} - \lambda_{11}$ is J mol$^{-1}$, then $R = 8.314$ J mol$^{-1}$ K$^{-1}$.

Table 1
Activity coefficient equations for binary system

| Name | Equation |
|---|---|
| Wilson (two-parameter) | $G^E/RT = -x_1 \ln(x_1 + \Lambda_{12}x_2) - x_2 \ln(x_2 + \Lambda_{21}x_1)$ |

$$\Lambda_{12} = \frac{V_2}{V_1}\exp\left[-\frac{(\lambda_{12} - \lambda_{11})}{RT}\right], \quad \Lambda_{21} = \frac{V_1}{V_2}\exp\left[-\frac{(\lambda_{21} - \lambda_{22})}{RT}\right]$$

Equation for component 1:

$$\ln\gamma_1 = -\ln(x_1 + \Lambda_{12}x_2) + x_2\left(\frac{\Lambda_{12}}{x_1 + \Lambda_{12}x_2} - \frac{\Lambda_{21}}{x_2 + \Lambda_{21}x_1}\right)$$

Equation for component 2:

$$\ln\gamma_1 = -\ln(x_2 + \Lambda_{21}x_1) + x_1\left(\frac{\Lambda_{21}}{x_2 + \Lambda_{21}x_1} - \frac{\Lambda_{12}}{x_1 + \Lambda_{12}x_2}\right)$$

| Name | Equation |
|---|---|
| NRTL (three-parameter) | $G^E = x_1 x_2\left(\dfrac{G_{21}\tau_{21}}{x_1 + G_{21}x_2} + \dfrac{G_{12}\tau_{12}}{x_2 + G_{12}x_1}\right)$ |

$$\tau_{21} = \frac{g_{21} - g_{11}}{RT}, \tau_{12} = \frac{g_{12} - g_{22}}{RT}$$

$$\ln G_{12} = -\alpha_{21}\tau_{21}, \quad \ln G_{21} = -\alpha_{12}\tau_{12}$$

$$\ln\gamma_1 = x_2^2\left[\frac{\tau_{21}G_{21}^2}{(x_1 + G_{21}x_2)^2} + \frac{\tau_{12}G_{12}}{(x_2 + G_{12}x_1)^2}\right]$$

$$\ln\gamma_2 = x_1^2\left[\frac{\tau_{12}G_{12}^2}{(x_2 + G_{12}x_1)^2} + \frac{\tau_{21}G_{21}}{(x_1 + G_{21}x_2)^2}\right]$$

| Name | Equation |
|---|---|
| UNIQUAC (two-parameter) | $G^E = G^E$ (combinatorial contribution) $+ G^E$ (residual contribution) |

$$\phi_1 = \frac{x_1 r_1}{x_1 r_1 + x_2 r_2}, \phi_2 = \frac{x_2 r_2}{x_1 r_1 + x_2 r_2}$$

$$\theta_1 = \frac{x_1 q_1}{x_1 q_1 + x_2 q_2}, \theta_2 = \frac{x_2 q_2}{x_1 q_1 + x_2 q_2}$$

$$\tau_{21} = \exp\left(-\frac{u_{21} - u_{11}}{RT}\right), \tau_{12} = \exp\left(-\frac{u_{12} - u_{22}}{RT}\right)$$

Table 1
Activity coefficient equations for binary system

| UNIQUAC (two-parameter) | $\ln \gamma_1 = \ln \gamma_1^C + \ln \gamma_1^R$ |
|---|---|

$$\ln \gamma_1^C = \ln \frac{\phi_1}{x_1} + \frac{z}{2} q_1 \ln \frac{\theta_1}{\phi_1} + \phi_2 (l_1 - \frac{r_1}{r_2} l_2)$$

$$\ln \gamma_1^R = -q_1 \ln(\theta_1 + \tau_{21}\theta_2) + \theta_2 q_1 (\frac{\tau_{21}}{\theta_1 + \tau_{21}\theta_2} - \frac{\tau_{12}}{\theta_2 + \tau_{12}\theta_1})$$

$$\ln \gamma_2 = \ln \gamma_2^C + \ln \gamma_2^R$$

$$\ln \gamma_2^C = \ln \frac{\phi_2}{x_2} + \frac{z}{2} q_2 \ln \frac{\theta_2}{\phi_2} + \phi_1 (l_2 - \frac{r_1}{r_2} l_1)$$

$$\ln \gamma_2^R = -q_2 \ln(\theta_2 + \tau_{12}\theta_1) + \theta_1 q_2 (\frac{\tau_{12}}{\theta_2 + \tau_{12}\theta_1} - \frac{\tau_{21}}{\theta_1 + \tau_{21}\theta_2})$$

$$l_1 = \frac{z}{2}(r_1 - q_1) - (r_1 - 1), \quad l_2 = \frac{z}{2}(r_2 - q_2) - (r_2 - 1)$$

z = lattice coordination number set equal to 10.

Table 2
Activity coefficient equations for multi-component system

| Name | Equation |
|---|---|
| Wilson | $G^E / RT = -\sum_i x_i \ln(\sum_j x_j \Lambda_{ij})$ |
| | $\Lambda_{ij} = \frac{V_j}{V_i} \exp\left[-\frac{(\lambda_{ij} - \lambda_{ii})}{RT}\right],$ |
| | Equation for component $i$: |
| | $\ln \gamma_i = 1 - \ln(\sum_j x_j \Lambda_{ij}) - \sum_k (\frac{x_k \Lambda_{ki}}{\sum_j x_j \Lambda_{kj}})$ |
| NRTL | $\dfrac{G^E}{RT} = \sum_i x_i \dfrac{\sum_j \tau_{ji} G_{ji} x_j}{\sum_k G_{ki} x_k}$ |
| | $\tau_{ji} = \dfrac{g_{ji} - g_{ii}}{RT}$ |

Table 2
Activity coefficient equations for multi-component system

| | |
|---|---|
| | $G_{ji} = \exp(-\alpha_{ji}\tau_{ji})$, $\tau_{ii} = \tau_{jj} = 0$, $G_{ii} = G_{jj} = 1$, $\alpha_{ij} = \alpha_{ji}$ $$\ln\gamma_i = \frac{\sum_j \tau_{ji}G_{ji}x_i}{\sum_k G_{ki}x_k} + \sum_j \frac{x_j G_{ij}}{\sum_k G_{kj}x_k}\left[\tau_{ij} - \frac{\sum_l x_l \tau_{lj}G_{lj}}{\sum_k G_{kj}x_k}\right]$$ |
| UNIQUAC | $G^E = G^E$ (combinatorial contribution) $+ G^E$ (residual contribution) $$\phi_i = \frac{r_i x_i}{\sum_j r_j x_j}, \quad \theta_i = \frac{q_i x_i}{\sum_j q_j x_j}$$ $$\tau_{ji} = \exp\left(-\frac{u_{ji} - u_{ii}}{RT}\right)$$ $$\ln\gamma_i = \ln\gamma_i^C + \ln\gamma_i^R$$ $$\ln\gamma_i^C = \ln\frac{\phi_i}{x_i} + \frac{z}{2}q_i \ln\frac{\theta_i}{\phi_i} + l_i - \frac{\phi_i}{x_i}\sum_j x_j l_j$$ $$\ln\gamma_i^R = q_i\left[1 - \ln(\sum_j \theta_j \tau_{ji}) - \sum_j \frac{\theta_j \tau_{ij}}{\sum_k \theta_k \tau_{kj}}\right]$$ $$l_i = \frac{z}{2}(r_i - q_i) - (r_i - 1)$$ z = lattice coordination number set equal to 10. |

The value of $\alpha_{ji}$ in NRTL equation for different systems is:

(1) $\alpha_{ji} = 0.20$ for mixtures of saturated hydrocarbons and polar non-associated component (e.g. n-hexane/acetone, iso-octane/nitroethane).

(2) $\alpha_{ji} = 0.30$ for mixtures of nonpolar compounds (e.g. benzene/n-heptane), except fluorocarbons and paraffins; mixtures of nonpolar and polar non-associated components (e.g. benzene/acetone); mixtures of polar components that exhibit negative deviations from Raoult's law (e.g. acetone/chloroform) and moderate positive deviations (e.g. ethanol/water); mixtures of water and polar non-associated components (e.g. water/acetone).

(3) $\alpha_{ji} = 0.40$ for mixtures of saturated hydrocarbons and homolog perfluorocarbons (e.g. n-hexane/perfluoro-n-hexane).

(4) $\alpha_{ji} = 0.47$ for mixtures of an alcohol or other strongly self-associated components with nonpolar components (e.g. ethanol/benzene); mixtures of carbon tetrachloride with either acetonitrile or nitromethane; mixtures of water with either butyl glycol, organic acids or pyridine.

It may be possible that the equation forms we have obtained from references are somewhat different from those given in Tables 1 and 2. So the binary-pair parameters of liquid-phase activity coefficient equation can't be directly used in the simulation software programs, and must be transformed into "standard form" (the equations listed in Tables 1 and 2). Only can the binary-pair parameters of "standard form" be directly input in some commercial softwares. As a result, the following transformation procedure is recommended:

(1) The activity coefficients for a binary system at finite dilution over the range of temperature investigated are calculated from the original equation. For instance, if $x_1 = 0$ (at the same time $x_2 = 1$), then $\gamma_1$ at infinite dilution can be solved by using the original equation.

(2) By using of the data of activity coefficients at infinite dilution, the binary-pair parameters of the "standard form" are correlated, and this step is reverse to the above step.

The following example is given to clarify the transformation procedure in detail.

*Example:* In the production of acetic acid, acetic acid often exists with some water. Acetic acid and water may be separated by special distillation processes, e.g. azeortopic distillation, membrane distillation, extractive distillation, etc. The experimental data of activity coefficients of acetic acid and water are available, and have been utilized to correlate the binary-pair parameters by the maximum likelihood regression. The objective function $Q$ is:

$$Q = \sum_{i=1}^{N} \left| \gamma_i^{exp} - \gamma_i^{cal} \right| \tag{70}$$

where $N$ is the number of data points, $\gamma_i^{exp}$ is the activity coefficient obtained from the experimental data and $\gamma_i^{cal}$ is the activity coefficient calculated from the thermodynamic equation. Perhaps, the readers would like to ask why the following objective function $Q$ isn't chosen as usual:

$$Q = \sum_{i=1}^{N} \left( \gamma_i^{exp} - \gamma_i^{cal} \right)^2 \tag{71}$$

It is primarily due to the fact that many values of $\left| \gamma^{exp} - \gamma^{cal} \right|$ are often in the range of 0-1 and the average relative deviation (ARD), $\sum\limits_{i=1}^{N} \left| \left( \gamma_i^{exp} - \gamma_i^{cal} \right) / \gamma_i^{exp} \right| / N$, may be magnified after

square. Suppose that the binary-pair parameters of Wilson equation are known, but what we need is the binary-pair parameters of NRTL equation. Please compute the binary-pair parameters of NRTL equation ($\alpha_{ij} = 0.47$). The symbols of "1" and "2" represent water and

acetic acid, respectively. (In the Wilson equation, $\Lambda_{12} = 3766.34$ J mol$^{-1}$ and

$\Lambda_{21} = -3301.30$ J mol$^{-1}$. The temperature is in the range of 80 - 150°C.)

*Solution:* According to Wilson equation, $\gamma_1^{\infty}$ at $x_1 \to 0$ and $\gamma_2^{\infty}$ at $x_2 \to 0$ at different

temperatures are calculated. By using the obtained activity coefficients at infinite dilution, the binary-pair parameters of NRTL equation are correlated by the Marquardt method with the

result that $\Lambda_{12} = 329.76$ J mol$^{-1}$, $\Lambda_{21} = 131.41$ J mol$^{-1}$, and the average relative deviation is

0.42%. The comparison of activity coefficients at infinite dilution of Wilson and NRTL

equations is given in Table 3, from which it can be seen that the values of $\gamma_1^{\infty}$ and $\gamma_2^{\infty}$ are in

good agreement between these two models.

Apart from the Wilson, NRTL and UNIQUAC equations in which the experimental data must be given to correlate the binary-pair parameters, there is a widely used prediction model, i.e. UNIFAC (UNIQUAC function-group activity coefficients) group contribution method. This method, first presented by Franklin [18], is currently very popular and can be used to predict the liquid-phase activity coefficient of binary or multi-component systems even when experimental phase equilibrium data are unavailable. It has several advantages over the Wilson, NRTL, UNIQUAC equations: (1) size and binary interaction parameters are available for a wide range of types of function groups (more than 100 function groups); (2) extensive comparisons with experimental data are available; (3) it is an open system, and more function groups and more parameters will be filled in the UNIFAC list in the future. But it still has a pity, that is, the ions (cation and anion groups) aren't complete in the UNIFAC menu. In particular, in the recent years, ionic liquids have attracted more attention in special distillation processes, which will be emphasized in later chapters; (4) Experimental measurements of vapor-liquid phase equilibrium are very time-consuming and therefore expensive. For example, if measurements are performed for a 10-component system at just one constant pressure (e.g. atmospheric pressure) in 10% mole steps and an average number of 10 data points can be experimentally determined daily, the measurements (in total 92378 data points) will take more than 37 years [19]. Therefore, with the view to multi-component mixtures, UNIFAC model is more advantage than Wilson, NRTL and UNIQUAC equations in shortening the measurement time. That is why it was very popular and desirable in the synthesis, design and optimization of separation processes in the past 25 years.

Table 3
Comparison of activity coefficients at infinite dilution of Wilson and NRTL equations

| | $T\,/\,^\circ C$ | | | | | | | |
|---|---|---|---|---|---|---|---|---|
| | 80 | 90 | 100 | 110 | 120 | 130 | 140 | 150 |
| Wilson equation | | | | | | | | |
| $\gamma_1^\infty$ | 1.1712 | 1.1647 | 1.1576 | 1.1501 | 1.1422 | 1.1342 | 1.1261 | 1.1179 |
| $\gamma_2^\infty$ | 1.1617 | 1.1611 | 1.1593 | 1.1565 | 1.1529 | 1.1484 | 1.1433 | 1.1377 |
| NRTL equation | | | | | | | | |
| $\gamma_1^\infty$ | 1.1633 | 1.1587 | 1.1543 | 1.1501 | 1.1462 | 1.1424 | 1.1389 | 1.1355 |
| $\gamma_2^\infty$ | 1.1690 | 1.1640 | 1.1593 | 1.1549 | 1.1507 | 1.1467 | 1.1429 | 1.1393 |

UNIFAC model is still developing, and there are four versions until now:
(1) The original UNIFAC model that can be applied for infinite dilution and finite concentration (thereof called model 1)

The original UNIFAC model [20-26] was widely used before. The activity coefficient is expressed as functions of composition and temperature. The model has a combinatorial contribution to the activity coefficient, i.e. $\ln\gamma_i^C$, essentially due to differences in size and shape of the molecules, and a residual contribution, i.e. $\ln\gamma_i^R$, essentially due to energetic interactions.

$$\ln\gamma_i = \ln\gamma_i^C + \ln\gamma_i^R \tag{72}$$

I . Combinatorial part.

$$\ln\gamma_i^C = \ln\frac{\phi_i}{x_i} + \frac{z}{2}q_i\ln\frac{\theta_i}{\phi_i} + l_i - \frac{\phi_i}{x_i}\sum_j x_j l_j \tag{73}$$

$$l_i = \frac{z}{2}(r_i - q_i) - (r_i - 1); \quad z = 10 \tag{74}$$

$$\theta_i = \frac{q_i x_i}{\sum_j q_j x_j}; \quad \phi_i = \frac{r_i x_i}{\sum_j r_j x_j} \tag{75}$$

Pure component parameters $r_i$ and $q_i$ are, respectively, relative with molecular van der Waals volumes and molecular surface areas. They are calculated as the sum of the group volume and group area parameters, $R_k$ and $Q_k$,

$$r_i = \sum_k v_k^{(i)} R_k; \quad q_i = \sum_k v_k^{(i)} Q_k \tag{76}$$

where $v_k^{(i)}$, always an integer, is the number of groups of type $k$ in molecule $i$. Group parameters $R_k$ and $Q_k$ are normally obtained from van der Waals group volumes and surface areas, $V_k$ and $A_k$, given by Bondi [27],

$$R_k = \frac{V_k}{15.17}; \ Q_k = \frac{A_k}{2.5 \times 10^9} \tag{77}$$

II. Residual Part.

$$\ln \gamma_i^R = \sum_k v_k^{(i)} [\ln \Gamma_k - \ln \Gamma_k^{(i)}] \tag{78}$$

$\Gamma_k$ is the group residual activity coefficient, and $\Gamma_k^{(i)}$ is the residual activity coefficient of group $k$ in a reference solution containing only molecules of type $i$.

$$\ln \Gamma_k = Q_k [1 - \ln(\sum_m \theta_m \psi_{mk}) - \sum_m (\theta_m \psi_{km} / \sum_n \theta_n \psi_{nm})] \tag{79}$$

$$\theta_m = \frac{Q_m X_m}{\sum_n Q_n X_n}; X_m = \frac{\sum_i v_m^{(i)} x_i}{\sum_i \sum_k v_k^{(i)} x_i} \tag{80}$$

$X_m$ is the fraction of group $m$ in the mixture.

$$\psi_{nm} = \exp[-(a_{nm}/T)] \tag{81}$$

Parameter $a_{nm}$ characterizes the interaction between groups $n$ and $m$. For each group-group interaction, there are two parameters: $a_{nm} \neq a_{mn}$.

Eqs. (79) and (80) also hold for $\ln \Gamma_k^{(i)}$, except that the group composition variable, $\theta_k$, is now the group fraction of group $k$ in pure fluid $i$. In pure fluid, $\ln \Gamma_k = \ln \Gamma_k^{(i)}$, which means that as $x_i \rightarrow 1$, $\gamma_i^R \rightarrow 1$. $\gamma_i^R$ must be close to unity because as $x_i \rightarrow 1$, $\gamma_i^C \rightarrow 1$ and $\gamma_i \rightarrow 1$.

(2) The modified UNIFAC model that can be applied for infinite dilution and finite concentration (thereof called model 2)

In the modified UNIFAC(Dortmund) model [28-33], as in the original UNIFAC model, the activity coefficient is also the sum of a combinatorial and a residual part:

$$\ln \gamma_i = \ln \gamma_i^C + \ln \gamma_i^R \tag{82}$$

The combinatorial part is changed in an empirical way to make it possible to deal with compounds very different in size:

$$\ln \gamma_i^C = 1 - V_i' + \ln V_i' - 5q_i(1 - \frac{V_i'}{F_i} + \ln(\frac{V_i'}{F_i})) \tag{83}$$

The parameter $V_i'$ can be calculated by using the relative van der Waals volumes $R_k$ of the different groups.

$$V_i' = \frac{r_i^{3/4}}{\sum_j x_j r_j^{3/4}} \tag{84}$$

All other parameters are calculated in the same way as in the original UNIFAC model:

$$V_i = \frac{r_i x_i}{\sum_j x_j r_j} \tag{85}$$

$$r_i = \sum_k v_k^{(i)} R_k \tag{86}$$

$$F_i = \frac{q_i x_i}{\sum_j x_j q_j} \tag{87}$$

$$q_i = \sum_k v_k^{(i)} Q_k \tag{88}$$

The residual part can be obtained by using the following relations:

$$\ln \gamma_i^R = \sum_k v_k^{(i)} (\ln \Gamma_k - \ln \Gamma_k^{(i)}) \tag{89}$$

$$\ln \Gamma_k = Q_k (1 - \ln(\sum_m \theta_m \psi_{mk}) - \sum_m \frac{\theta_m \psi_{km}}{\sum_n \theta_n \psi_{nm}}) \tag{90}$$

whereby the group area fraction $\theta_m$ and group mole fraction $X_m$ are given by the following equations:

$$\theta_m = \frac{Q_m X_m}{\sum_n Q_n X_n} \tag{91}$$

$$X_m = \frac{\sum_j v_m^{(j)} x_j}{\sum_j \sum_n v_n^{(j)} x_j} \tag{92}$$

In comparison to the original UNIFAC model, the van der Waals properties are changed slightly, and at the same time temperature-dependent parameters are introduced to permit a better description of the real behavior (activity coefficients) as a function of temperature.

$$\psi_{nm} = \exp(-\frac{a_{nm} + b_{nm}T + c_{nm}T^2}{T}) \tag{93}$$

Thus, in order to calculate the activity coefficient, such parameters as $R_k$, $Q_k$, $a_{nm}, b_{nm}, c_{nm}, a_{mn}, b_{mn}, c_{mn}$ should be pre-determined. Since the existing parameters of the modified UNIFAC model (model 2) are extended with the help of the Dortmund Data Bank (DDB) and the integrated fitting routines, even the values of $R_k$ and $Q_k$ are possibly different from those in the model 1. As the group parameters are replenished step by step, this model has the tendency to substitute the original UNIFAC model because of its better predictions of the real behavior of non-electrolyte systems. The present status of all research concerning the modified UNIFAC model (Dortmund) is always available via the internet: http://www.uni-oldenburg.de/tchemie/consortium.

(3) The UNIFAC model that is appropriate for infinite dilution concentration (thereof called model 3 or in the reference $\gamma^{\infty}$-based UNIFAC)

An UNIFAC parameter table exclusively based on $\gamma^{\infty}$ data is presented by Bastos et al. [34]. It aims at the improvement of the general accuracy and range of applicability of the UNIFAC model as far as the calculation for $\gamma^{\infty}$ and $S^{\infty}$ values is involved. Therefore, it can be regarded as an useful supplement to the existing vapor-liquid equilibrium (VLE) and liquid-liquid equilibrium (LLE) parameters.

The 190 pairs of parameters of 40 different groups have been estimated from about 8000 data points based on experimental $\gamma^{\infty}$ data with an average relative error of 20%. The equation forms in model 3, as well as the values of $R_k$ and $Q_k$, are the same as in model 1. The difference between these two models is only the interaction parameters of the UNIFAC groups.

(4) The UNIFAC model that can be applied for prediction of vapor-liquid equilibrium in mixed solvent-salt system (thereof called model 4)

The activity coefficient $\gamma_i$ of a solvent $i$ in a liquid-solvent mixture is calculated as [35-37]:

$$\ln \gamma_i = \ln \gamma_i^{D-H} + \ln \gamma_i^C + \ln \gamma_i^R \tag{94}$$

where $\gamma_i^{D-H}$ is the Debye-Huckel term and $\gamma_i^C$ and $\gamma_i^R$ represent the UNIFAC combinatorial and residual contributions.

The Debye-Huckel term is calculated from the following equation as described by

Macedo et al [38]:

$$\ln \gamma_i^{D-H} = \frac{2AM_i d_s}{b^3 d_i}\left[1 + b\sqrt{I} - \frac{1}{1+b\sqrt{I}} - 2\ln(1+b\sqrt{I})\right] \tag{95}$$

where $M_i$ is the molecular weight of solvent $i$, $I$ is the ionic strength, $d_s$ and $d_i$ are the densities of salt and solvent, respectively, $A$ and $b$ are expressed as the function of dielectric constant $\varepsilon$ of the solvent mixture:

$$A = 1.327757 \times 10^5 d_s^{1/2} /(\varepsilon T)^{3/2} \tag{96}$$

$$b = 6.359696 d_s^{1/2} /(\varepsilon T)^{1/2} \tag{97}$$

It is shown that in order to determine the activity coefficient we need to know not only the molecular structure of solvent and salt, but also the properties of density and dielectric constant, which, however, sometimes is difficult to obtain and thus casts a shade to its wide application.

In another way, one similar model [39] is developed, which consists of three terms for the excess Gibbs energy: (1) a Debye-Huckel term which represents the long-range(LR) interactions; (2) a virial term which accounts for the middle-range(MR) interactions caused by the ion-dipole effects; (3) a UNIFAC term which represents the short-range(SR) interaction. Therefore, the excess Gibbs energy is calculated as the sum of three contributions:

$$G^E = G_{LR}^E + G_{MR}^E + G_{SR}^E \tag{98}$$

That is to say,

$$\ln \gamma_i = \ln \gamma_i^{D-H} + \ln \gamma_i^{MR} + \ln \gamma_i^C + \ln \gamma_i^R \tag{99}$$

Compared with Eq. (94), only one term, $\ln \gamma_i^{MR}$, is added.

*Example:* 1,3-butadiene is a basic organic raw material for synthesizing rubber [40-42]. Separation of 1,3-butadiene from C4 mixture (consisting of n-butane, isobutene, 1-butene, trans-2-butene, cis-2-butene, 1,3-butadiene, 1,2-butadiene, methyl acetylene, butyne-1, butyne-2, VAC (vinylacetylene), etc.) is an important separation process in the petrochemical industry. A special distillation process, i.e. extractive distillation with DMF (N,N-dimethylformamide) as the separating agent, is often used to extract 1,3-butadiene from C4 mixture. In order to investigate the effect of DMF on improving the relative volatility of C4 mixture, use the UNIFAC models 1, 2 and 3 to estimate the liquid-phase activity coefficients of the key components of n-butane, 1-butene and 1,3-butadiene at infinite dilution. And compare the estimated and experimental values at different temperatures.

*Solution:* The volume and area parameters, $R_k$ and $Q_k$, and the interaction parameters,

$a_{nm}$ (K), $b_{nm}$, $c_{nm}$ (K$^{-1}$) of the UNIFAC groups ($n, m$) for the DMF/C4 system in the models

1, 2 and 3 are respectively collected, and listed in Table 4 through Table 8. The experimental data of activity coefficients at infinite dilution at temperatures of 323.15K, 333.15K, 353.15K,

373.15K and 393.15K are listed in Table 9. By using the models 1, 2 and 3, activity coefficients at infinite dilution are calculated, which are listed in Table 9 where "Exp." denotes experimental values, and "Cal.1", "Cal.2" and "Cal.3" denote calculated values from models 1, 2 and 3, respectively. It manifests that the calculated values of models 1 and 2 correspond well with the experimental data with the average relative deviation (ARD) of 8.86% and 6.72%, respectively. However, the average relative deviation (ARD) of model 3 is up to 19.24%. But it doesn't mean that model 3 isn't more accurate than models 1 and 2 in predicting the activity coefficient at infinite dilution, although it is set up just for this purpose. Their more generalization comparison will be discussed afterwards.

Table 4

The volume and area parameters, $R_k$ and $Q_k$ of the UNIFAC groups for the DMF/C4 system in models 1 and 3

|  | $R_k$ | $Q_k$ |
|---|---|---|
| DMF | 3.0856 | 2.736 |
| CH$_2$ | 0.6744 | 0.540 |
| CH$_3$ | 0.9011 | 0.848 |
| CH$_2$ = CH | 1.3454 | 1.176 |

Table 5

The volume and area parameters, $R_k$ and $Q_k$ of the UNIFAC groups for the DMF/C4 system in model 2

|  | $R_k$ | $Q_k$ |
|---|---|---|
| DMF | 2.0000 | 2.0930 |
| CH$_2$ | 0.6325 | 0.7081 |
| CH$_3$ | 0.6325 | 1.0608 |
| CH$_2$ = CH | 1.2832 | 1.6016 |

Table 6

The interaction parameters, $a_{nm}$ (K) of the UNIFAC groups ($n,m$) for the DMF/C4 system in model 1

| $n$ | $m$ | | |
|---|---|---|---|
|  | DMF | CH$_2$ | C = C |
| DMF | 0.0 | 31.91 | -64.63 |
| CH$_2$ | 417.77 | 0.0 | 86.02 |
| C = C | 269.70 | -35.36 | 0.0 |

Table 7

The interaction parameters, $a_{nm}$ (K), $b_{nm}$, $c_{nm}$ (K$^{-1}$) of the UNIFAC groups ($n,m$) for the DMF/C4 system in model 2

| $n$ | | $m$ | | |
|---|---|---|---|---|
| | | DMF | CH$_2$ | C = C |
| $a_{nm}$ | | | | |
| | DMF | 0.0 | 151.00 | -152.20 |
| | CH$_2$ | 406.20 | 0.0 | 189.66 |
| | C = C | 388.40 | -95.418 | 0.0 |
| $b_{nm}$ | | | | |
| | DMF | 0.0 | -0.9023 | 0.0 |
| | CH$_2$ | 0.6526 | 0.0 | -0.2723 |
| | C = C | 0.0 | 0.06171 | 0.0 |
| $c_{nm}$ | | | | |
| | DMF | 0.0 | 0.0 | 0.0 |
| | CH$_2$ | 0.0 | 0.0 | 0.0 |
| | C = C | 0.0 | 0.0 | 0.0 |

Table 8

The interaction parameters, $a_{nm}$ of the UNIFAC groups ($n,m$) for the DMF/C4 system in model 3

| $n$ | $m$ | | |
|---|---|---|---|
| | DMF | CH$_2$ | C = C |
| DMF | 0.0 | 211.83 | 150.25 |
| CH$_2$ | 0.70 | 0.0 | -41.31 |
| C = C | 7.97 | 68.39 | 0.0 |

Table 9

Comparison of activity coefficients at infinite dilution among models 1, 2, and 3

| | $T/K$ | | | | |
|---|---|---|---|---|---|
| | 323.15 | 333.15 | 353.15 | 373.15 | 393.15 |
| n-butane | | | | | |
| Exp. | 8.593 | 8.061 | 7.268 | 6.590 | 6.037 |
| Cal.1 | 9.857 | 9.486 | 8.815 | 8.226 | 7.707 |
| Cal.2 | 8.451 | 7.881 | 6.992 | 6.151 | 5.520 |
| Cal.3 | 6.208 | 5.876 | 5.316 | 4.860 | 4.485 |

Table 9
Comparison of activity coefficients at infinite dilution among models 1, 2, and 3

| 1-butene | | | | | |
|----------|-------|-------|-------|-------|-------|
| Exp. | 4.495 | 4.316 | 4.068 | 3.825 | 3.619 |
| Cal.1 | 4.420 | 4.321 | 4.134 | 3.964 | 3.808 |
| Cal.2 | 4.382 | 4.271 | 4.066 | 3.881 | 3.713 |
| Cal.3 | 4.238 | 4.057 | 3.746 | 3.488 | 3.272 |
| 1,3-butadiene | | | | | |
| Exp. | 2.309 | 2.310 | 2.311 | 2.312 | 2.313 |
| Cal.1 | 2.347 | 2.319 | 2.266 | 2.215 | 2.167 |
| Cal.2 | 2.579 | 2.605 | 2.644 | 2.668 | 2.682 |
| Cal.3 | 3.136 | 3.030 | 2.845 | 2.688 | 2.556 |

Besides the UNIFAC model, there are still some other prediction models used for predicting activity coefficients at infinite dilution whenever experimental data are unavailable, e.g. MOSCED model, SPACE model and COSMO-RS model.

The MOSCED (modified separation of cohesive energy density) model is an extension of regular solution theory to mixtures that contain polar and hydrogen-bonding components [43-47]. The cohesive energy density is separated into dispersion forces, dipole forces and hydrogen bonding, with small corrections made for asymmetry. The dipolarity and hydrogen bond basicity and acidity parameters are correlated on the basis of a limited database of activity coefficients. By using the expression for the cohesive energy density and accounting for the asymmetry effect, the activity coefficient at infinite dilution for component 2 in solvent 1 is written as

$$\ln \gamma_2^{\infty} = \frac{v_2}{RT}\left[ (\lambda_1 - \lambda_2)^2 + \frac{q_1^2 q_2^2 (\tau_1 - \tau_2)^2}{\psi_1} + \frac{(\alpha_1 - \alpha_2)(\beta_1 - \beta_2)}{\xi_1} \right] + d_{12} \tag{100}$$

where $\lambda$ is a measure of a molecule's polarizability, $\tau$ represents its polarity, $\alpha$ and $\beta$ are respectively acidity and basicity parameters, $q$ is a measure of the dipole-induced dipole energy, $\psi$ and $\xi$ interpret the asymmetry effect, $v$ is molar volume, and $d_{12}$ is a Flory-Huggins term which is usually minor anyway. The outstanding characteristic of this model is that it can predict activity coefficients at infinite dilution using only pure component parameters which are available in the parameter table.

The SPACE (solvatochromic parameters for activity coefficients estimation) method, developed by Hait et al. [48, 49], uses a much larger database and recently established scales of solvent and solute dipolarity and hydrogen bonding. The SPACE equation assumes additivity and independence of the various contributions to the cohesive energy density: (1) dispersion, (2) dipolar interactions, (3) hydrogen-bonding interactions, and (4) size differences:

$$\ln \gamma_2^\infty = \frac{v_2}{RT}\left[(\lambda_1 - \lambda_2)^2 + (\tau_1 - \tau_{2eff})^2 + (\alpha_1 - \alpha_{2eff})(\beta_1 - \beta_{2eff})\right] + d_{12} \qquad (101)$$

The meanings of physical quantities in the above equation refer to Eq. (100). But the SPACE model uses effective values for solute parameters ($\tau_{2eff}$, $\alpha_{2eff}$, $\beta_{2eff}$), which are calculated by a linear interpolation of the SPACE solvent (1) and solute (2) parameters. Unfortunately, the complete SPACE parameters for all compounds studied are only provided in the supplementary material that must be ordered from the specified institution.

Table 10 gives the comparison of models 1,2, 3, as well as other liquid-phase activity coefficient models such as MOSCED and SPACE, in predicting the activity coefficient at infinite dilution from 11 alkanes in 67 solvents at 25℃.

It is shown that MOSCED and SPACE models with small average relative deviations (ARD) are more attractive among these five models. In the UNIFAC models (models 1, 2 and 3), the modified UNIFAC (model 2) shows substantial improvement, like the example given above with the most accuracy. Although the $\gamma^\infty$-based UNIFAC (model 3) is undesirable with the greatest average relative deviation of up to 20.7%, the maximum error (170.5%) may counteract the accuracy.

Table 10
Comparison of activity coefficient at infinite dilution for different models

| Parameters | Model 1 | Model 2 | Model 3 | MOSECD | SPACE |
|---|---|---|---|---|---|
| $E\%$ (error) | 20.0 | 9.8 | 20.7 | 8.8 | 8.1 |
| $N$ (data number) | 671 (91%) | 643 (87%) | 616 (84%) | 432 (59%) | 539 (73%) |
| Max. error % | 96.1 | 78.2 | 170.5 | 44.9 | 24.6 |
| Fraction with error > 20% | 40% | 21% | 41% | 7.2% | 2.6% |
| Fraction with error > 15% | 49% | 32% | 52% | 18% | 13% |

Until now, all of the liquid-phase activity coefficient models are only suitable for the components condensable and inorganic salts, and the organic ions (e.g. ionic liquids) aren't concerned. In the recent years, CSMs (Dielectric continuum solvation models) and their revised versions (i.e. COSMO, COSMO-RS) have been developed, and can extend to predict the activity coefficient in the mixtures containing ions. In particular, COSMO-RS (conductor-like screening model for real solvents) model is a novel and fruitful concept, and avoids the questionable dielectric approach [50-55]. It is capable of treating almost the entire equilibrium thermodynamics of fluid systems and should become a power alternative to fragment-based models like UNIFAC. But the reports on the actual application of this model in predicting activity coefficients aren't so many.

## 2. VAPOR-LIQUID-LIQUID PHASE EQUILIBRIUM

When components are notably dissimilar and activity coefficients are large, two and even more liquid phases may coexist at equilibrium, which then leads to the vapor-liquid-liquid phase equilibrium for two liquid phases. For example, consider the binary system of methanol (1) and cyclohexane (2) at 25℃ [1, 56], an equilibrium plot of $y_1$ (mole fraction in the vapor phase) against $x_1$ (mole fraction in the liquid phase) assuming an isotherm condition is drawn in Fig. 1. By Eq. (35),

$$y_i = \frac{\gamma_{iL} P_i^s}{P} x_i \tag{102}$$

Thus,

$$P = \sum_{i=1}^{N} x_i \gamma_{iL} P_i^s \tag{103}$$

Then, one obtains the following relation for computing $y_i$ from $x_i$:

$$y_1 = \frac{x_1 \gamma_1 P_1^s}{x_1 \gamma_1 P_1^s + x_2 \gamma_2 P_2^s} \tag{104}$$

where if one $x_1$ (or $x_2$) is given, only one $y_i$ is corresponding. But it isn't true for the inverse. As can be seen from Fig. 1, in some liquid-phase region, three values of $y_i$ exist corresponding to only one $x_1$. This indicates phase instability. Experimentally, single liquid phases can exist only for methanol-rich mixtures of $x_1 = 0.8248$ to 1.0 ($G$ - 1) and for cyclohexane-rich mixtures of $x_1 = 0.0$ to 0.1291 (0 - $A$). Since a coexisting vapor exhibits only a single composition, for instance, for line 2, two coexisting liquid phases ($B$ and $F$) prevail at opposite ends of the dashed line. Under this condition the liquid phases represent solubilities of methanol in cyclohexane and cyclohexane in methanol. However, in accordance with the lever-arm rule, one overall liquid composition for two coexisting liquid phases ($B$ and $F$) is represented by point $D$. So point $D$ is an unstable point, and points $C$ and $E$ are heterogeneous critical points. About how to judge the unstable point and calculate the composition of two coexisting liquid phases is discussed in chapter 3 (azeotropic distillation). However, as the temperature increases, the solubility generally becomes large. Therefore, at a certain temperature (such as 55℃), the coexisting liquid phases will disappear and become a homogeneous phase.

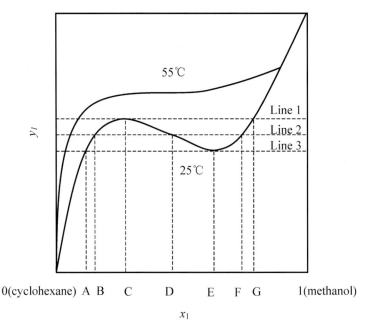

Fig. 1. Equilibrium curves for methanol (1) / cyclohexane (2) system.

## 3. SALT EFFECT

Recently, the study on solvent (or separating agent as called in special distillation processes) optimization has been very active in many unit operations of chemical engineering [57-59]. It has proved to be a very effective strategy to optimize the solvent with salt for vapor-liquid equilibrium and liquid-liquid equilibrium. Salt effect is commonly utilized in special distillation process [60-62], e.g. extractive distillation with salt as the separating agent, catalytic distillation with salt as catalyst, etc. This section tries to interpret the effect of salt on special distillation from thermodynamic viewpoints.

Nowadays there are many theories [63-66] about salt effect such as the electrostatic theory of Debye-McAulay in dilute electrolyte solutions, internal pressure theory of McDevit-Long, salt effect nature of Huangziqing, electrolyte solutions theory of Pitzer and scaled particle theory, in which the first four theories require the experimental data to correlate or make some simplification with a little accuracy or have no wide range to apply. But the scaled particle theory, which is deduced from thermodynamics and statistical physics, has defined physical meaning, and the required molecular parameters are readily available. Especially in the recent years, the study of scaled particle theory has been farther explored. Therefore it is advisable to deal with salt effect with this theory.

Herein, separation of C4 mixture by extractive distillation with DMF as an example is illustrated. The test system is composed of DMF, the salt NaSCN with weight fraction 10% in DMF, and C4 mixture. We use the subscript 1, 2, 3, 4 to represent one of the nonelectrolyte in

C4 mixture(solute), DMF(solvent), the cation $Na^+$ and the anion($SCN^{-1}$) respectively. The extension to other systems is an obvious one.

For a vapor and liquid phase in equilibrium, we know for the solute,

$$\mu_1^V = \mu_1^L \tag{105}$$

In terms of the grand canonical ensemble in statistical thermodynamics, the chemical potential of the solute in the gas phase is given by

$$\mu_1^V = kT \ln(\frac{\Lambda_1^3}{kT}) + kT \ln f_1^V \tag{106}$$

where

$\Lambda_1 = (\frac{h^2}{2\pi m_1 kT})^{1/2}$, and $f_1^V$ is the gas-phase fugacity of the solute.

The chemical potential of the solute in the liquid phase is given by

$$\mu_1^L = kT \ln \rho_1 \Lambda_1^3 + \overline{g}_1^h + \overline{g}_1^s \tag{107}$$

where $\overline{g}_1^h = \frac{\partial G_h}{\partial N_1}$, representing the free energy of introducing a hard sphere of diameter $\sigma_1$ into the solvent (electrolyte solution), and $\overline{g}_1^s = \frac{\partial G_s}{\partial N_1}$, representing the free energy needed to introduce the solute into the cavity due to the soft part of the chemical potential.

Putting Eqs. (106) and (107) in Eq. (105) and rearranging,

$$\ln(\frac{f_1^V}{\rho_1}) = \frac{\overline{g}_1^h}{kT} + \frac{\overline{g}_1^s}{kT} + \ln kT \tag{108}$$

The mole fraction of the solute gas in the solution is:

$$x_1 = \frac{\rho_1}{\sum_j \rho_j} \tag{109}$$

Therefore, Eq. (108) is written as

$$\ln(\frac{f_1^G}{x_1}) = \frac{\overline{g}_1^h}{kT} + \frac{\overline{g}_1^s}{kT} + \ln(kT\sum_j \rho_j) \tag{110}$$

At low pressure the fugacity in Eq. (110) can be replaced by partial pressure. According to Henry equation, the gas solubility at low solute concentration is described by the following equation:

$$P_1 = H_1 x_1 = H_1' c_1 \tag{111}$$

Combining Eqs. (110) and (111):

$$\ln H_1 = \frac{\overline{g}_1^{-h}}{kT} + \frac{\overline{g}_1^{-s}}{kT} + \ln(kT\sum_j \rho_j) \tag{112}$$

The solubility of a nonelectrolyte in a salt solution with low salt concentration is given by the Setschenow equation

$$\log\frac{c_0}{c} = k_s c_s \tag{113}$$

where $c_0$ is the solubility in pure solvent, $c$ is the solubility in a salt solution of concentration $c_s$, and $k_s$ is the salting coefficient which has a characteristic value for a given salt-nonelectrolyte pair. A positive value of $k_s$ corresponds to salting-out ($c_0 > c$), a negative value of $k_s$ salting-in ($c_0 < c$).

Differentiating Eq. (113) with respect to $c$, we can write in an ordinary expression:

$$\lim_{c_s \to 0} \log\frac{c_0}{c} = k_s c_s \tag{114}$$

Combining Eqs. (111), (112) and (114), the following equation is derived:

$$-\left(\frac{\partial \log c}{\partial c_s}\right) = k_s = \left[\frac{\partial(\overline{g}_1^h/2.3kT)}{\partial c_s}\right]_{c_s \to 0} + \left[\frac{\partial(\overline{g}_1^s/2.3kT)}{\partial c_s}\right]_{c_s \to 0} + \left[\partial(\ln\sum_{j=1}^m \rho_j)/\partial c_s\right]_{c_s \to 0}$$

$$= k_\alpha + k_\beta + k_\gamma \tag{115}$$

where $\overline{g}_1^h$ is the free energy change when a cavity large enough to hold the nonelectrolyte molecule is formed in the solution, $\overline{g}_1^s$ is the free energy change when the nonelectrolyte is introduced into the cavity, and $\rho_j$ is the number density of a solution species. $k_\alpha$, $k_\beta$ and $k_\gamma$ represent the contributions to the salting coefficient of each of the three terms on the right-hand side of Eq. (115). The problem now becomes one of deriving general expression for $k_\alpha, k_\beta$, and $k_\gamma$ in terms of parameters characteristic of the nonelectrolyte and the ions of the salts. We start with the expression for $k_\gamma$, since it is the easiest to deduce.

(1) Expression for $k_\gamma$

The total number density $\sum\rho_i$ is given by

$$\sum \rho_i = \rho_1 + \rho_2 + \rho_3 + \rho_4 \tag{116}$$

Because C4 mixture are at the infinite dilution, $\rho_1$ can be cancelled from the summation and we can write

$$\sum \rho_i = \rho_2 + \rho_3 + \rho_4 \tag{117}$$

For a 1:1 electrolyte such as $Na^+SCN^-$

$$\rho_3 = \rho_4 = \frac{N_0 c_s}{1000} \tag{118}$$

Applying the definition for the apparent molar volume, $\phi$, to DMF solution, it is readily shown that the variation of the number density of DMF, $\rho_2$, with concentration is given by the equation

$$\rho_2 = \frac{N_0 d_2}{M_2}(1 - \frac{c_s \phi}{1000}) \tag{119}$$

Substituting for $\rho_2$, $\rho_3$, and $\rho_4$ in Eq. (117), we obtain, after rearranging

$$\frac{1}{2.3} \ln \sum_{i=1}^{4} \rho_i = \frac{1}{2.3} \ln[\frac{N_0 d_2}{M_2}(1 - \frac{c_s \phi}{1000} + \frac{2c_s M_2}{1000 d_2})] \tag{120}$$

At low concentration, as $c_s \rightarrow 0$,

$$k_\gamma = [\frac{\partial \ln \sum \rho_i}{2.3 \partial c_s}]_{c_s \rightarrow 0} = \frac{2M_2}{2300 d_2} - \frac{\phi}{2300} \tag{121}$$

The values of molecular weight $M_2$, liquid density $d_2$ and apparent mole volume of salt $\phi$ are adapted from the references [67, 68], and are 73.10 g $mol^{-1}$, 0.944 g $mol^{-1}$, 34.49 ml $mol^{-1}$ for DMF, respectively. Substitute for these values in Eq. (121), and $k_\gamma$ is written as

$$k_\gamma = 0.0673 - 4.34 \times 10^{-4} \phi \tag{122}$$

(2). Expression for $k_\beta$

Shoor and Gubbins [64] derive the following expression for the interaction energy between a nonelectrolyte molecule and the solution species forming the cavity:

$$\frac{\overline{g}_1^s}{2.3kT} = -\frac{32\pi}{9000(2.3kT)}(\varepsilon_{13}^*\sigma_{13}^3 + \varepsilon_{14}^*\sigma_{14}^3) - \frac{4\pi N_0 d_2}{3M_2(2.3kT)}(1 - \frac{c_s\phi}{1000})(\frac{8\varepsilon_{12}^*\sigma_{12}^3}{3} + \frac{\mu_2^2\alpha_1}{\sigma_{12}^3}) \qquad (123)$$

Taking the derivative with respect to $c_s$ as $c_s \to 0$, and inserting numbers for the various constants ($T = 323.15\,K$, the dipole moment of DMF $\mu_2 = 3.86 \times 10^{-18}$ esu cm), we can write

$$k_\beta = [\frac{\partial(\overline{g}_1^s / 2.3kT)}{\partial c_s}]_{c_s \to 0}$$

$$= -\frac{9.05 \times 10^{18}}{k}(\varepsilon_{13}^*\sigma_{13}^3 + \varepsilon_{14}^*\sigma_{14}^3) + 1.168 \times 10^{17}\frac{\phi\varepsilon_{12}^*\sigma_{12}^3}{k} + 4.727 \times 10^{-3}\frac{\phi\alpha_1}{\sigma_{12}^3} \qquad (124)$$

To assign a numerical value to $k_\beta$, the mixture parameters $\varepsilon_{ij}^*$ and $\sigma_{ij}^*$ must be related to those of the pure species. We use the following mixing rules:

$$\varepsilon_{1j}^* = (\varepsilon_{11}^*\varepsilon_{jj}^*)^{1/2} \qquad (125)$$

$$\sigma_{1j} = \frac{1}{2}(\sigma_1 + \sigma_j) \qquad (126)$$

Values of energy parameters $\varepsilon_{jj}^*$ for this system are evaluated from the famous Mavroyannis-Stephen equation:

$$\varepsilon_{jj}^* = 3.146 \times 10^{-24}\frac{\alpha_j^{3/2}z_j^{1/2}}{\sigma_j^6} \qquad (127)$$

where $z_j$ is electron number of species $j$, $\sigma_j$ is its diameter (Å) and $\alpha_j$ is its polarizability (Å$^3$). Molecular polarizability $\alpha$ is obtained by Langevin-Debye equation:

$$\left(\frac{D-1}{D+2}\right)\frac{M}{d} = \frac{4\pi N_0}{3}\alpha \qquad (128)$$

where $D$ is dielectric constant (esu cm, "esu" is Electro Static Unit), $M$ is molecular weight (g mol$^{-1}$), $d$ is liquid density (g ml$^{-1}$) and $N_0$ is Avogardro's number, $6.023 \times 10^{23}$.

The data necessary for solving Eq. (125) through Eq. (128) are listed in Table 11, in which $v$ is molar volume(ml mol$^{-1}$).

Making these substitutions in Eq. (124), we arrive at a final expression for $k_\beta$:

$$k_\beta = -1.707 \times 10^{14} (\frac{\overset{*}{\varepsilon}_1}{k})^{1/2} \left[ \alpha_3^{3/4} z_3^{1/4} \frac{(\sigma_1 + \sigma_3)^3}{\sigma_3^3} + \alpha_4^{3/4} z_4^{1/4} \frac{(\sigma_1 + \sigma_4)^3}{\sigma_4^3} \right]$$

$$+ \frac{1}{8} \times 1.168 \times 10^{17} (\frac{\overset{*}{\varepsilon}_2}{k})^{1/2} (\frac{\overset{*}{\varepsilon}_1}{k})^{1/2} \phi(\sigma_1 + \sigma_2)^3 + 3.78 \times 10^{-2} \frac{\phi \alpha_1}{(\sigma_1 + \sigma_2)^3} \qquad (129)$$

Table 11
Data for calculating salting coefficients

| | $v$ | $\sigma$ | $\alpha$ | Electron Num | $\overset{*}{\varepsilon}/k$ |
|---|---|---|---|---|---|
| DMF | 79.0 | 5.16 | 79.1 | 40 | 169.8 |
| n-Butane | 101.4 | 5.52 | 81.8 | 34 | 108.8 |
| n-Butene | 95.3 | 5.42 | 85.8 | 32 | 127.3 |
| Trans-2-butene | 93.8 | 5.40 | 88.5 | 32 | 135.0 |
| Cis-2-butene | 91.2 | 5.36 | 85.8 | 32 | 134.4 |
| Butadiene-1,3 | 88.0 | 5.31 | 89.1 | 30 | 148.0 |
| $Na^+$ | - | 1.90 | 0.41 | - | - |
| $SCN^-$ | - | 4.26 | 2.30 | - | - |

(3) Expression for $k_\alpha$

The term for the free energy of cavity formation, according to scaled particle theory, is written in the form

$$\frac{\overline{g}_1^h}{kT} = -\ln(1 - y_3) + \frac{3y_2\sigma_1}{(1 - y_3)}[1 + \frac{y_1\sigma_1}{y_2} + \frac{3y_2\sigma_1}{2(1 - y_3)}] \qquad (130)$$

Neglecting the terms involving $\rho_1$, substituting for $\rho_2$, $\rho_3$, and $\rho_4$ from Eqs. (118) and (119), we arrive at the following expressions for $y_1$, $y_2$ and $y_3$:

$$y_1 = \frac{\pi}{6}[\frac{N_0 d_2}{M_2}(1 - \frac{c_s\phi}{1000})\sigma_2 + \frac{N_0 c_s}{1000}(\sigma_3 + \sigma_4)]$$

$$= 2.10 \times 10^{14} + c_s[3.15 \times 10^{20}(\sigma_3 + \sigma_4) - 2.10 \times 10^{11}\phi] \qquad (131)$$

$$y_2 = \frac{\pi}{6}[\frac{N_0 d_2}{M_2}(1 - \frac{c_s\phi}{1000})\sigma_2^2 + \frac{N_0 c_s}{1000}(\sigma_3^2 + \sigma_4^2)]$$

$$= 1.08 \times 10^7 + c_s[3.15 \times 10^{20}(\sigma_3^2 + \sigma_4^2) - 1.08 \times 10^4\phi] \qquad (132)$$

$$y_3 = \frac{\pi}{6}[\frac{N_0 d_2}{M_2}(1 - \frac{c_s\phi}{1000})\sigma_2^3 + \frac{N_0 c_s}{1000}(\sigma_3^3 + \sigma_4^3)]$$

$$=0.557+c_s[3.15\times10^{20}(\sigma_3^3+\sigma_4^3)-5.57\times10^{-4}\phi]$$ (133)

In order to determine $k_\alpha$, it is necessary to get an expression for the quantity $(1-y_3)$, which appears in Eq. (130). It is readily shown from Eq. (133) that

$$1-y_3=0.443\{1-c_s[7.11\times10^{20}(\sigma_3^3+\sigma_4^3)-12.57\times10^{-4}\phi]\}$$ (134)

We know that when $x\to0$, $\ln(1+x)$ is able to be extended and approximately equal

$x$, that is, $\ln(1+x)=x$ as $c_s\to0$:

$$\ln(1-y_3)=\ln0.443-c_s[7.11\times10^{20}(\sigma_3^3+\sigma_4^3)-12.57\times10^{-4}\phi]$$ (135)

In addition, as $x\to0$, $\dfrac{1}{1+x}\approx1-x$. Thus,

$$\frac{y_1}{y_2}=\frac{2.10\times10^{14}+c_s[3.15\times10^{20}(\sigma_3+\sigma_4)-2.10\times10^{11}\phi]}{1.08\times10^{7}+c_s[3.15\times10^{20}(\sigma_3^2+\sigma_4^2)-1.08\times10^{4}\phi]}$$

$$=1.94\times10^{7}+c_s[2.92\times10^{13}(\sigma_3+\sigma_4)-5.67\times10^{20}(\sigma_3^2+\sigma_4^2)]$$ (136)

$$\frac{y_2}{1-y_3}=2.44\times10^{7}+c_s[7.11\times10^{20}(\sigma_3^2+\sigma_4^2)+1.73\times10^{28}(\sigma_3^3+\sigma_4^3)-5.50\times10^{4}\phi]$$ (137)

$$1+\frac{y_1\sigma_1}{y_2}+\frac{3y_2\sigma_1}{2(1-y_3)}=1+5.60\times10^{7}\sigma_1+c_s\sigma_1[2.92\times10^{13}(\sigma_3+\sigma_4)+5.0\times10^{20}(\sigma_3^2+\sigma_4^2)$$
$$+2.60\times10^{28}(\sigma_3^3+\sigma_4^3)-8.25\times10^{4}\phi]$$ (138)

Obtain the expressions of Eq. (131) through Eq. (138), and we can write

$$\frac{\overline{g}_1^{-h}}{2.3kT}=0.3540+c_s[3.09\times10^{20}(\sigma_3^3+\sigma_4^3)-5.47\times10^{-4}\phi]+3.18\times10^{7}\sigma_1+1.78\times10^{15}\sigma_1^2$$

$$+c_s\sigma_1[9.27\times10^{20}(\sigma_3^2+\sigma_4^2)+2.26\times10^{28}(\sigma_3^3+\sigma_4^3)-7.17\times10^{4}\phi]+$$

$$c_s\sigma_1^2[9.29\times10^{20}(\sigma_3+\sigma_4)+6.78\times10^{28}(\sigma_3^2+\sigma_4^2)+2.09\times10^{36}(\sigma_3^3+\sigma_4^3)-6.64\times10^{12}\phi]$$ (139)

By virtue of the following equation:

$$k_\alpha=[\frac{\partial(\overline{g}_1^h/2.3kT)}{\partial c_s}]_{c_s\to0}$$ (140)

we arrive at the following expression for $k_\alpha$:

$$k_\alpha = 3.09 \times 10^{20} (\sigma_3^3 + \sigma_4^3) - 5.47 \times 10^{-4} \phi +$$

$$\sigma_1 \left[ 9.27 \times 10^{20} (\sigma_3^2 + \sigma_4^2) + 2.26 \times 10^{28} (\sigma_3^3 + \sigma_4^3) - 7.17 \times 10^4 \phi \right] +$$

$$\sigma_1^2 \left[ 9.27 \times 10^{20} (\sigma_3 + \sigma_4) + 6.78 \times 10^{28} (\sigma_3^2 + \sigma_4^2) + 2.09 \times 10^{36} (\sigma_3^3 + \sigma_4^3) - 6.64 \times 10^{12} \phi \right]$$

(141)

The salt coefficient $k_s$ can be calculated by adding $k_\gamma$ (Eq. (122)), $k_\beta$ (Eq. (129))

and $k_\alpha$ (Eq. (141)). In order to find $k_s$ for a particular system, it is necessary to know: (1)

the apparent molar volume, $\phi$, of the salt at infinite dilution; (2) the diameters ($\sigma_3$, $\sigma_4$) and

polarizabilities ($\alpha_3$, $\alpha_4$) of cation and anion; (3) the diameters ($\sigma_1$, $\sigma_2$), polarizabilities ($\alpha_1$,

$\alpha_2$) of the nonelectrolyte and solvent molecules.

*Example:* In terms of scaled particle theory, deduce the three terms, $k_\gamma$, $k_\beta$ and $k_\alpha$, of

salting coefficient $k_s$ of the concerned system composed of one nonelectrolyte as the solute

(1), water as the solvent (2) and a 1:1 type electrolyte of the cation (3) and the anion (4) at

low salt concentration ($T = 298.15$ K, the dipole moment of water $\mu_2 = 1.84 \times 10^{-18}$ esu cm).

*Solution:* According to the same way as the C4/DMF/NaSCN system, the three terms, $k_\gamma$,

$k_\beta$ and $k_\alpha$ can be deduced (see the results of Eq. (142) through Eq. (144)). Herein, the

procedures are omitted, but we encourage the readers to try deducing them.

$$k_\gamma = 0.016 - 4.34 \times 10^{-4} \phi$$

(142)

$$k_\beta = -1.85 \times 10^{14} \left( \frac{\varepsilon_1^*}{k} \right)^{1/2} \left[ \alpha_3^{3/4} z_3^{1/4} \frac{(\sigma_1 + \sigma_3)^3}{\sigma_3^3} + \alpha_4^{3/4} z_4^{1/4} \frac{(\sigma_1 + \sigma_4)^3}{\sigma_4^3} \right]$$

(143)

$$+ 6.26 \times 10^{17} \phi \left( \frac{\varepsilon_1^*}{k} \right)^{1/2} (\sigma_1 + \sigma_2)^3 + 4.00 \times 10^{-2} \frac{\phi \alpha_1}{(\sigma_1 + \sigma_2)^3}$$

$$k_\alpha = 2.15 \times 10^{20} (\sigma_3^3 + \sigma_4^3) - 2.47 \times 10^{-4} \phi +$$

$$\sigma_1 \left[ 6.45 \times 10^{20} (\sigma_3^2 + \sigma_4^2) + 1.34 \times 10^{28} (\sigma_3^3 + \sigma_4^3) - 4.23 \times 10^4 \phi \right] +$$

$$\sigma_1^2 \left[ 6.45 \times 10^{20} (\sigma_3 + \sigma_4) + 4.01 \times 10^{28} (\sigma_3^2 + \sigma_4^2) + 1.32 \times 10^{36} (\sigma_3^3 + \sigma_4^3) - 4.17 \times 10^{12} \phi \right]$$

(144)

## 4. NONEQUILIBRIUM THERMODYNAMIC ANALYSIS

Relative to equilibrium thermodynamics, nonequilibrium thermodynamics is less studied by researchers [54-58]. Especially, its application thereof in special distillation processes is a new thing. The aim of nonequilibrium thermodynamics analysis for special distillation processes is to interpret in what condition and/or why the process can take place, so as to substantially understand the essence and find the theoretical judging rule. At the same time, the mathematic model of entropy generation can be established for the guidance of energy saving, so as to realize decreasing entropy generation. As we know, in the actual operation, trays in the distillation column rarely, if ever, operate at equilibrium. That is to say, nonequilibrium thermodynamics in the most degrees reflects the real state. In the phase space, nonequilibrium thermodynamics is divided into the linear and non-linear ranges. The linear range is commonly studied because in this case the linear driving force equation is satisfied. But at some time for such special distillation processes as azeotropic distillation, adsorption distillation and extractive distillation, the separation ability of the separating agent is so weak as to make tray efficiency low due to the large liquid load on the tray. Undoubtedly, not only the vapor and liquid phases can't reach equilibrium, but also the actual composition is far away from the equilibrium composition. So under this condition the system isn't in the linear range of nonequilibrium thermodynamics, and the linear driving force equation isn't satisfied.

Here reactive extractive distillation with tray column is selected as an example. The results obtained may be applied to common extractive distillation. Moreover, it can deduce the similar results of nonequilibrium thermodynamics analysis for other special distillation processes in the same way.

In the reactive extractive distillation, it is assumed that there are four components $A$, $B$, $S$ and $E$ concerned, in which $A$ (the heavy component) and $B$ (the light component) are the components to be separated, $S$ is the separating agent, and $E$ is the resultant formed by the following reversible reaction and has a high boiling point:

$$A \ + \ S \ \rightleftharpoons \ E$$

According to the principle of nonequilibrium thermodynamics, entropy generation ratio per volume $\sigma$ can be written as

$$\sigma = J_q \nabla(\frac{1}{T}) + \sum_{i=1}^{4} J_i [-\nabla(\frac{\mu_i}{T}) + \frac{M_i F_i}{T}] - \frac{1}{T} \Pi \bullet \nabla u + \sum_i \frac{R_i}{T} \omega_i \qquad (145)$$

where the right terms mean the contributions due to the influences of heat conduction, diffusion (free diffusion and forced diffusion), viscosity flow, chemical reaction, respectively. Each term is composed of two factors: one is related with the irreversible ratio; another is related with the driving force resulting in the corresponding "flow".

In most cases, there is no apparent pressure gradient, temperature gradient and velocity gradient so that the three terms in the right side of Eq. (145) can be neglected. On the other hand, chemical reaction rate between $A$ and $S$ is generally quick and chemical reaction equilibrium can arrive in a very short time. Therefore, the term, $\sum_i \frac{R_i}{T}\omega_i$, can also be neglected. Thus, Eq. (145) is simplified as

$$\sigma = \sum_{i=1}^{4} J_i \nabla(-\frac{\mu_i}{T}) \tag{146}$$

where $\nabla\mu_i$ is the chemical potential gradient between the liquid bulk and vapor bulk.

In the reactive extractive distillation, the separating agent $S$ is generally high-efficiency, and can apparently improve the relative volatility of $B$ to $A$. Accordingly, the amount of the separating agent $S$ used in the distillation column is greatly reduced, which leads to a high tray efficiency. So it is reasonable to assume that the system concerned is in the linear range of nonequilibrium thermodynamics, and the linear driving force equation is satisfied. The mass transfer rate $J_i$ of the components $A$, $B$, $S$ and $P$ can be expressed as

$$J_A = L_{11}\nabla(-\frac{\mu_A}{T}) + L_{12}\nabla(-\frac{\mu_B}{T}) + L_{13}\nabla(-\frac{\mu_S}{T}) + L_{14}\nabla(-\frac{\mu_E}{T}) \tag{147}$$

$$J_B = L_{21}\nabla(-\frac{\mu_A}{T}) + L_{22}\nabla(-\frac{\mu_B}{T}) + L_{23}\nabla(-\frac{\mu_S}{T}) + L_{24}\nabla(-\frac{\mu_E}{T}) \tag{148}$$

$$J_S = L_{31}\nabla(-\frac{\mu_A}{T}) + L_{32}\nabla(-\frac{\mu_B}{T}) + L_{33}\nabla(-\frac{\mu_S}{T}) + L_{34}\nabla(-\frac{\mu_E}{T}) \tag{149}$$

$$J_P = L_{41}\nabla(-\frac{\mu_A}{T}) + L_{42}\nabla(-\frac{\mu_B}{T}) + L_{43}\nabla(-\frac{\mu_S}{T}) + L_{44}\nabla(-\frac{\mu_E}{T}) \tag{150}$$

By virtue of the characteristics of extractive distillation that the boiling point of component $S$ is generally far greater than that of component $A$ or $B$, the concentration of $S$ in vapor phase is minute and close to zero. On the other hand, $E$ is a resultant and thus the composition in vapor phase is supposed to be zero. So it may be thought that

$$J_S = 0 , \ J_E = 0 \tag{151}$$

When the system reaches the steady-state of nonequilibrium thermodynamics, one obtains:

$$J_A = C_1 , \ J_B = C_2 , \ J_S = 0 , \ J_E = 0 \tag{152}$$

According to the force balance equation $\sum_i J_i = 0$,

$$C_1 = -C_2 = C \tag{153}$$

where $C_1$, $C_2$ and $C$ are constant at steady-state.

When Eq. (152) is incorporated into Eqs. (147)-(150), the linear algebraic equation group describing the relationship of diffusion force and diffusion coefficient is:

$$L_{11}\nabla(-\frac{\mu_A}{T}) + L_{12}\nabla(-\frac{\mu_B}{T}) + L_{13}\nabla(-\frac{\mu_S}{T}) + L_{14}\nabla(-\frac{\mu_E}{T}) = C \tag{154}$$

$$L_{21}\nabla(-\frac{\mu_A}{T}) + L_{22}\nabla(-\frac{\mu_B}{T}) + L_{23}\nabla(-\frac{\mu_S}{T}) + L_{24}\nabla(-\frac{\mu_E}{T}) = -C \tag{155}$$

$$L_{31}\nabla(-\frac{\mu_A}{T}) + L_{32}\nabla(-\frac{\mu_B}{T}) + L_{33}\nabla(-\frac{\mu_S}{T}) + L_{34}\nabla(-\frac{\mu_E}{T}) = 0 \tag{156}$$

$$L_{41}\nabla(-\frac{\mu_A}{T}) + L_{42}\nabla(-\frac{\mu_B}{T}) + L_{43}\nabla(-\frac{\mu_S}{T}) + L_{44}\nabla(-\frac{\mu_E}{T}) = 0 \tag{157}$$

For the above linear algebraic equation group, the main matrix $D$ is:

$$D = \begin{vmatrix} L_{11} & L_{12} & L_{13} & L_{14} \\ L_{21} & L_{22} & L_{23} & L_{24} \\ L_{31} & L_{32} & L_{33} & L_{34} \\ L_{41} & L_{42} & L_{43} & L_{44} \end{vmatrix} \tag{158}$$

Thus, the chemical potential gradients for the components $A$, $B$, $S$ and $P$ are, respectively,

$$\nabla(-\frac{\mu_A}{T}) = \begin{vmatrix} C & L_{12} & L_{13} & L_{14} \\ -C & L_{22} & L_{23} & L_{24} \\ 0 & L_{32} & L_{33} & L_{34} \\ 0 & L_{42} & L_{43} & L_{44} \end{vmatrix} \bullet \frac{1}{D} \tag{159}$$

$$\nabla(-\frac{\mu_B}{T}) = \begin{vmatrix} L_{11} & C & L_{13} & L_{14} \\ L_{21} & -C & L_{23} & L_{24} \\ L_{31} & 0 & L_{33} & L_{34} \\ L_{41} & 0 & L_{43} & L_{44} \end{vmatrix} \bullet \frac{1}{D} \tag{160}$$

$$\nabla(-\frac{\mu_S}{T}) = \begin{vmatrix} L_{11} & L_{12} & C & L_{14} \\ L_{21} & L_{22} & -C & L_{24} \\ L_{31} & L_{32} & 0 & L_{34} \\ L_{41} & L_{42} & 0 & L_{44} \end{vmatrix} \bullet \frac{1}{D} \tag{161}$$

$$\nabla(-\frac{\mu_E}{T}) = \begin{vmatrix} L_{11} & L_{12} & L_{13} & C \\ L_{21} & L_{22} & L_{23} & -C \\ L_{31} & L_{32} & L_{33} & 0 \\ L_{41} & L_{42} & L_{43} & 0 \end{vmatrix} \bullet \frac{1}{D} \tag{162}$$

As a result, two requirements should be met in order to implement reactive extractive distillation process:

(1) In terms of the constraint for entropy generation in nonequilibrium thermodynamics, the entropy generation rate $\sigma$ is permanently positive. That is to say,

$$\sigma = \sum_j J_j X_j \geq 0 \qquad (163)$$

where $J_j = \sum_j \left( \frac{\partial J_j}{\partial X_j} \right)_0 X_j = \sum_j L_{ij} X_j$ $\qquad (164)$

In Eq. (164), $X_j$ is the thermodynamic force. It indicates that all the main sub-matrixes are always positive, or the self-coefficient $L_{ii}$ is positive while the inter-coefficient $L_{ij}$ may be either positive or negative.

(2) In terms of the separation principle of partial vaporization and partial condensation, component $A$ (as the heavy component) should be transferred, as a whole, from vapor phase to liquid phase in the distillation process. On the contrary, it should be from liquid phase to vapor phase for component $B$ (as the heavy component). So, if it is assumed that the positive direction is from vapor phase to liquid phase, then

$$J_A = C_1 = C > 0 \qquad (165)$$

The first requirement ensures that the separation process can occur, and the second requirement goes a further step to ensure that the direction of mass transfer is from vapor phase to liquid phase for component $A$ and from liquid phase to vapor phase for component $B$. Therefore, this separation process is able to be put into practice. For the common extractive distillation, the above analysis will certainly be applicable, but, of course, only three components, i.e. $A$, $B$ and $S$, are involved and there is no component $E$ formed by reaction.

The mathematic model of entropy generation in the common distillation process has been reported by Liu et al. [74], in which, however, it is only suitable for the vapor phase and not extended to the whole of the vapor and liquid phases on the tray. Moreover, it isn't suitable for special distillation processes.

For reactive extractive distillation, by using Eqs. (146) - (152), entropy generation is given by

$$\sigma = J_A \nabla \left[ -\frac{(\mu_A - \mu_B)}{T} \right] \qquad (166)$$

By using the linear driving force equation of nonequilibrium thermodynamics, one obtains:

$$J_A = L_{A-B} \nabla \left[ -\frac{(\mu_A - \mu_B)}{T} \right] = L_{A-B} \nabla (-\frac{\mu_{A-B}}{T}) \qquad (167)$$

where $L_{A-B}$ is the inter-coefficient at the driving force of the chemical potential difference between components $A$ and $B$. It is evident that $L_{A-B} > 0$.

The Fick's diffusion coefficient $D_{A-B}$ is written in another form by analogue with Eq. (167), as well as applying the Onsager inversion and Fick's diffusion laws,

$$D_{A-B} = -\frac{L_{A-B}}{T}\left(\frac{\partial \mu_{A-B}}{\partial c_{A-B}}\right)$$

(168)

By combining Eqs. (166) and (167), one obtains:

$$\sigma = \frac{J_A^2}{L_{A-B}}$$

(169)

Integrating Eq. (169) over the range of the whole tray volume, entropy generation $P$ on a tray is:

$$P = \int_V \sigma dV = \int_{V_{int\,erface}} \frac{J_A^2}{L_{A-B}} dV + \int_{V - V_{int\,erface}} \frac{J_A^2}{L_{A-B}} dV$$

(170)

where the interfacial volume $V_{int\,erface} = A \bullet \Delta x$, $A$ is the vapor-liquid phase contact area(i.e. bubble area), $\Delta x$ is the total height of vapor and liquid films and calculated by $\Delta x = D/k_c$, and $k_c$ is the total liquid-phase mass transfer coefficient. In terms of the assumption of thermodynamic interface, the energy consumption is mainly concentrated on the vapor and liquid films. So the second term in the right side of Eq. (170) can be omitted, and we have:

$$P = A\int_0^{\Delta x} \frac{J_A^2}{L_{A-B}} dx$$

(171)

Ignoring temperature gradient on the tray and substituting for $J_A$ in Eq. (167),

$$P = A\int_0^{\Delta x} L_{A-B} \frac{(\nabla \mu_{A-B})^2}{T^2} dx$$

(172)

At the steady-state of nonequilibrium thermodynamics, inter-coefficient $L_{A-B}$ and the corresponding chemical potential gradient remain constant, that is, $\nabla \mu_{A-B} = \frac{\Delta \mu_{A-B}}{\Delta x}$

and $\nabla c = \frac{\Delta c_{A-B}}{\Delta x}$. By taking Eq. (168) into Eq. (172), and integrating Eq. (172),

$$P = \frac{A}{L_{A-B}} D_{A-B}^2 \bullet (\Delta c_{A-B})^2 \bullet \Delta x = \frac{V}{L_{A-B}} D_{A-B}^2 \bullet (\Delta c_{A-B})^2 \qquad (173)$$

where $\Delta c_{A-B} = (c_A^V - c_B^V) - (c_A^* - c_B^*)$. $c_A^*$ and $c_B^*$ are mole concentration equilibrium with components $A$ and $B$ in the vapor phase respectively. From Eq. (173), entropy generation rate per unit interfacial volume $\sigma_V$ is obtained:

$$\sigma_V = \frac{P}{V} = \frac{D_{A-B}^2 \bullet (\Delta c_{A-B})^2}{L_{A-B}} \qquad (174)$$

It is concluded from Eq. (173) that
(1) Entropy generation $P$ at nonequilibrium state is always greater than zero. When the system reaches equilibrium, $P = 0$ because $\Delta c_{A-B} = 0$.

(2) At the steady-state of nonequilibrium thermodynamics, $\dfrac{dP}{dt} = 0$ due to constant values of $V$, $L_{A-B}$, $D_{A-B}$ and $\Delta c_{A-B}$.

At equilibrium state $P = 0$ and at the linear range of nonequilibrium thermodynamics, the nonequilibrium steady-state is stable and doesn't spontaneously form the regular structure in time and space. The results obtained from this text are consistent with those from classic nonequilibrium thermodynamics, which means that the deducing results are reasonable. Evidently, the above analysis is also suitable for common extractive distillation with tray column. Of course, in this case, only three components of $A$, $B$ and $P$ excluding the new component $S$ formed by reaction, are concerned.

Furthermore, the results can also be extended to other special distillation processes such as adsorption distillation, catalytic distillation, etc. Let us see the following example of nonequilibrium thermodynamic analysis for adsorption distillation.

*Example:* In the adsorption distillation column, the tiny solid particles are used as the adsorption agent, and blended with liquid phase on the tray. There exists four components: $M$ (the heavy key component), $N$ (the light key component), $P$ (the carrier for carrying $S$) and $S$ (adsorption agent) involved [75]. The following reversible chemical reaction may take place on the tray,

$$M + S \rightleftarrows M^S$$
$$N + S \rightleftarrows N^S$$

It is required to set up the condition that the separation process can occur from the viewpoint of nonequilibrium thermodynamics.

*Solution:* The similar procedure as reactive extractive distillation can be deduced for this situation, and the results similar as Eqs. (163)- (165) and Eq. (173) will be obtained. The interested reader can try to do it.

## 5. MULTI-COMPONENT MASS TRANSFER

The most difference between equilibrium thermodynamics and nonequilibrium thermodynamics is that mass transfer (as well as heat transfer) should be considered in the latter. In the linear range of nonequilibrium thermodynamics, for binary mixtures or for diffusion of dilute species $i$ in a $c$-component mixture, Fick's law of diffusion postulates a linear dependence of the flux $J_i$ with respect to the molar average mixture velocity $u$ and its composition gradient $\nabla x_i$ :

$$J_i = C_i(u_i - u) = -C_t D_i \nabla x_i \tag{175}$$

$$u = \frac{\sum_{i=1}^{c} C_i u_i}{C_t} = \frac{\sum_{i=1}^{c} C_i u_i}{\sum_{i=1}^{c} C_i} \tag{176}$$

The molar flux $N_i$ with respect to a laboratory-fixed coordinate reference frame is given by

$$N_i = C_i u_i = C_t x_i u_i = J_i + x_i N_t = -C_t D_i \nabla x_i + x_i N_t \tag{177}$$

$$N_t = \sum_{i=1}^{c} N_i \tag{178}$$

If one takes the review that Eq. (175) provides a definition of the effective Fick diffusivity $D_{\mathit{eff}}$ of component $i$ in a $c$-component mixture, then this parameter shows a complicated, often unpredictable behaviour because the fundamental driving force for diffusion is the gradient of chemical potential rather than mole fraction or concentration gradient.

For multi-component mass transfer, it is common to use the generalized Maxwell-Stefan equation [76-80], which is divided into implicit and explicit forms. The implicit formulation of Maxwell-Stefan equation is related to the chemical potential gradients, and the molar transfer rate $N_i^L$ in the liquid phase and $N_i^V$ in the vapor phase are, respectively,

$$\frac{x_i^L}{RT^L} \frac{\partial \mu_i^L}{\partial \eta} = \sum_{k=1}^{c} \frac{x_i^L N_k^L - x_k^L N_i^L}{C_t^L \kappa_{i,k}^L a} \tag{179}$$

and

$$\frac{y_i^V}{RT^V} \frac{\partial \mu_i^V}{\partial \eta} = \sum_{k=1}^{c} \frac{x_i^V N_k^V - x_k^V N_i^V}{C_t^V \kappa_{i,k}^V a} \tag{180}$$

where $a$ is the effective interfacial area per unit volume (m$^2$ m$^{-3}$).

The $\kappa_{i,k}^L$ and $\kappa_{i,k}^V$ represent the mass transfer coefficients of the $i-k$ pair in the liquid and vapor phases, and how to calculate them will be discussed in what follows. Undoubtedly, it is very difficult to directly obtain $N_i^L$ and $N_i^V$ from Eqs. (179) and (180).

However, the explicit formulation of Maxwell-Stefan equation is more preferred to use. Moreover, for the convenience, $J_i^L$, $J_i^V$, $N_i^L$ and $N_i^V$ are determined as mass transfer rates instead of mass transfer fluxes. For the liquid phase, by using $(c-1)\times(c-1)$ matrix notation,

$$[k_{ik}^L a] = [B_{ik}^L]^{-1} a[\Gamma_{ik}^L] \quad (i, k = 1, 2, \ldots, c\text{-}1.) \tag{181}$$

and the elements of the matrix $[B_{ik}^L]$ are:

$$B_{ii}^L = \frac{x_i}{\kappa_{ic}^L} + \sum_{\substack{k=1 \\ k \neq i}}^{c} \frac{x_k}{\kappa_{ik}^L} \tag{182}$$

$$B_{ik(i \neq k)}^L = -x_i \left( \frac{1}{\kappa_{ik}^L} - \frac{1}{\kappa_{ic}^L} \right) \quad (i, k = 1, 2, \ldots, c\text{-}1.) \tag{183}$$

where $x_i$ is the molar fraction of $i$-component in the liquid phase.

The elements of the $(c-1)\times(c-1)$ matrix of thermodynamic factors $[\Gamma_{ik}^L]$ are:

$$\Gamma_{ik} = \delta_{ik} + x_i \frac{\partial \ln \gamma_i}{\partial x_i} \quad (i, k = 1, 2, \ldots, c\text{-}1.) \tag{184}$$

As we know, $\delta$ function (the Kronecker delta) is defined as: $\delta_{ik} = 1$ as $i = k$; otherwise $\delta_{ik} = 0$ $(i \neq k)$.

Similarly, for the vapor phase,

$$[k_{ik}^V a] = [B_{ik}^V]^{-1} a \tag{185}$$

Comparing to Eq. (181), the thermodynamic factors $[\Gamma_{ik}^V]$ don't appear in Eq. (185). It is due to the assumption of ideal gas at or below middle pressure in the vapor phase such that in this case $[\Gamma_{ik}^V] = I$ (unit matrix).

The elements of the matrix $[B_{ik}^V]$ are:

$$B_{ii}^V = \frac{y_i}{\kappa_{ic}^V} + \sum_{\substack{k=1 \\ k \neq i}}^{c} \frac{y_k}{\kappa_{ik}^V} \tag{186}$$

$$B_{ik(i \neq k)}^V = -y_i \left( \frac{1}{\kappa_{ik}^V} - \frac{1}{\kappa_{ic}^V} \right) \quad (i, k = 1, 2, \ldots, c\text{-}1.) \tag{187}$$

where $y_i$ is the molar fraction of $i$-component in the vapor phase.

Thus, the molar rates $J$ with respect to the molar average mixture velocity in matrix equation form is:

$$[J_i^V] = [k_{ik}^V a](Y - Y^I) \tag{188}$$

$$[J_i^L] = [k_{ik}^L a](X - X^I) \tag{189}$$

$$Y = (y_1, y_2, \ldots, y_{c-1})^T, \quad X = (x_1, x_2, \ldots, x_{c-1})^T \tag{190}$$

where the superscript "$I$" denotes the phase interface.

The flux of the last $c$ component, $J_c^V$ and $J_c^L$, isn't independent, but is obtained by

$$\sum_{i=1}^{c} J_i^V = 0, \quad \sum_{i=1}^{c} J_i^L = 0 \tag{191}$$

The molar rates $N$ with respect to a laboratory-fixed coordinate reference frame in matrix equation form is:

$$[N_i^V] = [J_i^V] + [\bar{y}_i N_T^V] \tag{192}$$

$$[N_i^L] = [J_i^L] + [\bar{x}_i N_T^L] \tag{193}$$

where $\bar{y}_i$ and $\bar{x}_i$ are the average mole fraction, in general, $\bar{y}_i = (y_i + y_i^I)/2$ and $\bar{x}_i = (x_i + x_i^I)/2$. $N_T^V$ and $N_T^L$ are the total mass transfer rates, and are written as:

$$N_T^V = \sum_{i=1}^{c} N_i^V, \quad N_T^L = \sum_{i=1}^{c} N_i^L \tag{194}$$

For the assumption of constant molar flow in a distillation column, $N_T^V = 0$ and $N_T^L = 0$, which, however, at most cases isn't valid. A schematic diagram of the flow of liquid and vapor phases in tray $n$ is shown in Fig. 2, which consists of three trays or a segment of a packing column possibly converted according to the concept of HETP (height equivalent of theoretical plate).

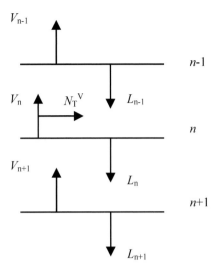

Fig. 2. Schematic diagram of the flow of liquid and vapor phases in tray $n$; the direction of mass transfer is assumed from vapor phase to liquid phase.

It is assumed that the direction of mass transfer is from vapor phase to liquid phase. Thus,

$$N_T^V = V_{n+1} - V_n, \quad N_T^L = L_n - L_{n-1} \tag{195}$$

At the steady-state of nonequilibrium thermodynamics (i.e. in the linear range), we obtain:

$$N_i^V = N_i^L, \quad N_T^V = N_T^L \tag{196}$$

Some questions may arise in solving the explicit formulation of Maxwell-Stefan equation, that is,

(1) how to calculate $\dfrac{\partial \ln \gamma_i}{\partial x_i}$ in Eq. (184) ?

The general way to calculate $\dfrac{\partial \ln \gamma_i}{\partial x_i}$ is by the so-called perturbation method. For instance,

provided a minute perturbation( $x_i \rightarrow x_i + 0.0001$ ), then

$$\frac{\partial \ln \gamma_i}{\partial x_i} \approx \frac{\ln \gamma_i (T, x_i + 0.0001) - \ln \gamma_i (T, x_i)}{(x_i + 0.0001) - (x_i)} \tag{197}$$

The activity coefficient, $\gamma_i$, can be derived, as shown in Table 2.

(2) how to calculate the $(c\text{-}1)\times(c\text{-}1)$ inverse matrix $[B_{ik}^L]^{-1}$ and $[B_{ik}^V]^{-1}$?

This is essentially a mathematical problem, and can be solved by the following computer program (see Table 12), which is programmed with Visual Basic Language and can be called by the interested readers.

Table 12

The computer program of solving the $(c\text{-}1)\times(c\text{-}1)$ inverse matrix of $[B_{ik}]$

```
Sub inverse(B() as double, c%)
Dim k%, i%, j%, alfa#
For k = 1 To c-1
alfa = 1# / B(k, k)
B(k, k) = alfa
For i = 1 To c-1
If i <> k Then
B(i, k) = -alfa * B(i, k)
End If
Next i
For i = 1 To c-1
For j = 1 To c-1
If i <> k And j <> k Then
B(i, j) = B(i, j) + B(i, k) * B(k, j)
End If
Next j
Next i
For j = 1 To c-1
If j <> k Then
B(k, j) = alfa * B(k, j)
End If
Next j
Next k
End sub
```

(3) how to calculate $\kappa_{i,k}^L a$ and $\kappa_{i,k}^V a$ ?

The binary mass transfer coefficients of $\kappa_{i,k}^L a$ and $\kappa_{i,k}^V a$ (mol s$^{-1}$) are the most important in the Maxwell-Stefan equation. For packing distillation column, these are obtained from the experimental data, depending on physical properties of the mixture, flow pattern of vapor and liquid, packing configuration and so on. For tray distillation column, often with sieve tray, it is believed that $\kappa_{i,k}^L a$ and $\kappa_{i,k}^V a$ can be predicted from the empirical correlation of the AIChE method [81]. However, in the AIChE method, only estimation of the numbers of mass transfer unit in the liquid and vapor sides, i.e. $N_L$ and $N_G$, are given. In terms of the famous two-film theory, the relationship between mass transfer coefficient and number of mass transfer unit is derived as follows:

For the liquid phase,

$$\delta_L = H_L N_L \tag{198}$$

$$\delta_L = \frac{L/A}{\kappa_{i,k}^{L0} a} \tag{199}$$

$$\kappa_{i,k}^{L} a = \kappa_{i,k}^{L0} a \bullet (A\delta_L) = LN_L \tag{200}$$

and for the vapor phase,

$$\delta_V = H_V N_V \tag{201}$$

$$\delta_V = \frac{V/A}{\kappa_{i,k}^{V0} a} \tag{202}$$

$$\kappa_{i,k}^{V} a = \kappa_{i,k}^{V0} a \bullet (A\delta_V) = VN_V \tag{203}$$

where $\delta_L$ and $\delta_V$ are respectively the height of liquid and vapor films (m), $A\delta_L$ and $A\delta_V$ are the corresponding volume of liquid and vapor film (m$^3$), respectively, and $L$ and $V$ are the flow of liquid and vapor phases (mol s$^{-1}$), respectively. $\kappa_{i,k}^{L0} a$ and $\kappa_{i,k}^{V0} a$ are the binary mass transfer coefficients, but their units are mol m$^{-3}$ s$^{-1}$.

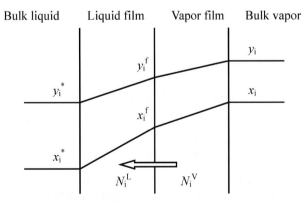

Fig. 3. Schematic diagram of two-film theory; the direction of mass transfer is assumed from vapor phase to liquid phase.

The schematic diagram of two-film theory is shown in Fig. 3. The entire resistance to mass transfer in a given turbulent phase is in a thin, stagnant region of that phase at the interface, called a film. This film is similar to the laminar sublayer that forms when a fluid flows in the turbulent regime parallel to a flat plate.

So we can calculate $[B_{ik}^{L}]^{-1}a$ and $[B_{ik}^{V}]^{-1}a$ on the condition that $\kappa_{i,k}^{L}a$ and $\kappa_{i,k}^{V}a$ are known, and then $[k_{ik}^{L}a]$ and $[k_{ik}^{V}a]$ are determined by Eqs. (181) and (185). Let us see one example about how to obtain $\kappa_{i,k}^{L}a$ and $\kappa_{i,k}^{V}a$ by means of the AIChE method.

*Example:* In a distillation column with sieve tray, the operation conditions are listed in Table 13. This table is designed for the convenience to be dealt with by MS Excel software. The mixture on the sieve tray is composed of 50 mol% benzene (1) and 50 mol% propylene (2) for the liquid phase, and 10 mol% benzene (1) and 90 mol% propylene (2) for the vapor phase ($T = 80\,^\circ\text{C}$, $P = 700\,\text{kPa}$). Compute $\kappa_{i,k}^{L}a$ and $\kappa_{i,k}^{V}a$ (kmol s$^{-1}$) on the sieve tray.

Table 13
The operation conditions on the sieve tray

| Notation | Physical quantity | Value |
|---|---|---|
| A1 | The vapor volume flowrate (m$^3$ h$^{-1}$) | 359.7 |
| A2 | The liquid volume flowrate (m$^3$ h$^{-1}$) | 10.5 |
| A3 | Vapor density (kg m$^{-3}$) | 1.172 |
| A4 | Liquid density (kg m$^{-3}$) | 750 |
| A5 | Surface tension (dyn cm$^{-1}$) | 20.0 |
| A6 | Column diameter (m) | 0.8 |
| A7 | Tray spacing (m) | 0.3 |
| A8 | Slant-hole area ratio (%) | 8.0 |
| A9 | Overflow type | Single |
| A10 | Weir height (m) | 0.04 |
| A11 | Weir length (m) | 0.6 |
| A12 | Downcomer gap (m) | 0.03 |

*Solution:* The calculated procedure is listed in Table 14.

Table 14

The calculated procedure of $[k_{ik}^{L}]a$ and $[k_{ik}^{V}]a$

| Notation | Physical quantity | Equation | Value |
|---|---|---|---|
| A13 | Weir area (m$^2$) | A13 = 0.25*ASIN(A11/A6)*A6^2 - 0.25*A11*SQRT(A6^2-A11^2) | 0.056 |
| A14 | Column area (m$^2$) | A14 = 0.25*3.1415926*A6^2 | 0.503 |

Table 14 (continued)

The calculated procedure of $[k_{ik}^L]a$ and $[k_{ik}^V]a$

| | | | |
|---|---|---|---|
| A15 | Gas velocity relative to bubble area (m s$^{-1}$) | A15 = A1/3600/(A14-A13) | 0.224 |
| A16 | Vapor viscosity (Pa s) | $\mu_1 = \mu_{01}(\dfrac{T}{T_0})^m = 6.985 \times 10^{-6}(\dfrac{273.15+80}{273.15})^{1.00}$ | 9.459 $\times 10^{-6}$ |
| | | $\mu_2 = \mu_{02}(\dfrac{T}{T_0})^m = 7.505 \times 10^{-6}(\dfrac{273.15+80}{273.15})^{0.92}$ | |
| | | $\mu_m = \sum\limits_{i=1}^{2} y_i\mu_i = (0.1 \times 9.031 + 0.9 \times 9.506) \times 10^{-6}$ | |
| A17 | Vapor diffusion coefficient (m$^2$ s$^{-1}$) | $D_{12} = \dfrac{0.00143T^{1.75}}{PM_{12}^{0.5}[(\sum v)_1^{1/3} + (\sum v)_2^{1/3}]}$ | 9.553 $\times 10^{-7}$ |
| | | $M_{12} = \sum\limits_{i=1}^{2} y_i M_i = (0.1 \times 42 + 0.9 \times 78) = 74.4$ | |
| | | $(\sum v)_1 = 90.96, \quad (\sum v)_2 = 61.56$ | |
| A18 | Sc dimensionless number | A18 = A16/(A17*A3) | 8.448 |
| A19 | Vapor F-factor (kg$^{0.5}$ m$^{-0.5}$ s$^{-1}$) | A19 = A15*A3^0.5 | 0.242 |
| A20 | Average liquid flow path width (m) | A20 = 0.5*(A6+A11) | 0.700 |
| A21 | Number of mass transfer unit in the vapor phase | A21 = (0.776+4.58*A10-0.24*A19+105*A2/3600/A20) *A18^(-0.5) | 0.460 |
| A22 | Average molecular weight in the vapor phase (kg kmol$^{-1}$) | $M_{12} = \sum\limits_{i=1}^{2} y_i M_i = (0.1 \times 42 + 0.9 \times 78)$ | 74.4 |
| A23 | Binary mass transfer coefficient in the vapor side, $\kappa_{i,k}^V a$ (kmol s$^{-1}$) | A22 = A21*A1/3600*A3/A22 | 7.240 $\times 10^{-4}$ |
| A24 | Average liquid flow path length (m) | A24 = 2*((A6/2)^2-(A11/2)^2)^0.5 | 0.529 |

Table 14 (continued)

The calculated procedure of $[k_{ik}^L]a$ and $[k_{ik}^V]a$

| A25 | Liquid hold-up (m³ m⁻²) | A25 = 0.0061+0.725*A10-0.006*A19 +1.23*A2/3600/A20 | 0.0388 |
|---|---|---|---|
| A26 | Liquid diffusion coefficient (m² s⁻¹) | $D_{12}^0 = \dfrac{(8.52\times10^{-8})T}{\mu_2 V_2^{1/3}}[1.40(\dfrac{V_2}{V_1})^{1/3}+(\dfrac{V_2}{V_1})]$ | $1.839 \times 10^{-8}$ |

$$D_{21}^0 = \frac{(8.52\times10^{-8})T}{\mu_1 V_1^{1/3}}[1.40(\frac{V_1}{V_2})^{1/3}+(\frac{V_1}{V_2})]$$

(at infinite dilution)

$$V_1 = 88.26 \times 10^{-6} \text{ m}^3 \text{ mol}^{-1}$$
$$V_2 = 68.76 \times 10^{-6} \text{ m}^3 \text{ mol}^{-1}$$

$$\mu_1 = 3.081\times10^{-4} \text{ Pa s}$$

$$\mu_2 = 0.496\times10^{-4} \text{ Pa s}$$

$$D_{12}^0 = 3.062\times10^{-8}, \quad D_{21}^0 = 6.154\times10^{-9}$$

$$x_{12} = (1+x_2-x_1)/2, \quad x_{21} = (1+x_1-x_2)/2$$

$$D_{12} = D_{21} = D_{12}^0 x_{12} + D_{21}^0 x_{21}$$

(at finite concentration)

| A27 | Average liquid resident time (s) | A27 = A24*A25/(A2/3600/A20) | 4.924 |
|---|---|---|---|
| A28 | Number of mass transfer unit in the liquid phase | A28 = 197*A26^0.5*(0.394*A19+0.17)*A27 | 3.492 |
| A29 | Average molecular weight in the liquid phase (kg kmol⁻¹) | $M_{12} = \sum_{i=1}^{2} x_i M_i = (0.5\times42+0.5\times78)$ | 60.0 |
| A30 | Binary mass transfer coefficient in the liquid side, $\kappa_{i,k}^L a$ (kmol s⁻¹) | A22 = A28*A2/3600*A4/A29 | 0.127 |

Although it is able to derive the $(c-1)\times(c-1)$ matrix of $[k_{ik}^L a]$ and $[k_{ik}^V a]$ on the basis

of $\kappa_{i,k}^L a$ and $\kappa_{i,k}^V a$, the mass transfer rate $N_i$ (kmol s⁻¹) (it should be cautious that in Eqs.

(177) and (178), the unit of $N_i$ is kmol m$^2$ s$^{-1}$, called mass transfer flux) is more important because it is needed only in the nonequilibrium model. So the next step for us is to explore how to calculate $N_i$. This is illustrated by the following example, which is similar to the source from Example 12.1 on page 665 of Seader and Henley [1] and Example 11.5.1 on page 283 of Taylor and Krishna [82].

*Example:* In a distillation column with sieve tray, the operation conditions are listed in Table 15, which is obtained from tray $n$ from a rate-based calculation of a ternary distillation at about 100 kPa, involving acetone (1), methanol (2) and water (3). In Table 15, $K_n^I$ is the equilibrium ration at the phase interface (it implies here the two-film theory is adopted). Vapor and liquid phases are assumed to be completely mixed, and the number of tray is counted from the top to the bottom.

Table 15
The operation conditions on the sieve tray

| Component | $y_n$ | $y_{n+1}$ | $y_n^I$ | $K_n^I$ | $x_n$ |
|---|---|---|---|---|---|
| 1 | 0.2971 | 0.1700 | 0.3521 | 2.7590 | 0.1459 |
| 2 | 0.4631 | 0.4290 | 0.4677 | 1.2250 | 0.3865 |
| 3 | 0.2398 | 0.4010 | 0.1802 | 0.3673 | 0.4676 |
| Total | 1.0000 | 1.0000 | 1.0000 | - | 1.0000 |

The computed products of the gas-phase binary mass transfer coefficients and interfacial area, $\kappa_{i,k}^V a$, are (in kmol s$^{-1}$):

$$\kappa_{12}a = \kappa_{21}a = 1955 ; \quad \kappa_{13}a = \kappa_{31}a = 2407 ; \quad \kappa_{23}a = \kappa_{32}a = 2797$$

(a) Compute the molar diffusion rates $J_i^V$ ;

(b) Computer the mass transfer rates $N_i^V$ ( $N_T^V = -54$ kmol s$^{-1}$).

*Solution:* (a) Compute the reciprocal rate function, i.e. the $(c-1)\times(c-1)$ matrix of $[B_{ik}^V a]$

$(c=3)$, from Eqs. (186) and (187), assuming linear mole fraction gradients such as $\bar{y}_i$ can be replaced by $(y_i + y_i^I)/2$.

Thus:

$$\bar{y}_1 = (0.2971 + 0.3521)/2 = 0.3246$$

54

$$\bar{y}_2 = (0.4631 + 0.4677)/2 = 0.4654$$

$$\bar{y}_3 = (0.2398 + 0.1802)/2 = 0.2100$$

From Eq. (186),

$$\frac{B_{11}}{a} = \frac{\bar{y}_1}{\kappa_{13}a} + \frac{\bar{y}_2}{\kappa_{12}a} + \frac{\bar{y}_3}{\kappa_{13}a} = \frac{0.3246}{2407} + \frac{0.4654}{1955} + \frac{0.2100}{2407} = 0.000460$$

$$\frac{B_{22}}{a} = \frac{\bar{y}_2}{\kappa_{23}a} + \frac{\bar{y}_1}{\kappa_{21}a} + \frac{\bar{y}_3}{\kappa_{23}a} = \frac{0.4654}{2797} + \frac{0.3246}{1955} + \frac{0.2100}{2797} = 0.000408$$

$$\frac{B_{12}}{a} = -\bar{y}_1 \left( \frac{1}{\kappa_{12}a} - \frac{1}{\kappa_{13}a} \right) = -0.3246 \left( \frac{1}{1955} - \frac{1}{2407} \right) = -0.0000312$$

$$\frac{B_{21}}{a} = -\bar{y}_2 \left( \frac{1}{\kappa_{21}a} - \frac{1}{\kappa_{23}a} \right) = -0.4654 \left( \frac{1}{1955} - \frac{1}{2797} \right) = -0.0000717$$

In matrix form:

$$[\frac{B_{ik}^V}{a}] = \begin{vmatrix} \dfrac{B_{11}}{a} & \dfrac{B_{12}}{a} \\ \dfrac{B_{21}}{a} & \dfrac{B_{22}}{a} \end{vmatrix} = \begin{vmatrix} 0.000460 & -0.0000312 \\ -0.0000717 & 0.000408 \end{vmatrix}$$

By using the computer program given in Table 12,

$$[B_{ik}^V]^{-1}a = [\frac{B_{ik}^V}{a}]^{-1} = \begin{vmatrix} 2200 & 168.2 \\ 386.6 & 2480 \end{vmatrix} = [k_{ik}^V a]$$

From Eq. (188),

$$\begin{bmatrix} J_1^V \\ J_2^V \end{bmatrix} = \begin{bmatrix} k_{11}^V a & k_{12}^V a \\ k_{21}^V a & k_{22}^V a \end{bmatrix} \begin{bmatrix} (y_1 - y_1^I) \\ (y_2 - y_2^I) \end{bmatrix}$$

$$J_1^V = k_{11}^V a(y_1 - y_1^I) + k_{12}^V a(y_2 - y_2^I) = -121.8 \ \text{kmol s}^{-1}$$

$$J_2^V = k_{21}^V a(y_1 - y_1^I) + k_{22}^V a(y_2 - y_2^I) = -32.71 \ \text{kmol s}^{-1}$$

From Eq. (191),

$$J_3^V = -J_1^V - J_2^V = 154.51 \ \text{kmol s}^{-1}$$

(b). From Eq. (192),

$$N_1^V = J_1^V + \bar{y}_1 N_T^V = -121.8 + 0.3246 \times (-54) = -139.4 \ \text{kmol s}^{-1}$$

Similarly,

$$N_2^V = -57.81 \ \text{kmol s}^{-1}, \ N_3^V = 143.2 \ \text{kmol s}^{-1}.$$

In this example, the values of $y_i^I$ are predetermined. Sometimes the actual case is just inverse. Before calculating $N_i^V$ and $N_i^L$, unfortunately, $y_i^I$ and $x_i^I$ are unknown. But at the steady state of nonequilibrium thermodynamics,

$$N_i^V = N_i^L, \quad N_T^V = N_T^L = N_T \tag{204}$$

From Eq. (204),

$$J_i^V + \bar{y}_i N_T^V = J_i^L + \bar{x}_i N_T^L \tag{205}$$

$$J_i^V + \frac{(y_i + y_i^I)}{2} N_T = J_i^L + \frac{(x_i^I + x_i)}{2} N_T \tag{206}$$

$$y_i^I = K_i x_i^I \tag{207}$$

From Eqs. (206) and (207), $x_i^I$ and $y_i^I$ are obtained, under the condition that the net mass transfer rate $N_T$ is known, by continuous iteration because $J_i^V$ and $J_i^L$ are the functions of $y_i^I$ and $x_i^I$, respectively. Subsequently, $N_i^V$ and $N_i^L$ are solved.

In the multi-component system of nonequilibrium thermodynamics, not only mass transfer should be considered, but also heat transfer is indispensable to establish the energy balance equation. To determine the component heat transfer coefficient, the simplest way is to use the analogous expressions among momentum, heat and mass transfer, such as Reynolds analogy, Chilton-Colburn analogy and Prandtl analogy. The details about these analogies are discussed in standard texts [1-12]. Therefore, the heat transfer coefficient is convenient to be derived on the basis of the preceding mass transfer coefficient.

## REFERENCES

[1] J.D. Seader and E.J. Henley (eds.), Separation Process Principles, Wiley, New York, 1998.
[2] J.F. Richardson, J.H. Harker and J.R. Backhurst (eds.), Chemical Engineering Particle Technology and Separation Process, Butterworth-Heinemann, Oxford, 2002.
[3] J.S. Tong (ed.), The Fluid Thermodynamics Properties, Petroleum Technology Press, Beijing, 1996.
[4] R.C. Reid, J.M. Prausnitz and B.E. Poling (eds.), The Properties of Gases and Liquids, McGraw-Hill, New York, 1987.
[5] B.E. Poling, J.M. Prausnitz and J.P. Oconnell (eds.), The Properties of Gases and Liquids (fifth edition), McGraw-Hill, New York, 2000.
[6] J.S. Tong, G.H. Gao and Y.P. Liu (eds.), Chemical Engineering Thermodynamics, Tsinghua University Press, Beijing, 1995.
[7] Z.Q. Zhu (ed.), Supercritical Fluids Technology, Chemical Industry Press, Beijing, 2001.

[8] J.M. Prausnitz (ed.), Molecular Thermodynamics of Fluid-Phase Equilibria, Prentice-Hall, New Jersey, 1969.

[9] J.M. Smith and H.C. Van Ness (eds.), Introduction to Chemical Engineering Thermodynamics, McGraw-Hill, New York, 1975.

[10] S. Malanowski and A. Anderko (eds.), Modelling Phase Equilibria: Thermodynamic Background and Practical Tools, Wiley, New York, 1992.

[11] S.I. Sandler (ed.), Chemical Engineering Thermodynamics, Wiley, New York, 1989.

[12] T.M. Reed and K.E. Gubbins (eds.), Applied Statistical Mechanics: Thermodynamic and Transport Properties of Fluids, McGraw-Hill, New York, 1973.

[13] C. Perego and P. Ingallina, Catal. Today, 73 (2002) 3-22.

[14] G. Buelna, R.L. Jarek, S.M. Thornberg and T.M. Nenoff, J. Mol. Catal. A-Chem., 198 (2003) 289-295.

[15] A. Geatti, M. Lenarda, L. Storaro, R. Ganzerla and M. Perissinotto, J. Mol. Catal. A-Chem., 121 (1997) 111-118.

[16] C. Perego, S. Anarilli, R. Millini, G. Bellussi, G. Girotti and G. Terzoni, Microporous Mat., 6 (1996) 395-404.

[17] J.D. Shoemaker and E.M. Jones, Hydro. Proc., 66 (1987) 57-58.

[18] J.L. Franklin, Ind. Eng. Chem., 41 (1949) 1070-1076.

[19] J.P. Novak, J. Matous and J. Pick (eds.), Liquid-liquid Equilibria, Elsevier, Amsterdam, 1987.

[20] S.S. Jorgensen, B. Kolbe, J. Gmehling and P. Rasmussen, Ind. Eng. Chem. Proc. Des. Dev., 18 (1979) 714-722.

[21] J. Gmehling, P. Rasmussen and A. Fredenslund, Ind. Eng. Chem. Proc. Des. Dev., 21 (1982) 118-127.

[22] E.A. Macedo, U. Weidlich, J. Gmehling and P. Rasmussen, Ind. Eng. Chem. Proc. Des. Dev., 22 (1983) 676-678.

[23] D. Tiegs, J. Gmehling, P. Rasmussen and A. Fredenslund, Ind. Eng. Chem. Res., 26 (1987) 159-161.

[24] H.K. Hansen, P. Rasmussen and A. Fredenslund, M. Schiller and J. Gmehling, Ind. Eng. Chem. Res., 30 (1991) 2352-2355.

[25] J. Gmehling, Fluid Phase Equilib., 144 (1998) 37-47.

[26] J. Gmehling, Fluid Phase Equilib., 107 (1995) 1-29.

[27] A. Bondi, Physical Properties of Molecular Liquids, Crystals and Glasses, Wiley, New York, 1968.

[28] J. Gmehling, J. Li and M. Schilleer, Ind. Eng. Chem. Res., 32 (1993) 178-193.

[29] J. Lohmann and J. Gmehling, J. Chem. Eng. Japan, 34 (2001) 43-54.

[30] H. K. Hansen, P. Rasmussen, A. Fredenslund, M. Schiller and J. Gmehling, Ind. Eng. Chem. Res., 30 (1991) 2355-2358.

[31] J. Gmehling, J. Lohmann, A. Jakob, J. Li and R. Joh, Ind. Eng. Chem. Res., 37 (1998) 4876-4882.

[32] J. Gmehling, R. Wittig, J. Lohmann and R. Joh, Ind. Eng. Chem. Res., 41 (2002) 1678-1688.

[33] R. Wittig, J. Lohmann and J. Gmehling, Ind. Eng. Chem. Res., 42 (2003) 183-188.

[34] J.C. Bastos, M.E. Soares and A.G. Medina, Ind. Eng. Chem. Res., 27 (1988) 1269-1277.

[35] I. Kikic, M. Fermeglia and P. Rasmussen, Chem. Eng. Sci., 46 (1991) 2775-2780.

[36] C. Achard, C.G. Dussap and J.B. Gros, Fluid Phase Equilib., 98 (1994) 71-89.

[37] M. Aznar and A.S. Telles, Braz. J. Chem. Eng., 18 (2001) 127-137.

[38] E.A. Macedo, P. Skovborg and P. Rasmussen, Chem. Eng. Sci., 45 (1990) 875-882.

[39] W. Yan, M. Topphoff, C. Rose and J. Gmehling, Fluid Phase Equilib., 162 (1999) 97-113.

[40] Z.G. Lei, H.Y. Wang, Z. Xu, R.Q. Zhou and Z. T. Duan, Chemical Industry and Engineering Progress (China), 20 (2001) 6-9.

[41] H. Asatani and W. Hayduk, Can. J. Chem. Eng., 61 (1983) 227-232.

[42] R.R. Bannister and E. Buck, Chem. Eng. Prog., 65 (1969) 65-68.

[43] E.R. Thomas and L.R. Eckert, Ind. Eng. Chem. Process Des. Dev., 23 (1984) 194-209.

[44] Z.S. Yang, C.L. Li and J.Y. Wu, Petrochemical Technology (China), 30 (2001) 285-288.

[45] Y.Z. Wu, S.K. Wang, D.R. Hwang and J. Shi, Can. J. Chem. Eng., 70 (1992) 398-402.

[46] W.J. Howell, A.M. Karachewski, K.M. Stephenson, C.A. Eckert, J.H. Park, P.W. Carr and S.C. Rutan, Fluid Phase Equilib., 52 (1989) 151-160.

[47] J.H. Park and P.W. Carr, Anal. Chem., 59 (1987) 2596-2602.

[48] M.J. Hait, C.A. Eckert, D.L. Bergmann, A.M. Karachewski, A.J. Dallas, D.I. Eikens, J.J.J. Li, P.W. Carr, R.B. Poe and S.C. Rutan, Ind. Eng. Chem. Res., 32 (1993) 2905-2914.

[49] R.B. Poe, S.C. Rutan, M.J. Hait, C.A. Eckert and P.W. Carr, Anal. Chim. Acta, 277 (1993) 223-238.

[50] A. Klamt, J. Phys. Chem., 99 (1995) 2224-2235.

[51] A. Klamt, V. Jonas, T. Burger and J.C.W. Lohrenz, J. Phys. Chem. A, 101 (1998) 5074-5085.

[52] A. Klamt and F. Eckert, Fluid Phase Equilib., 172 (2000) 43-72.

[53] W. Arlt, O. Spuhl and A. Klamt, Chem. Eng. Process., 43 (2004) 221-238.

[54] M. Seiler, D. Kohler and W. Arlt, Sep. Purif. Technol., 29 (2002) 245-263.

[55] M. Seiler, W. Arlt, H. Kautz and H. Frey, Fluid Phase Equilib., 201 (2002) 359-379.

[56] K. Strubl, V. Svoboda, R. Holub and J. Pick, Collect. Czech. Chem. Commun., 35 (1970) 3004-3019.

[57] M. Agarwal and V.G. Gaikar, Chem. Eng. Commun, 115 (1992) 83-94.

[58] W. Wardencki and A.H.H. Tameesh, J. Chem. Technol. Biotechnol., 31 (1981) 86-92.

[59] B.S. Rawat and I.B. Gulati, J. Chem. Technol. Biotechnol., 31 (1981) 25-32.

[60] Z. Duan, L. Lei and R. Zhou, Petrochemical Technology (China), 9 (1980) 350-353.

[61] L. Lei, Z. Duan and Y. Xu, Petrochemical Technology (China), 11 (1982) 404-409.

[62] Q. Zhang, W. Qian and W. Jiang, Petrochemical Technology (China), 13 (1984) 1-9.

[63] Y. Li, Thermodynamics of Metal Extraction, Tsinghua University Press, Beijing, 1988.

[64] S.K. Shoor and K.E. Gubbins, J. Phys. Chem., 73 (1969) 498-505.

[65] W.L. Masterton and T.P. Lee, J. Phys. Chem., 74 (1970) 1776-1782.

[66] R.A. Pierotti, Chem. Rev., 76 (1976) 717-726.

[67] F.J. Millero, Chem. Rev., 71 (1971) 147-176.

[68] L. Chen (ed.), Handbook of Solvent, Chemical Industry Press, Beijing, 1994.

[69] R.S. Li (ed.), Irreversible Thermodynamics and Dissipation, Beijing, Tsinghua University Press, 1986.

[70] K. Miyazaki, K. Kitahara and D. Bedeaux, Physica A, 230 (1996) 600-630.

[71] J.M. Rubi and P. Mazur, Physica A, 276 (2000) 477-488.

[72] Z.W. Chen, Z.Y. Jian and W.Q. Jie, Journal of Xi'an Institute of Technology, 21 (2001) 142-149.

[73] H. Liang, Z.G. Wang, B.C. Lin, C.G. Xu and R.N. Fu, J. Chromatogr. A, 763 (1997) 237-251.

[74] Q.L. Liu, P. Li and Z.B. Zhang, Chem. J. Chinese Univ., 22 (2001) 1209-1212.

[75] H. Cheng, M. Zhou, X.Y. Tang, C.J. Xu and G.C. Yu, Journal of TianJin University (China), 33 (2000) 11-13.

[76] R. Krishna and J.A. Wesselingh, Chem. Eng. Sci., 52 (1997) 861-911.

[77] R. Baur, R. Taylor and R. Krishna, Chem. Eng. Sci., 56 (2001) 2085-2102.

[78] H.A. Kooijman and R. Taylor, Ind. Eng. Chem. Res., 30 (1991) 1217-1222.

[79] R. Baur, A.P. Higler, R. Taylor and R. Krishna, Chem. Eng. J., 76 (2000) 33-47.

[80] A.P. Higler, R. Taylor and R. Krishna, Chem. Eng. Sci., 54 (1999) 2873-2881.

[81] Committee of chemical engineering handbook (eds.), Chemical Engineering Handbook 2nd ed., Chemical Industry Press, Beijing, 1996.

[82] R. Taylor and R. Krishna (eds.), Multicomponent Mass Transfer, John Wiley and Sons, New York, 1993.

# Chapter 2. Extractive distillation

Extractive distillation is more and more commonly applied in industry, and is becoming an important separation method in chemical engineering. Separation sequence of the columns, combination with other separation processes, tray configuration and operation policy are included in the process section. Since the solvent plays an important role in the design of extractive distillation, such conventional and novel separating agents as solid salt, liquid solvent, the combination of liquid solvent and solid salt, and ionic liquid are covered. The prominent characteristics of extractive distillation is that one new solvent (i.e. separating agent) with a high boiling point, is added to the components to be separated so as to increase their relative volatility. Selection of a suitable solvent is fundamental to ensure an effective and economical design. CAMD is a useful tool and is applied for screening the solvents and thus reducing the experimental work. Molecular thermodynamic theories, which can interpret the microscale mechanism of selecting the solvents, are also presented. To accurately describe the extractive distillation process, mathematical model is indispensable. There are two types of mathematical models to simulate extractive distillation process, i.e. equilibrium (EQ) stage model and non-equilibrium (NEQ) stage model. The EQ stage model, an old method, is widely used even until now, but the NEQ stage model should receive more attention. The model equations of NEQ stage are given in chapter 4 (catalytic distillation).

## 1. INTRODUCTION

Extractive distillation is commonly applied in industry, and is becoming an increasingly important separation method in petrochemical engineering [1]. The production scale in industrial equipment is diverse, from several kilotons (column diameter about 0.5 m) to hundred kilotons (column diameter about 2.5 m) per year. It is mainly used in the following cases [1-5]: one application is separating hydrocarbons with close boiling point, such as C4, C5, C6 mixtures and so on; the other is the separation of mixtures which exhibit an azeotrope, such as alcohol/water, acetic/water, acetone/methanol, methanol/methyl acetate, ethanol/ethyl acetate, acetone/ethyl ether and so on.

In extractive distillation, an additional solvent (i.e. separating agent) is used to alter the relative volatility of the components to be separated. In this way, it is possible to obtain one pure component at the top of one column and the other, together with the solvent at the bottom, which may be separated easily in a secondary distillation column due to the high boiling point of the solvent. The solvent doesn't need to be vaporized in the extractive distillation process. However, in azeotropic distillation (discussed in chapter 3), which is also often used for the separation of close boiling point or azetropic mixtures, both the solvent and components must be vaporized into the top of an azeotropic distillation column. Moreover, the amount of azetropic solvent is usually large, which leads to large energy consumption compared to extractive distillation. For this reason, extractive distillation is used more often than azeotropic distillation [6]. In recent years, an interesting special distillation method, i.e. adsorption distillation, has been proposed and will be compared with extractive distillation (see chapter 5).

The ease of separation of a given mixture with key components $i$ and $j$ is given by the relative volatility:

$$\alpha_{ij} = \frac{y_i / x_i}{y_j / x_j} = \frac{\gamma_i P_i^0}{\gamma_j P_j^0} \tag{1}$$

where $x$ is molar fraction in the liquid phase, $y$ is molar fraction in the vapor phase, $\gamma$ is the activity coefficient, and $P_i^0$ is the pure component vapor pressure.

The solvent is introduced to change the relative volatility as far as away from unity as possible. Since the ratio of $P_i^0 / P_j^0$ is almost no change in a narrow temperature range, the only way to affect the relative volatility is by introducing a solvent which can change the ratio $\gamma_i / \gamma_j$. This ratio, in the presence of the solvent, is called selectivity $S_{ij}$:

$$S_{ij} = (\frac{\gamma_i}{\gamma_j})_s \tag{2}$$

In some cases a significant change in operating pressure and hence temperature, alters $a_{ij}$ enough to eliminate an azeotrope.

Besides altering the relative volatility, the solvent also should be easily separated from the distillation products, that is to say, high boiling point difference between the solvent and the components to be separated is desirable. Other criteria, e.g. corrosion, prices, sources, etc. should also be taken into consideration. However, the relative volatility (which is consistent with selectivity) is the most important. When solvents are ranked in the order of relative volatility (or selectivity), the solvent with the highest relative volatility is always considered to be the most promising solvent for a given separation task. This may indicate that, from the viewpoint of economic consideration, the use of the solvent with the highest relative volatility (or selectivity) will always give the lowest total annual cost (TAC) of the extractive distillation process [7]. The economic evaluations were carried out by using many different solvents for separating three different binary mixtures: 2-methyl-butene / isoprene (mixture A), n-butane / trans-2-butene (mixture B), and n-hexane / benzene (mixture C). For each case a complete design and costing of the process was done by using cost estimating computer programs. Figures 1, 2 and 3 show, respectively, the ranking of solvent selectivity at infinite dilution with the TAC of extractive distillation process for separating 2-methyl-butene / isoprene, n-butane / trans-2-butene, and n-hexane / benzene mixtures by various potential solvents.

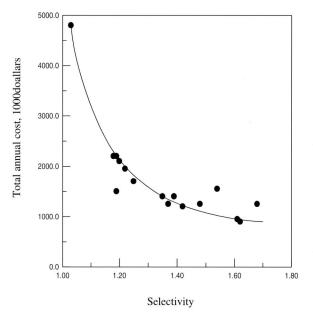

Fig. 1. The effect of solvent selectivity on the total annual cost of an extractive distillation operation (2-methyl-1-butene/isoprene); adapted from the source [7].

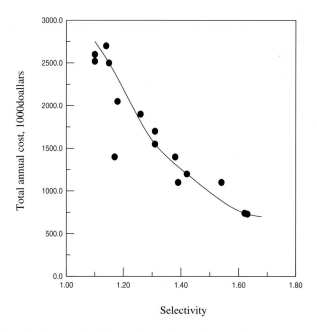

Fig. 2. The effect of solvent selectivity on the total annual cost of an extractive distillation operation (n-butane/trans-2-butene); adapted from the source [7].

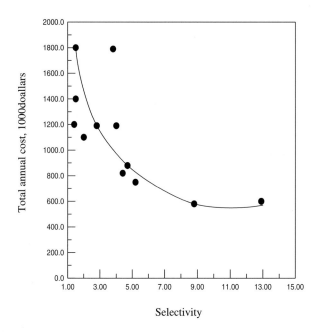

Fig. 3. The effect of solvent selectivity on the total annual cost of an extractive distillation operation (n-hexane/benzene); adapted from the source [7].

It can be seen from Figs. 1, 2, and 3 that as the solvent selectivity increases, the total annual cost of the extractive distillation tends to decrease. This indicates that the solvent with the highest selectivity has a potential as the best solvent.

Berg [8] classified the liquid solvents into five groups according to their potential for forming hydrogen bonds. However, it is found by many experiences that Class I and Class II, which are a highly hydrogen bonded liquids, are successful solvents in most cases.

Class I : Liquids capable of forming three dimensional networks of strong hydrogen bonds, e.g. water, glycol, glycerol, amino alcohols, hydroxylamine, hydroxyacids, polyphenols, amides, etc. Compounds such as nitromethane and acetonitrile also form three dimensional hydrogen bond networks, but the bonds are much weaker than those involving OH and $NH_2$ groups. Therefore, these types of compounds are placed in Class II.

Class II: Other liquids composed of molecules containing both active hydrogen atoms and donor atoms (oxygen, nitrogen and fluorine), e.g. alcohols, acids, phenols, primary and secondary amines, oximes, nitro-compounds with alpha-hydrogen atoms, nitriles with alpha-hydrogen atoms, ammonia, hydrazine, hydrogen fluoride, hydrogen cyanide, etc.

Table 1 illustrates some examples of the single liquid solvents commonly used in the extractive distillation [9-11], and proves the idea of Berg to be reliable.

For simplification, it is noted that here the words "separating agents" and "solvents" are confused and have the same meaning in what follows.

Table 1
Examples of the single liquid solvents commonly used in the extractive distillation

| No. | Components to be separated | Solvents as the separating agents |
|-----|----------------------------|-----------------------------------|
| 1 | alcohol (ethanol, isopropanol, tert-butanol) and water | ethylene glycol |
| 2 | acetic acid and water | tributylamine |
| 3 | acetone and methanol | water, ethylene glycol |
| 4 | methanol/methyl acetate | Water |
| 5 | propylene and propane | ACN |
| 6 | C4 hydrocarbons | acetone, ACN (acetonitrile), DMF (N,N-dimethylformamide), NMP (N-methyl-2-pyrrolidone), NFM (N-formylmorpholine) |
| 7 | alcohol (ethanol, isopropanol) and water | DMF |
| 8 | C5 hydrocarbons | DMF |
| 9 | aromatics and non-aromatics | DMF, NMP, NFM |

## 2. PROCESS OF EXTRACTIVE DISTILLATION

In this section, process of extractive distillation is discussed in four aspects: column sequence, combination with other separation processes, tray configuration and operation policy, which relate to the process design.

### 2.1. Column sequence

In general, for a two-component system the extractive distillation process is made up of two columns, i.e. an extractive distillation column and a solvent recovery column. The two column process is diagrammed in Fig. 4 where components A and B to be separated are respectively obtained from the top of two columns, and the solvent S is recovered in the solvent recovery column and recycled [12, 13].

For a multi-component system, the arrangement of column sequence may be complicated, but is developed from the above double-column process. An interesting example is for the separation of C4 mixture [14-20].

Among C4 mixture (butane, butene-1, trans-butylene-2, cis-butylene-2, 1,3-butadiene and vinylacetylene (VAC)), 1,3-butadiene is a basic organic raw material. In industry, 1,3-butadiene can be separated from C4 mixtures by extractive distillation with the solvent acetronitile (ACN). On the basis of the double-column process, the extractive distillation process for separating 1,3-butadiene is designed. The first separation sequence is: extractive distillation (column 1) – solvent recovery (column 2) – extractive distillation (column 3) – solvent recovery (column 4), and is illustrated in Fig. 5. In this case, butane and butene are regarded as one part and used for fuel.

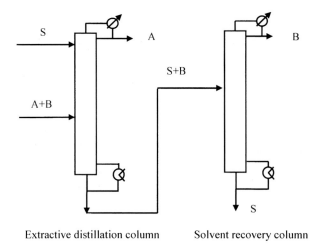

Fig. 4. The double-column process for extractive distillation.

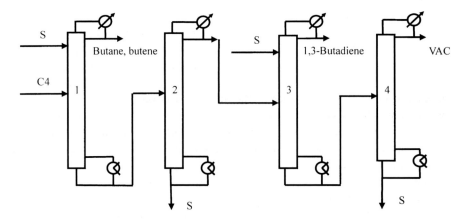

Fig. 5. The extractive distillation process designed for separating C4 mixture: 1. the first extractive distillation column; 2. solvent recovery column; 3. the second extractive distillation column; 4. flash column.

In Fig. 5 the feedstock flows into the first extractive distillation column 1. The solvent S enters the top sections of columns 1 and 3. The mixtures of butane and butene are escaped from the top of column 1 while VAC from the top of column 4. The product, 1,3-butadiene is obtained from the top of the second extractive distillation column 3. The solvent S is recovered in columns 2 and 4 and recycled. It can be seen that column 2 is unnecessary and can be eliminated because the solvent is still needed in the subsequent column 3 and needn't be separated. At the same time, much energy is consumed because butadiene-1,3 goes through the unnecessary repeated evaporation and condensation. The current process for separating C4 mixture with ACN is shown in Fig. 6.

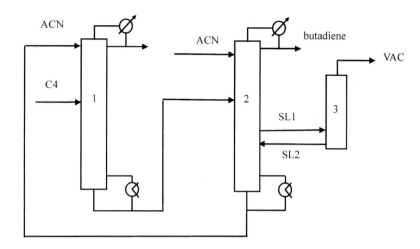

Fig. 6. The current process of extractive distillation with ACN method for separating C4 mixture: 1. the first extractive distillation column; 2. the second extractive distillation column; 3. flash column.

The product, 1,3-butadiene, is obtained from the top of the second extractive distillation column 2. Alkyne hydrocarbons mainly containing VAC and drawn out through stream SL1, are removed in flash column 3. But SL2 stream is also returned into column 2. This is to say, column 2 and column 3 are thermally coupled. There exists an obvious disadvantage in the current extractive distillation process. The liquid load is very high in the lower section of column 2, which has negative influence on the tray efficiency and decreasing production capacity of 1,3-butadiene. It is the current case that the demand for 1,3-butadiene goes up gradually year by year. Moreover, the thermal coupling between columns 2 and 3 leads to the difficulty in operation and control.

To solve this problem, a new optimization process was put forward as shown in Fig. 7. The vapor stream SV is drawn from the lower section of the first extractive distillation column 1 into the second extractive distillation column 2. SL1 mainly containing ACN and alkyne hydrocarbons is flashed in column 3. It can be seen that the thermal coupling in the optimization process is eliminated and operation and control would be more convenient.

By process simulation, it is found that the liquid load in the lower section of the second extractive distillation column 2 is greatly reduced from 51943 kg h$^{-1}$ to 12584 kg h$^{-1}$, about 75.8% of the original process. This demonstrates that it is possible to farther increase the yield of butadiene on the basis of old equipment. The vapor and liquid loads of the old and new processes are respectively illustrated in Figs. 8 and 9 [21].

On the other hand, comparison of heat duties on reboilers $Q_R$ and condensers $Q_C$ for Fig. 5 through Fig. 7 is made and the result is listed in Table 2, where the heat duties of the column used for separating VAC and solvent are neglected due to the much smaller values than those of other columns.

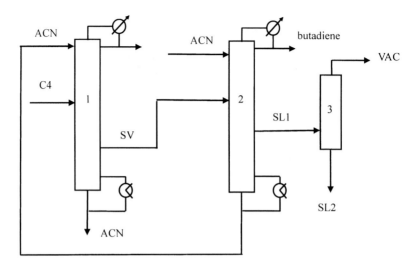

Fig. 7. The optimum process of extractive distillation with ACN method for separating C4 mixture: 1. the first extractive distillation column; 2. the second extractive distillation column; 3. flash column.

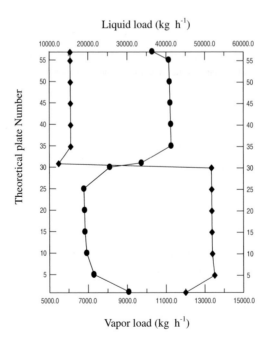

Fig. 8. The vapor and liquid load distribution diagram of the second extractive distillation column for the original process. ◆— liquid load along the column; ●— vapor load along the column.

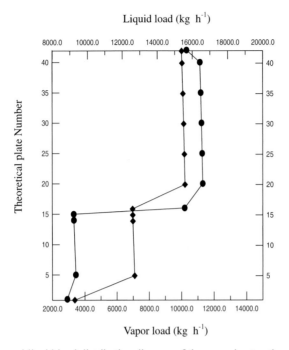

Liquid load (kg h$^{-1}$)

Theoretical plate Number

Vapor load (kg h$^{-1}$)

Fig. 9. The vapor and liquid load distribution diagram of the second extractive distillation column for the optimum process. ◆— liquid load along the column ; ●— vapor load along the column.

Table 2
Comparison of heat duties of these three processes

| No. | The Fig. 5 process | | The Fig. 6 process | | The Fig. 7 process | |
|---|---|---|---|---|---|---|
| | $Q_C$ | $Q_R$ | $Q_C$ | $Q_R$ | $Q_C$ | $Q_R$ |
| Column1 | 1161.5 | 2936.3 | 1161.5 | 2936.3 | 1163.6 | 4693.0 |
| Column 2 | 1183.3 | 1984.7 | 1122.6 | 2545.1 | 1122.3 | 1178.0 |
| Column 3 | 1122.0 | 1728.1 | 0.0 | 0.0 | 0.0 | 0.0 |
| Total duty | 3466.8 | 6649.1 | 2284.1 | 5481.4 | 2285.9 | 5871.0 |

It can be seen from Table 2 that heat duties on reboilers $Q_R$ and condensers $Q_C$ are the largest in the Fig. 5 process, and those are approximately equal in the processes of Figs. 6 and 7, where 1,3-butadiene goes through the minimum number of evaporation and condensation. As a result, it indicates that the Fig. 7 process is the best from the aspects of equipment investment, energy consumption and production capacity.

However, formerly we only paid more attention on extracting 1,3-butadiene from C4 mixture. Indeed 1,3-butadiene is an important raw material, which is used for the preparation of synthetic rubber, e.g. the copolymers of butadiene and styrene. But it is a pity that the residual C4 mixture has always been burned as fuel. Until recently, many plants require the residual C4 mixture to be utilized more reasonably. Among the residua, n-butene, i.e. the mixture of 1-butene, trans-2-butene and cis-2-butene, is able to be used as one monomer of synthesizing polymers and one reactant of producing acetone by reacting with water. Isobutene can be selectively combined with methanol to make methyl tert-butyl ether (MTBE) [22-25]. Butane is still used as fuel. Therefore, the residual C4 mixture can be used in a more reasonable way, not just as fuel. But only are the useful C4 components separated from mixtures, the reasonable utilization is possible.

C4 mixture can be divided into four groups: butane (n-butane and isobutane), butene (n-butene, isobutene), 1,3-butadiene and VAC. N-butene and isobutene are regarded as one group, which is based on the consideration that isobutene is easy to be removed from n-butene by reacting with methanol. These four groups have different fluidity of electron cloud, the sequence being butane < butene < 1,3-butadiene < VAC. Therefore, the interaction force between the polar solvent and C4 mixture is different, and the corresponding order is that butane < butene < 1,3-butadiene < VAC. Thus C4 mixture can be separated in series in the presence of solvent by extractive distillation.

In terms of the above optimum process of separating 1,3-butadiene from C4 mixture, it is straightforward to deduce another new process for separating butane, butene, 1,3-butadiene and VAC in series, as diagrammed in Fig. 10.

It is evident that the new process has the advantages in equipment investment, energy consumption and production capacity. That is to say, for a very complicated separation of C4 mixture, we can still find out the most optimum process step by step from the simplest double-column process of extractive distillation as shown in Fig. 4.

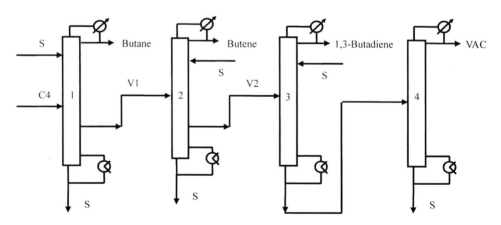

Fig. 10. A new process for separating butane, butene, 1,3-butadiene and VAC.

## 2.2. Combination with other separation processes

As mentioned in the foregoing introduction section, extractive distillation is used more often than azeotropic distillation because of low energy consumption and flexible selection of the possible solvents. However, anything has its own defects, and extractive distillation isn't exceptive. For instance, comparing to azeotropic distillation, extractive distillation can't obtain a very high purity of product because the solvent coming from the bottom of solvent recovery column more or less contains some impurities which may influence the separation effect. Moreover, comparing with liquid-liquid extraction, extractive distillation consumes much more energy. So it is advisable to combine extractive distillation with other separation process such as azeotropic distillation, liquid-liquid extraction and so on.

(1) Combination of extractive distillation and azeotropic distillation

Extractive distillation can't obtain a very high purity of product, but consumes less energy; however, azeotropic distillation can obtain a very high purity of product, but consumes much more energy. Therefore, a separation method, i.e. the combination of extractive distillation and azeotropic distillation is proposed [26], which eliminates the disadvantages of both extractive distillation and azeotropic distillation, and retains the advantages of them. This method has been used for the separation of 2-propanol (IPA) and water while the high purity of 2-propanol over 99.8%wt is required. The process of this method is diagrammed in Fig. 11, where column 1, column 2, column 3 and column 4 are extractive distillation column, solvent recovery column (separating agent $S_1$ for extractive distillation), azeotropic distillation column and solvent recovery column (separating agent $S_2$ for azeotropic distillation), respectively.

This process is designed according to the sequence, firstly extractive distillation which ensures low energy consumption, and then azeotropic distillation which ensures high purity of 2-propanol as product. But this sequence can't be in reverse. Otherwise, high purity of 2-propanol is difficult to be obtained.

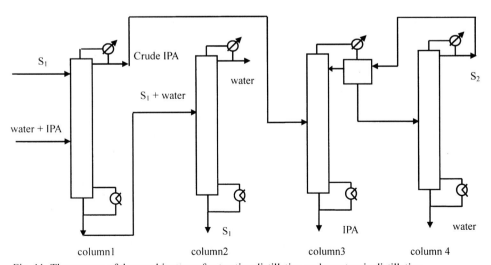

Fig. 11. The process of the combination of extractive distillation and azeotropic distillation.

In addition, the energy consumption of this process is greater than that of extractive distillation, but less than that of extractive distillation. So it is concluded that this process is especially suitable for the separation of requiring high purity of product. For instance, in the case of separating acetic acid and water, the concentration of acetic acid in water below 20ppm is required in industry.

(2) Combination of extractive distillation and liquid-liquid extraction

This method has been sucessfully used for the recovery of benzene and toluene from pyrolysis hydrogenation gasoline fraction [27, 28]. The mixture of benzene and toluene is firstly separated into two parts, i.e. one mainly containng benzene and the other mainly containing toluene. Then the rich-benzene mixture is dealt with by extractive distillation, while the rich-toluene mixture is dealt with by liquid-liquid extraction. The solvent used in these two processes is the same, sulfolane.

The reason why extractive distillation is selected for the separation of the rich-benzene mixture is that its boiling-point is lower than that of rich-toluene mixture, and thus the boiling-point difference between the rich-benzene mixture and the solvent is great enough, which facilitates the solvent recovery. However, the boiling-point difference between the rich-toluene mixture and the solvent is relatively small, which leads to some impurities accumulated in the solvent. Therefore, it is wise to select liquid-liquid extraction for the separation of the rich-toluene mixture.

The practical result from the Aromatics Plant of Yangzi Petrochemical Company with a capacity of 360kt a year showed that the the freezing point of benzene obtained by this method was 5.48 ℃, the sulfur content in benzene was less than 0.5μg g$^{-1}$, and the non-aromatics content in toluene was less than 1000μg g$^{-1}$. The recovery yield of benzene and toluene was 99.9% and 99.1% respectively. This proves that the combination of extractive distillation and liquid-liquid extraction is effective for the recovery of benzene and toluene from pyrolysis hydrogenation gasoline fraction.

## 2.3. Tray configuration

One prominent characteristics of extractive distillation is that the solvent ratio (namely the mass ratio of solvent and feed) is very high, generally 5-8, which restricts increasing capacity. The liquid load is very large in the extractive distillation column, but the vapor load is relatively small. So when we design the extractive distillation column, it should be paid more attention to the channels of passing liquid phase.

In most cases, either tray column or packing column can be adopted. However, if the extractive distillation column is operated under the middle or even high pressure such as the separation of C3, C4 and C5 mixtures, or if the components to be separated are easy to polymerize such as the separation of C5 mixture, then it is better to choose tray column. Herein, two types of plate trays used in the extractive distillation column are introduced [29-35].

Under the middle or even high pressure, the plate trays, especially double overflow trays, are generally used as internal fittings. Both double overflow valve trays and double overflow slant-hole trays have ever been adopted. However, it is reported that, if double overflow valve trays are replaced by double overflow slant-hole trays in a column for separating propane and propylene, the feeds to be treated are raised above 50% with a tray efficiency similar to or higher than that of the valve trays and an energy save of 10% by decreasing pressure drop to about 1/3 of the original.

Slant-hole trays, very excellent and extensively applied in the industry, has opposite stagger arrangement of slant-holes, which causes rational flowing of vapor and liquid phases, level blowing of vapor, permission of a large vapor speed, no mutual interference, steady liquid level and high tray efficiency. On the basis of studying and analyzing the columns with multiple downcomer (MD) trays, a new-type multi-overflow compound slant-hole trays were invented, which adopt the downcomer similar to MD trays. The number of downcomers used is not too many, but only two. The downcomer has the feature of simple structure, longer flowing distance of liquid, higher capacity of column and high efficiency of trays. The configuration of the double overflow slant-hole trays is diagrammed in Fig. 12. In terms of the vapor and liquid load of the extractive distillation column, the tray parameters can be determined often by a computer program for tray design. Of course, it should be mentioned that the designed values must be within the range of normal operation condition.

Under the condition that the components to be separated are easy to polymerize, the big-hole sieve trays are desirable. For instance, for the separation of the C5 mixture, the key components being pentene and isoprene, it is known that the polymerization reactions among unsaturated hydrocarbons take place and hinder the normal operation of the column while the traditional valve trays are employed. However, in this case the big-hole sieve trays with the hole diameter of 10-15 mm are effective because the holes with large diameter prevent the trays from jam due to the formation of polymer. In addition, this type of trays eliminates the gradient of liquid layer and causes the rational flowing of vapor and liquid phases on the tray, with the help of directed holes arranged for decreasing the radial mix and bubble promoter installed in the outlet of the downcomer.

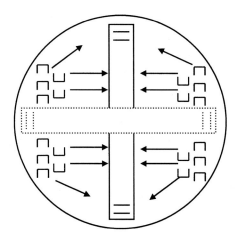

Fig. 12. Configuration of multi-overflow(double flow) slant-hole tray.

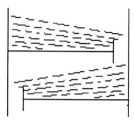

Fig. 13. The gradient of liquid layer on the traditional trays.

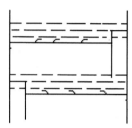

Fig. 14. The gradient of liquid layer on the big-hole trays.

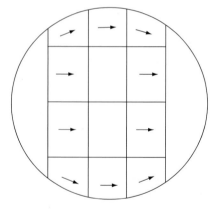

Fig. 15. The flowing direction of liquid phases on the big-hole trays.

Figs. 13 and 14 illustrate the gradient of liquid layer on the traditional trays and the improved gradient of liquid layer on the big-hole sieve trays, respectively. Fig. 15 shows the flowing direction of liquid phases on the big-hole sieve trays. This manifests that the flowing of liquid phases on the big-hole sieve tray is indeed rational.

## 2.4. Operation policy

We know that the simplest extractive distillation process is made up of two columns, i.e. an extractive distillation column and a solvent recovery column. Herein, we call it double-column process. However, at few times, only one column, which is either as extractive distillation column or as solvent recovery column, is employed in the extractive distillation process. Herein, we call it single column process.

(1) Double-column process

For extractive distillation, double-column process is often adopted. The double-column process can be operated either in the batch way or in the continuous way. In the continuous way, two columns are operated at the same time, while in the batch way, there is still one column left not run.

Frankly speaking, we would like the continuous way than the batch way because the batch operation is tedious and the production efficiency is very low. However, the reason why the batch way is carried out arises from the solvent recovery column. Due to the very high boiling point of the solvent, the solvent recovery column has to be operated at vacuum pressure in order to decrease the temperature of the bottom and thus prevent the solvent from decomposition. Moreover, at some time there may be a strict requirement for the purity of the recycled solvent. All these lead to the difficulty of the continuous operation in the solvent recovery column. A strategy to solve them is by decreasing the boiling point of the solvent and canceling the vacuum system (see the next section 3.2. Extractive distillation with liquid solvent).

However, in fine chemical engineering, the batch way is often used. Although it is complicated, it is flexible and the amount of treated feed is not so many.

(2) Single column process

For the single column process, there are two cases: batch mode and semi-continuous mode [36-41]. In the batch mode, the solvent is charged in the reboiler with the feed mixture at the beginning of operation. Therefore, it limits the amount of feed mixture to be processed, as the reboiler has a limited capacity. This increases the number of batches to be processed in a campaign mode operation. In this mode, finding the optimum feed charge to solvent ratio is an important factor to maximize the productivity.

In the semi-continuous mode, the solvent is fed to the column in a semi-continuous fashion at some point of the column. There may be two strategies in this mode of operation while charging of the initial feed mixture is concerned:

(a) Full charge. In this strategy the feed mixture is charged in the reboiler to its maximum capacity at the beginning of operation. For a given condenser vapor load, if the reflux ratio and the solvent feed rate aren't carefully controlled, the column will possibly be flooded.

(b) Fractional charge. "Full charge" isn't a suitable option to achieve the required product specifications for many azeotropic mixtures in a given sized column, operating even at the maximum reflux ratio, because the amount of solvent required in such separations is more than those that the column can accommodate without flooding. Mujtaba [40] proposed a fractional feed charge strategy for this type of mixture. In this strategy the feed mixture is charged to a certain fraction of the maximum capacity. The column can operate at a reflux ratio greater than maximum reflux ratio for some period until the reboiler level reaches to its maximum capacity.

However, it can be deduced that comparing with batch mode and semi-continuous mode, the latter is more promising because this mode can deal with much more products in the same column. A complete batch extractive distillation of semi-continuous mode consists of the following steps:

(a) Operation under total reflux without solvent feeding.

(b) Operation under total reflux with solvent feeding. For decreasing the concentration of the less volatile component (LVC) in the distillate, this step isn't absolutely necessary because the first part of the distillate containing too much of the LVC can be recycled to the next charge.

(c) Operation under finite reflux ratio with solvent feeding (production of the more volatile component (MVC)).

(d) Operation under finite reflux ratio without solvent-feeding (separation of the LVC from the solvent). Before the start of the production of LVC an off-cut can be taken.

It should be noted that the holdup on the reboiler must not exceed the maximum capacity at any time within the entire operation period to avoid column flooding.

Many researchers focus on exploring the semi-continuous mode in such aspects as simulations, improving equipment with an external middle vessel and so on, among which batch extractive distillation with an external idle vessel is an interesting topic. The typical process and apparatus are illustrated in Fig. 16.

In one example [37], the packed column with a 38 mm diameter was composed of three parts, i.e. scrubbing section, rectifying section and stripping section. The scrubbing section was packed to a height of 150 mm with 2.5 × 2.5 Dixon rings, and the rectifying and stripping were, respectively, packed to a height of 500 mm with 2.5 × 2.5 Dixon rings. The external middle vessel is a 2000 ml agitated flat bottom flask. The materials to be separated were charged to the middle vessel. The liquid flow from the middle vessel to the stripping section was controlled by an electric-magnetic device. The column was operated under the following steps:

(a) Operation under total reflux without solvent feeding;

(b) Operation under total reflux with solvent feeding;

(c) Operation under finite reflux ratio and total reboiler ratio with solvent feeding (production of the more volatile component from the top of the column, while no production of solvent from the bottom);

(d) Operation under finite reflux ratio and finite reboiler ratio with solvent feeding (production of more volatile component from the top and solvent from the bottom of the column simultaneously);

(e) Operation under finite reflux ratio and finite reboiler ratio without solvent feeding (production of slop cut or less volatile component from the top while solvent from the bottom);

(f) When the less volatile component or solvent in the middle vessel reaches the required purity or there is only a little liquid in the middle vessel, the operation is stopped.

The operation and simulation of this process in a batch column with a middle vessel is very complex, but it has many advantages such as flexibility, multi-component separation in one column and so on. It is especially suitable for the batch extractive distillation with large solvent ratio. However, when the boiling point of solvent is very high while the boiling point of the components to be separated are low, it is inconvenient to use the column, since the reboiler needs a high temperature heat resource or vacuum system.

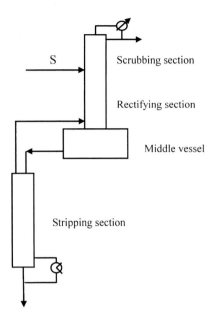

Fig. 16. Single column process with an external middle vessel.

## 3. SOLVENT OF EXTRACTIVE DISTILLATION

The two factors that influence the extractive distillation process are separation process and solvents (or separation agents). Assuming that the separation process is determined, the task is to select the basic solvent with high separation ability. When a basic solvent is found, this solvent should be further optimized to improve the separation ability and to decrease the solvent ratio and liquid load of the extractive distillation column.

The solvent is the core of extractive distillation. It is well-known that selection of the most suitable solvent plays an important role in the economical design of extractive distillation. Excellent solvents should at least decrease the solvent ratio and the liquid load of the extractive distillation column, and make the operation easily implemented. Up to date, there are four kinds of solvents used in extractive distillation, i.e. solid salt, liquid solvent, the combination of liquid solvent and solid salt, and ionic liquid, which are discussed in this text.

### 3.1. Extractive distillation with solid salt
*3.1.1. Definition of extractive distillation with solid salt*
In certain systems where solubility permits, it is feasible to use a solid salt dissolved into the liquid phase, rather than a liquid additive, as the separating agent for extractive distillation. The extractive distillation process in which the solid salt is used as the separating agent is called extractive distillation with solid salt. However, herein, the ionic liquids aren't included in the solid salts although they are a kind of salts. Ionic liquid as a special separating agent

will be discussed separately.

The so-called "salt effect in vapor-liquid equilibrium (VLE)" refers to the ability of a solid salt which has been dissolved into a liquid phase consisting of two or more volatile components to alter the composition of the equilibrium vapor without itself being present in the vapor. The feed component in which the equilibrium vapor is enhanced is thought to have been "salting-out" by the salt; otherwise, if the equilibrium vapor is declined, then "salting-in". This phenomenon can be described by the following famous Setschenow equation which expresses the solubility of a non-electrolyte in a solid salt solution with low salt concentration [42]:

$$\log \frac{c_0}{c} = k_s c_s \tag{3}$$

The notation meaning of the above equation refers to Eq. (113) in chapter 1.

### 3.1.2. The process of extractive distillation with salt

The process of extractive distillation with salt is somewhat different from the process shown in Fig. 4, in that the salt isn't recovered by means of distillation. Anyway, any extractive distillation process can be taken on as consisting of one extractive distillation column, and one solvent recovery equipment but sometimes maybe not distillation column [43-45].

Fig. 17 demonstrates a typical flowsheet for salt-effect extractive distillation. The solid salt, which must be soluble to some extent in both feed components, is fed at the top of the column by dissolving it at a steady state into the boiling reflux just prior to entering the column. The solid salt, being nonvolatile, flows entirely downward in the column, residing solely in the liquid phase. Therefore, no scrubbing section is required above the situation of feeding separating agent to strip agent from the overhead product. Recovery of the salt from the bottom product for recycle is by either full or partial drying, rather than by the subsequent distillation operation for recovering the liquid separating agents.

Several variations on the Fig. 17 process are possible. For example, an azeotrope-containing system could be separated by first taking it almost to the azeotrope by ordinary distillation without adding any separating agent, then across the former azeotrope point by extractive distillation with solid salt present, usually containing a little components to be separated, and then to final purity by further distillation without solid salt. The solid salt is generally recovered by evaporation.

In this case, the flowsheet of extractive distillation with salt is shown in Fig. 18. One advantage of this process is that the salt isn't difficult to be recovered, only by evaporation. Thus the process is convenient to operate.

### 3.1.3. Case studies
(1) Separation of ethanol and water

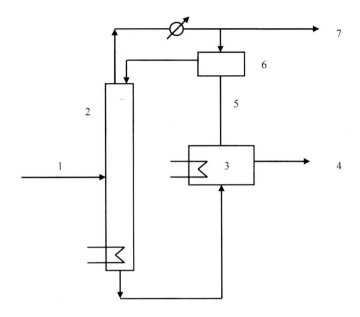

Fig. 17. One flowsheet of extractive distillation with salt. 1- feed stream; 2- extractive distillation column; 3- salt recovery equipment; 4- bottom product; 5- the salt recovered; 6- reflux tank; 7- overhead product.

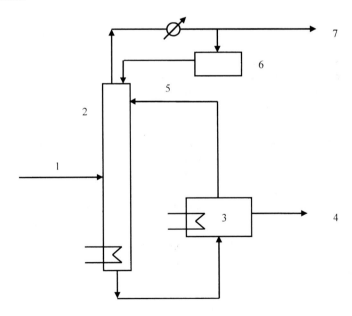

Fig. 18. One flowsheet of extractive distillation with salt. 1- feed stream; 2- extractive distillation column; 3- evaporation tank; 4- bottom product; 5- salt solution; 6- reflux tank; 7- overhead product.

Ethanol is a basic chemical material and solvent used in the production of many chemicals and intermediates. Especially in the recent year, ethanol has received increasing attention because it is an excellent alternative fuel and has a virtually limitless potential for growth. However, ethanol are usually diluted and must be separated from water. It is known that ethanol forms azeotrope with water, and can't be extracted to a high concentration from the aqueous solutions by ordinary distillation methods.

A process known as HIAG (Holz Industrie Acetien Geselleschaft), licensed by DEGUSSA and based on patents registered by Adolph Gorhan, employed extractive distillation using a 70/30 mixture of potassium and sodium acetates as the separating agent, and produced above 99.8% ethanol completely free of separating agent directly from the top of the column [44, 45].

The separation of ethanol and water is the most important application of extractive distillation with solid salt. The influence of various salts on the relative volatility of ethanol and water was investigated by Duan et al. [46-48], and the results are listed in Table 3, where the volume ratio of the azeotropic ethanol-water mixture to the separating agent is 1.0, and the concentration of salt in the separating agent is 0.2 g (salt)/ml (solvent).

Table 3
The influence of various solid salts and liquid solvents on the relative volatility of ethanol and water

| No. | Separating agent | Relative volatility |
|-----|------------------|---------------------|
| 1 | no solvent | 1.01 |
| 2 | ethylene glycol | 1.85 |
| 3 | calcium chloride saturated | 3.13 |
| 4 | potassium acetate | 4.05 |
| 5 | ethylene glycol+NaCl | 2.31 |
| 6 | ethylene glycol+$CaCl_2$ | 2.56 |
| 7 | ethylene glycol+$SrCl_2$ | 2.60 |
| 8 | ethylene glycol+$AlCl_3$ | 4.15 |
| 9 | ethylene glycol+$KNO_3$ | 1.90 |
| 10 | ethylene glycol+$Cu(NO_3)_2$ | 2.35 |
| 11 | ethylene glycol+$Al(NO_3)_3$ | 2.87 |
| 12 | ethylene glycol+$CH_3COOK$ | 2.40 |
| 13 | ethylene glycol+$K_2CO_3$ | 2.60 |

It can be seen from the above that the higher the valence of metal ion is, the more obvious the salt effect is. That is to say, the order of salt effect is: $AlCl_3 > CaCl_2 > NaCl_2$; $Al(NO_3)_3 > Cu(NO_3)_2 > KNO_3$. Besides, the effects of acidic roots are different, in an order of $Ac^- > Cl^- > NO^-_3$.

(2) Separation of isopropanol and water

Azeotropic distillation with benzene as the entrainer is also a conventional process for the isopropanol-water separation. The Ishikawajima-Harima Heavy industries (IHI) company in Japan has developed a process for isopropanol production from aqueous solution using the solid salt, calcium chloride, as the azeotrope-breaking agent. In a 7300 tonnes per year plant, IHI reported a capital cost of only 56%, and an energy requirement of only 45% of those for the benzene process [44, 45].

(3) Separation of nitric acid and water

A process in North America using extractive distillation by salt effect is the production of nitric acid from aqueous solution using a solid salt, magnesium nitrate as the separating agent. Hercules Inc. has been operating such plants since 1957, and reported lower capital costs and lower overall operating costs than for the conventional extractive distillation process which uses a liquid solvent, sulfuric acid, as the separating agent [44, 45].

*3.1.4. The advantages and disadvantages of extractive distillation with solid salt*

In the systems where solubility considerations permit their use, solid salt as the separating agents have major advantages. The ions of a solid salt are typically capable of causing much large effects than the molecules of a liquid agent, both in the strength of attractive forces they can exert on feed component molecules, and in the degree of selectivity exerted. This means that the salt is of good separation ability. In the extractive distillation process, the solvent ratio is much smaller than that of the liquid solvent (mentioned afterwards), which leads to high production capacity and low energy consumption. Moreover, since solid salt isn't volatile, it can't be entrained into the product. So no salt vapor is inhaled by operators. From this viewpoint, it is environment-friendly.

However, it is a pity that when solid salt is used in industrial operation, dissolution, reuse and transport of salt is quite a problem. The concurrent jam and erosion limit the industrial value of extractive distillation with solid salt. That is why this technique, extractive distillation with solid salt, isn't widely used in industry.

## 3.2. Extractive distillation with liquid solvent

*3.2.1. Definition of extractive distillation with liquid solvent*

Like the extractive distillation with solid salt, in certain systems where solubility permits, it is feasible to use a liquid solvent dissolved into the liquid phase, rather than salt, as the separating agent for extractive distillation. Therefore, the extractive distillation process in which the liquid solvent is used as the separating agent is called extractive distillation with liquid solvent. Note that, herein, the ionic liquids aren't included in the liquid solvents although they are in the liquid phase from room temperature to a higher temperature. The contents about ionic liquid as a special separating agent will be discussed afterwards.

*3.2.2. The process of extractive distillation with liquid solvent*

See Fig. 4 (the process of extractive distillation with liquid solvent).

*3.2.3. Case studies*

(1) Reactive extractive distillation

There are many examples for extractive distillation with a single liquid solvent as the separating agent, as shown in Table 1. Herein, this example is concerned with the separation of acetic acid and water by reactive extractive distillation in which chemical reaction is involved.

Acetic acid is an important raw material in the chemical industries. But in the production of acetic acid, it often exists with much water. Because a high-purity of acetic acid is needed in industry, the problem of separating acetic acid and water is an urgent thing. By now, there are three methods commonly used for this separation, i.e. ordinary distillation, azetropic distillation and extractive distillation [49-54]. Although ordinary distillation is simple and easy to be operated, its energy consumption is large and a lot of trays are required. The number of trays for azeotropic distillation is fewer than that for ordinary distillation. But the amount of azeotropic agent is large, which leads to much energy consumption because the azeotropic agents must be vaporized in the column. However, in the extractive distillation process the separating agents aren't vaporized and thus the energy consumption is relatively decreased. Therefore, extractive distillation is an attractive method for separating acetic acid and water, and has been studied by Berg [49, 50].

The reported separating agents in extractive distillation are sulfolane, adiponitrile, pelargonic acid, heptanoic acid, isophorone, neodecanoic acid, acetophenone, nitrobenzene and so on. It is evident that the interaction between acetic acid or water and these separating agents is mainly physical force including the van der Waals bonding and hydrogen bonding.

Recently, a new term, reactive extractive distillation, has been put forward [149-150], and a single solvent, tributylamine, is selected as the separating agent.

If we select the solvent tributylamine (b.p. 213.5℃) as the separating agent, then the following reversible chemical reaction may take place:

$$HAc \quad + \quad R_3N \quad \rightleftharpoons \quad R_3NH^+ \cdot {}^-OOCCH_3$$

where HAc, $R_3N$ and $R_3NH^+ \cdot {}^-OOCCH_3$ represent acetic acid, tributylamine and the salt formed by reaction, respectively. This reaction may be reversible because weak acid (acetic acid) and weak base (tributylamine) are used as reactants. That is to say, for the extractive distillation process the forward reaction occurs in the extractive distillation column and the reverse reaction occurs in the solvent recovery column. Therefore, this new separation method is different from traditional extractive distillation with liquid solvent, and based on the reversible chemical interaction between weak acid (acetic acid) and weak base (separating agent). So we call it reactive extractive distillation.

A new substance, $R_3NH^+ \cdot {}^-OOCCH_3$, is produced in this reaction, which can be verified by infrared spectra (IR) technique. It can be seen from the IR diagrams that a new characteristic peak in the range of 1550 cm$^{-1}$ to 1600 cm$^{-1}$ appears in the mixture of acetic acid and tributylamine, and is assigned to the carboxylic-salt function group, —COO$^-$. This indicates that chemical reaction between HAc and $R_3N$ indeed takes place.

On the other hand, this chemical reaction is reversible, which can be verified by mass chromatogram (MS) technique. Two peaks denoting acetic acid and tributylamine respectively can be found for the mixture of acetic acid 10 %wt and tributylamine 90 %wt.

So it indicates that HAc, $R_3N$ and the product produced by them all can be detected by the combination of IR and MS techniques. In general, the reaction rate between weak acid and weak base is very quick. So the chemical reaction between HAc and $R_3N$ may be reversible. The further proof of reversible reaction is supported by investigating the chemical equilibrium constant.

In order to verify the effect of tributylamine as the separating agent, the vapor liquid equilibrium (VLE) was measured by experiment. Fig. 19 shows the equilibrium data for the ternary system of water (1) + acetic acid (2) + tributylamine (3), plotted on a tributylamin free basis. It may be observed that the solvent, tributylamin, enhances the relative volatility of water to acetic acid in such a way that the composition of the more volatile component (water) is higher in the liquid than in the vapor phase. The reason may be that the interaction forces between acetic acid and tributylamin molecules are stronger than those between water and tributylamin molecules because in the former reversible chemical reaction takes place. As a result, water would be obtained as the overhead product in the extractive distillation column, being acetic acid and the solvent, tributylamine, the bottom product.

By analyzing the reaction system, it is found that the new group –NH is formed and the old group –OH is disappeared during the reaction. This reaction is exothermic and the heat generated can be obtained from the reference [54], i.e. -2.17 kJ mol$^{-1}$. In addition, the ionization constant $PK_a$ of acetic acid and tributylamine at 25℃ can be found from the reference [55], i.e. 4.76 and 10.87 respectively. Therefore, the chemical equilibrium constant $K_p$ at 25℃ can be deduced, and thus the relationship of chemical equilibrium constant with temperature is given by:

$$K_p = \frac{C_{salt}^2}{C_{HAc}C_{R_3N}} = \exp(-4.6284 + \frac{261.01}{T}) \tag{4}$$

where $T$ is the absolute temperature (K), and $C$ is the mole concentration (kmol m$^{-3}$).

It is found that chemical equilibrium constant is small, about 0.02 under the operation condition of the extractive distillation column. This means that this reaction is reversible, and the chemical interaction between acetic acid and tributylamine is weak.

In terms of the mechanism of reactive extractive distillation, the following criteria should be satisfied in order to make ensure that this method can be implemented.

(a) The chemical reaction is reversible. The reaction product (here as $R_3NH^+ \cdot {}^-OOCCH_3$) is used as carrier to carry the separated materials back and forth.

(b) One of the reactants (here as acetic acid) is a low boiling-point component. It can be easily removed by distillation so that the separating agent (here as tributylamine) can be regenerated and recycled.

(c) There are no other side reactions between the separating agent and the component to be separated. Otherwise, the separation process will be complicated and some extra equipment may be added, which results in no economy of this technique.

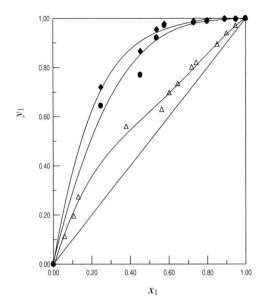

Fig. 19. VLE curves on the solvent free basis for the ternary system of water (1) + acetic acid (2) + tributylamine (3) at 101.33 kPa. ◆ — solvent/feed volume ratio 2:1; ● — solvent/feed volume ratio 1:1; △ — no solvent.

It is obvious that the system consisting of water, acetic acid and tributylamine meets these requirements. So it seems to be advisable to separate water and acetic acid by reactive extractive distillation. However, it should be cautious that one solvent with a tri-amine $R_3N$, not di-amine $R_2NH$ or mono-amine $RNH_2$, can be selected as the separating agent because the reaction between solvent with two or single amine groups and acetic acid is irreversible, and some amides will be produced.

(2) The mixture of liquid solvents

Compared with the single solvent, the mixture of the liquid solvents as the separating agents is complicated and interesting. However, in most cases the number of the liquid solvents in the mixture is just two. According to the aims of adding one solvent to another, it can be divided into two categories: increasing separation ability and decreasing the boiling point of the mixture.

(a) Increasing separation ability

As we know, selection of the most suitable solvent plays an important role in the economical design of extractive distillation. Moreover, the most important factor in selecting the solvents is relative volatility. Increasing separation ability of the solvent means increasing the relative volatility of the components to be separated. That is to say, when one basic solvent is given, it is better to find the additive to make into a mixture to improve the relative volatility, and thus to decrease the solvent ratio and liquid load of the extractive distillation column. It has been verified by many examples that adding a little of extra solvent (additive) to one basic solvent will increase the separation ability greatly.

For example, adding small amounts of water has improved the separation ability of acetonitrile (ACN) in separating C4 mixture [56, 57]. Some results are listed in Table 4, where the subscript (1 - 13) represents n-butane, isobutane, isobutene, 1-butene, trans-2-butene, propadiene, cis-2-butene, 1,3-butadiene, 1,2-butadiene, methyl acetylene, butyne-1, butyne-2 and vinylacetylene (VAC) respectively, and 100%, 80% and 70% are the solvent weight fraction in the mixture of ACN and C4.

Table 4
The relative volatility of C4 mixture to 1,3-butadiene at 50 ℃

| | ACN | | | ACN + 5 %wt water | | | ACN + 10 %wt water | | |
|---|---|---|---|---|---|---|---|---|---|
| | 100% | 80% | 70% | 100% | 80% | 70% | 100% | 80% | 70% |
| $\alpha_{18}$ | 3.01 | 2.63 | 2.42 | 3.46 | 2.94 | 2.66 | 3.64 | 3.11 | 2.75 |
| $\alpha_{28}$ | 4.19 | 3.66 | 3.37 | 4.95 | 4.11 | 3.72 | 5.40 | 4.35 | 3.82 |
| $\alpha_{38}$ | 1.92 | 1.79 | 1.72 | 2.01 | 1.84 | 1.75 | 2.09 | 1.89 | 1.78 |
| $\alpha_{48}$ | 1.92 | 1.79 | 1.72 | 2.01 | 1.84 | 1.75 | 2.09 | 1.89 | 1.78 |
| $\alpha_{58}$ | 1.59 | 1.48 | 1.42 | 1.59 | 1.54 | 1.46 | 1.78 | 1.58 | 1.49 |
| $\alpha_{68}$ | 2.09 | 2.14 | 2.17 | 1.97 | 2.04 | 2.08 | 1.88 | 1.97 | 2.02 |
| $\alpha_{78}$ | 1.45 | 1.35 | 1.30 | 1.48 | 1.36 | 1.30 | 1.51 | 1.37 | 1.30 |
| $\alpha_{88}$ | 1.00 | 1.00 | 1.00 | 1.00 | 1.00 | 1.00 | 1.00 | 1.00 | 1.00 |
| $\alpha_{98}$ | 0.731 | 0.720 | 0.712 | 0.750 | 0.733 | 0.722 | 0.755 | 0.738 | 0.728 |
| $\alpha_{10.8}$ | 1.00 | 1.09 | 1.16 | 0.947 | 1.060 | 1.130 | 0.897 | 1.050 | 1.120 |
| $\alpha_{11.8}$ | 0.481 | 0.499 | 0.508 | 0.456 | 0.478 | 0.490 | 0.435 | 0.462 | 0.476 |
| $\alpha_{12.8}$ | 0.296 | 0.303 | 0.308 | 0.278 | 0.288 | 0.293 | 0.267 | 0.279 | 0.287 |
| $\alpha_{13.8}$ | 0.389 | 0.413 | 0.430 | 0.364 | 0.400 | 0.414 | 0.344 | 0.379 | 0.403 |

It can be seen from Table 4 that adding a little water to ACN indeed increases the relative volatilities of $\alpha_{18}$, $\alpha_{28}$, $\alpha_{38}$, $\alpha_{48}$, $\alpha_{58}$, $\alpha_{68}$ and $\alpha_{78}$ in which 1,3-butadiene is as the heavy component and decreases the relative volatilities of $\alpha_{98}$, $\alpha_{10,8}$, $\alpha_{11,8}$, $\alpha_{12,8}$ and

$\alpha_{13,8}$ in which 1,3-butadiene is as the light component. So the mixture of water and ACN can

enhance the separation ability. However, ACN is blended with water to separate C4 mixture with the defect that ACN is prone to hydrolyze, which leads to equipment corrosion and operation difficulty. From this viewpoint, it is thought that water may be not the best additive. Which additive is the best? This question will be explored in the 2.5. CAMD of extractive distillation section.

Other examples in which the new additive is added to improve the separation ability of the original solvent are illustrated in Tables 5 and 6 for separating cyclopentane (1) / 2,2-dimethylbutane (2) and n-pentane (1) / 1-pentene (2), respectively. In Table 5, NMP, CHOL and NMEP stand for N-methyl-2-pyrrolidone, cyclohexanol and N- ( β -mercaptoethyl)-2-pyrrolidone, respectively [58-60].

Table 5

Relative volatility $\alpha_{12}$ of the mixture of the solvents for separating cyclopentane (1) and

2,2-dimethylbutane (2); adapted from the reference [60]

| No. | Solvent/feed mass ratio | Mixture of the solvents | $\alpha_{12}$ |
|---|---|---|---|
| 1 | 3 : 1 | NMEP | 1.28 |
| 2 | 3 : 1 | A: NMEP+CHOL | 1.40 |
| 3 | 3 : 1 | B: NMEP+CHOL | 1.42 |
| 4 | 3 : 1 | C: NMEP+NMP | 1.32 |
| 5 | 5 : 1 | NMEP | 1.48 |
| 6 | 5 : 1 | CHOL | 1.32 |
| 7 | 5 : 1 | NMP | 1.37 |
| 8 | 5 : 1 | A: NMEP+CHOL | 1.64 |
| 9 | 5 : 1 | B: NMEP+CHOL | 1.57 |
| 10 | 5 : 1 | C: NMEP+NMP | 1.54 |
| 11 | 7 : 1 | NMEP | 1.71 |
| 12 | 7 : 1 | CHOL | 1.26 |
| 13 | 7 : 1 | NMP | 1.33 |
| 14 | 7 : 1 | A: NMEP+CHOL | 1.70 |
| 15 | 7 : 1 | B: NMEP+CHOL | 1.64 |
| 16 | 7 : 1 | C: NMEP+NMP | 1.74 |

(b) Decreasing boiling point

In general, the solvent has a much higher boiling point than the components to be separated in order to ensure the complete recovery of the solvent. It indicates that at the bottom of the solvent recovery column, a water stream with a higher temperature than the boiling point of the solvent is needed to heat the solvent to reach its boiling point at normal pressure. This means that much energy with high level has to be consumed. If the solvent recovery column is operated at below normal pressure, the degree of vacuum pressure must not be too high due to the restriction of the temperature of condensing water at the top. Therefore, the best strategy is to find the suitable additive to make into a mixture to decrease the boiling point. It is evident that the boiling point of the additive is relatively lower than the basic solvent, but higher than the components to be separated. Table 7 illustrates the examples of the mixture of the solvents, in which the function of the additive is to decrease the boiling point [61, 62].

Table 6

Selectivity $S_{12}$ of the mixture of the solvents for separating n-pentane (1) and 1-pentene (2); adapted from the reference [60]

| No. | Mixture of the solvents | | Concentration | $S_{12}$ |
|---|---|---|---|---|
| | A | B | Volume fraction of B, % | |
| 1 | methyl cellosolve | nitromethane | 0 | 1.69 |
| 2 | methyl cellosolve | nitromethane | 5 | 1.70 |
| 3 | methyl cellosolve | nitromethane | 100 | 2.49 |
| 4 | pyridine | butyrolactone | 0 | 1.60 |
| 5 | pyridine | butyrolactone | 32.1 | 1.79 |
| 6 | pyridine | butyrolactone | 100 | 2.17 |
| 7 | ethyl methyl ketone | butyrolactone | 0 | 1.62 |
| 8 | ethyl methyl ketone | butyrolactone | 50 | 1.79 |
| 9 | ethyl methyl ketone | butyrolactone | 100 | 2.17 |

Table 7
The examples of the mixture of the solvents for decreasing the boiling point

| No. | Systems | Mixture of the solvents |
|---|---|---|
| 1 | C4 hydrocarbons | ACN + water |
| 2 | aromatics and non-aromatics | NMP + water |
| 3 | benzene and non-aromatics | NFM + an additive |

However, one question arises, that is, the separation ability of the mixture of the solvents is strengthened or declined? The fact is that, at some time, the separation ability of the mixture of the solvents is improved as the example of Table 4, where the mixture of ACN and water not only improves the separation ability, but also decreases the boiling point. At other time, the separation ability of the mixture of the solvents indeed declines, but not obviously. The reason is that as the temperature decreases, there is a positive effect on the separation ability of the mixture of the solvents. The temperature effect on selectivity is given by

$$\frac{d(\lg S_{12}^0)}{d(1/T)} = \frac{L_1^0 - L_2^0}{2.303R} \tag{5}$$

where $L_i^0$ is partial molar heat of solution, component $i$ at infinite dilution in the solvent, $T$ is the absolute temperature (K), and $R$ is gas constant (8.314 J mol$^{-1}$ K)).

It is shown that $\lg S_{12}^0$ is proportional to the reciprocal absolute temperature. This means that although the separation ability of the additive is weaker than that of the basic solvent, the relatively low boiling point of the mixture of the solvents in some degrees offsets the negative influence of the separation ability of the additive, which leads to no apparent decrease of the separation ability of the mixture of the solvents. That is a reason why some researchers are willing to improve the solvent by decreasing the boiling point.

In addition, the mixture of the solvents can raise the yield ratio of the product. Zhu et al. [61] studied an extractive distillation process for recovering high purity of benzene from pyrolysis gasoline. The selected solvent is the mixture of N-formylmorpholine (NFM) and an additive. When the weight fraction of the additive is in the range of 5% - 20%, the yield ratio of benzene increases from 99.0% to 99.8% and the bottom temperature of the solvent recovery column is less than 190℃.

*3.2.4. The advantages and disadvantages of extractive distillation with liquid solvent*

In most cases, the solvent and feed mass ratio (i.e. solvent ratio) in the extractive distillation with liquid solvents is very high, up to 5-8. For example, for the separation of C4 hydrocarbon using N,N-dimethylformamide (DMF) or acetonitrile (ACN) as the solvents, the solvent ratio is 7-8 in industry, which leads to much consumption of energy. However, in the systems where solubility considerations permit their use, the separation ability of the solid salts is much greater than that of the liquid solvents, and thus the solvent ratio is generally low. The reason why extractive distillation with liquid solvent is more widely used in industry rather than extractive distillation with solid salt is that there is no problems of dissolution, reuse and transport for the liquid solvent because it is in the liquid phase under the operation condition of extractive distillation process. In summary, the advantages of extractive distillation with liquid solvent predominate over its disadvantages. On the contrary, the disadvantages of extractive distillation with solid salt predominate over its advantages.

**3.3. Extractive distillation with the combination of liquid solvent and solid salt**

*3.3.1. Definition of extractive distillation with liquid solvent and solid salt*

Like the extractive distillation with solid salt or liquid solvent, in certain systems where solubility permits, it is feasible to use a combination of liquid solvent and solid salt dissolved into the liquid phase, rather than only salt or liquid solvent, as the separating agent for extractive distillation. Therefore, the extractive distillation process in which the combination of liquid solvent and solid salt is used as the separating agent is called extractive distillation with the combination of liquid solvent and solid salt.

### 3.3.2. The process of extractive distillation with the combination of liquid solvent and solid salt

See Fig. 4 (the process of extractive distillation with liquid solvent). That is to say, the processes of extractive distillation with liquid solvent and with the combination of liquid solvent and solid salt, are identical.

### 3.3.3. Case studies

Extractive distillation with the combination of liquid solvent and solid salt as the separating agent is a new process for production of high-purity products. This process integrates the advantages of liquid solvent (easy operation) and solid salt (high separation ability). In industrial operation, when only salt is used, dissolution, reuse and transport of solid salt is quite a problem. The concurrent jam and erosion limits the industrial value of extractive distillation with solid salt only. However, the mixture of liquid solvent and solid salt can avoid the defects and realize continual production in industry.

Extractive distillation with the combination of liquid solvent and solid salt can be suitable either for the separation of polar systems or for the separation of non-polar systems, respectively.

Herein, we select aqueous alcohol solutions, ethanol/water and isopropanol/water, as the representatives of polar systems and C4 mixture as the representatives of non-polar systems. It is known that both anhydrous alcohol and 1,3-butadiene are basic chemical raw materials. Anhydrous alcohol is not only used as chemical reagent and organic solvent, but also used as the raw material of many important chemical products and intermediates; 1,3-Butadiene mainly comes from C4 mixture and is utilized for the synthesis of polymers on a large scale. It has been reported [13, 63-64] that the systems of aqueous alcohol and C4 mixture are able to be separated by extractive distillation. Since these two materials are very important in industry, the separation of them by extractive distillation is interesting.

(1) Separation of the systems of ethanol /water and isopropanol/water

Firstly the equilibrium data of the ethanol (1) - water (2) system, which corresponded well with the reference data [65], were measured. It is verified that the experimental apparatus was reliable. Then the measurements were respectively made for the system ethanol (1) - water (2) - ethylene glycol (solvent/feed volume ratio is 1:1) and the system ethanol (1) - water (2) - ethylene glycol - $CaCl_2$ (solvent/feed volume ratio is 1:1 and the concentration of salt is 0.1 g/ml solvent) at normal pressure.

On the other hand, measurements were also made for the system of isopropanol (1) / water (2) / ethylene glycol / glycollic potassium at normal pressure [13]. Isopropanol (1) / water (2) and the mixture of ethylene glycol and glycollic potassium were blended with the feed/solvent volume ratio of 1:1. The mixture of ethylene glycol and glycollic potassium were prepared from ethylene glycol and potassium hydroxide with the weight ratio 5:1 and 4:1 respectively. During preparation, the mixture of ethylene glycol and potassium hydroxide were fed into a distillation column and the water produced was removed. Afterwards, the experimental VLE data were measured.

The experimental results from both systems show that under the same liquid composition the mole fraction of alcohol in the vapor phase with salt is higher than that without salt. It means that adding salt to ethylene glycol is advisable for improving the separation ability of the solvent, while alcohol as a light component and water as a heavy component.

(2) Separation of C4 mixture

When acetonitrile (ACN) is regarded as a basic solvent, organic solvents including water and salts will be added [64]. The aim is to explore the effect of them on $\alpha_{ij}^{\infty}$. Among organic solvents, water and ethylenediamine are better additives than other solvents. However, it is found that a little of solid salt added to ACN can effectively improve the relative volatility and the effect of solid salt is close to water but stronger than ethylenediamine. However, formerly ACN was mixed with water to separate C4 mixture with the defect that ACN is prone to hydrolyze, which leads to equipment corrosion and operation difficulty.

On the other hand, N,N-dimethylformamide (DMF) is another solvent commonly used to separate C4 mixture. Due to the same reason as ACN, DMF is used as a single solvent. It is expected to modify it with solid salt. Many substances are strongly soluble in DMF including many kinds of solid salts. The influence of solid salts and organic additives on the separation ability of DMF is tested [63, 64]. The same phenomenon as ACN exists. Salts added to DMF also improve the relative volatilities of C4 to some extent, and at the same additive concentration the effect of the salts is more apparent than that of organic solvents. Moreover, if some factors such as relative volatilities, price, erosion, source and so on are considered, the salts NaSCN and KSCN are the best additives.

*3.3.4. The advantages and disadvantages of extractive distillation with the combination of liquid solvent and solid salt*

As mentioned above, extractive distillation with the combination of liquid solvent and solid salt as the separating agent integrates the advantages of liquid solvent (easy operation) and solid salt (high separation ability). From the above discussion, we know that whether it is the separation of polar or non-polar systems, extractive distillation with the combination of liquid solvent and solid salt may be feasible. So it is concluded that extractive distillation with the combination of liquid solvent and solid salt is a promising separation method. If we meet with the problems about extractive distillation in the future, it is wise to try to improve it by adding solid salt.

Unfortunately, many solid salts are corrosive to the equipment and easies to decompose at a high temperature. In some cases the kinds of solid salts that we can select are a few. An economic calculation must be made in determining the final salts. The benefit from adding salts in the production should exceed the price of salts and other charges.

On the other hand, since the amount of solid salts added to the liquid solvents is often small, the role of solid salts in improving the separation ability is doomed to be limited.

Besides, liquid solvents are volatile, which inevitably pollutes the top product of the extractive distillation column. Accordingly, new separating agents should be sought to avoid these problems brought on by liquid solvent and solid salt.

### 3.4. Extractive distillation with ionic liquid

*3.4.1. Definition of extractive distillation with ionic liquid*

Like the extractive distillation with solid salt or liquid solvent or the combination, in certain systems where solubility permits, it is feasible to use an ionic liquid dissolved into the liquid phase, rather than only salts or liquid solvents, as the separating agent for extractive distillation. Therefore, the extractive distillation process where ionic liquid is used as the separating agent is called extractive distillation with ionic liquid.

Then, what are ionic liquids? Ionic liquids are salts consisting entirely of ions, which exist in the liquid state at ambient temperature, i.e. they are salts that don't normally need to be melted by means of an external heat source.

Ionic liquids are often used in the chemical reaction [66-71]. The cases of applications on the separation process are a few, especially rarely reported on the extractive distillation (only one patent found) [72].

Ionic liquids are called "green" solvents. When they are used in the extractive distillation, the "green" is brought out in the following aspects:

(1) Negligible vapor pressure, which means that ionic liquids don't pollute the product at the top of the column. However, when the liquid solvent or the combination of liquid solvent and solid salt is used as the separating agent, the solvent may be entrained more or less into the product. In the case of strict restriction for the impurity of the product, it is advisable to use ionic liquids as the separating agents.

(2) A wide liquid range of about $300\,°C$ with a melting point around room temperature. This temperature range in most cases corresponds with the operation condition of the extractive distillation.

(3) A wide range of materials including inorganic, organic and even polymeric materials are soluble in ionic liquids, which ensures that the ionic liquids have an enough solubility for the components to be separated, and can play a role in increasing the relative volatility in the liquid phase.

(4) Potential to be reused and recycled. Due to their non-volatility, ionic liquids are easy to be recovered from the components to be separated, and the simplest way is by evaporation in a tank.

(5) Many ionic liquids are high thermal and chemical stability with or without water under the operation temperature of extractive distillation. However, some excellent liquid solvents, which are commonly used until now, such as acetontrile (ACN), dimethylformamide (DMF), N-methyl-pyrrolidone (NMP), etc., are not thermal and chemical stability with or without water, and are easy to be decomposed.

Fig. 20. Typical 1-alkyl-3-methylimidazolium cations and the abbreviations used to refer to them. R = Me, $R^1$ = Et: [emim]⁺; R = Me, $R^1$= n-Bu: [bmim]⁺; R = Me, $R^1$ = n-hexyl: [hmim]⁺; R = Me, $R^1$ = n-octyl: [omim]⁺.

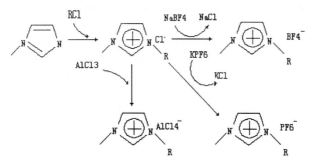

Fig. 21. One synthesis route for ionic liquids.

The most commonly used ionic liquids are those with alkylammonium, alkylphosophonium, N-alkylpyridinium, and N, N-dialkylimidazolium cations. However, much more ionic liquids are synthesized based on 1,3- dialkylimidazolium cations with 1-butyl-3-methylimidazolium [bmim]⁺ being probably the most common cation (See Fig. 20). The most common anions are [PF₆]⁻, [BF₄]⁻, [SbF₆]⁻, [CF₃SO₃]⁻, [CuCl₂]⁻, [AlCl₄]⁻, [AlBr₄]⁻, [AlI₄]⁻, [AlCl₃Et]⁻, [NO₃]⁻, [NO₂]⁻ and [SO₄]²⁻. Ionic liquids of this type have displayed the useful combination of low melting point, along with high thermal and chemical stability necessary for extractive distillation process.

The most widely used methodology in the synthesis of ionic liquids is metathesis of a halide salt of the organic cation with a group 1 or ammonium salt containing the desired anion. Alternatively, the halide salt of the organic cation may be reacted with a Lewis acid. Fig. 21 illustrates a synthesis route, in which synthesis of one ionic liquid may be carried out under an inert atmosphere because such substances as AlCl₃, [AlCl₄]⁻ are sensitive to air and moisture.

### 3.4.2. The process of extractive distillation with ionic liquid

The process of extractive distillation with ionic liquid may be same to the process of extractive distillation with solid salt (see Figs. 17 and 18).

### 3.4.3. Case study

Extractive distillation with ionic liquid as the separating agent is a very new process for producing high-purity products. This process integrates the advantages of liquid solvent (easy operation) and solid salt (high separation ability). But compared with extractive distillation with the combination of liquid solvent and solid salt, it has no problem of entrainment of the solvent into the top product of the column.

Extractive distillation with ionic liquids can be suitable either for the separation of polar systems or for the separation of non-polar systems. The following polar and non-polar systems have been investigated [72]: ethanol/water, acetone/methanol, water/acetic acid, tetrahydrofuran (THF)/water and cyclohexane/benzene. It is found that the separation effect is very apparent for these systems when the corresponding ionic liquids are used. However, it is cautious that the suitable ionic liquid should have enough solubility for the given components to be separated. The reason why ionic liquid can greatly raise the relative volatility is the same as that of using solid salt, that is, due to the salt effect.

Recently, we measured the equilibrium data of the cyclohexane (1)-toluene (2) system in which three kinds of ionic liquids are selected as the separating agents. Ionic liquids were prepared according to the synthesis route shown in Fig. 21, and [bmim]$^+$ was the cation. Their molecular structures have been verified by HNMR technique. The experimental VLE data at normal pressure are listed in Table 8, in which the mole fractions are on solvent free basis.

Table 8
Vapor-liquid equilibrium data of cyclohexane (1) and toluene (2) at normal pressure, temperature $T$ / K, liquid phase $x_1$, and vapor phase $y_1$, mole fraction for all cases, solvent/feed volume ratio is 1:1.

| No. | $T$ /$°C$ | $x_1$ | $x_2$ | $y_1$ | $y_2$ |
|---|---|---|---|---|---|
| (1) No separating agent | | | | | |
| 1 | 97.3 | 0.1267 | 0.8733 | 0.2468 | 0.7532 |
| 2 | 99.3 | 0.2614 | 0.7386 | 0.4488 | 0.5512 |
| 3 | 96.0 | 0.3534 | 0.6466 | 0.5852 | 0.4148 |
| 4 | 89.0 | 0.5021 | 0.4979 | 0.7167 | 0.2833 |
| 5 | 89.1 | 0.6516 | 0.3484 | 0.7604 | 0.2396 |
| 6 | 88.9 | 0.7738 | 0.2262 | 0.8336 | 0.1664 |
| (2) [AlCl4]$^-$ as the anion | | | | | |
| 1 | 105.5 | 0.1267 | 0.8733 | 0.2989 | 0.7011 |
| 2 | 98.0 | 0.2614 | 0.7386 | 0.5592 | 0.4408 |
| 3 | 94.0 | 0.3534 | 0.6466 | 0.6608 | 0.3392 |
| 4 | 86.8 | 0.5021 | 0.4979 | 0.7644 | 0.2356 |
| 5 | 86.0 | 0.6516 | 0.3484 | 0.8541 | 0.1459 |
| 6 | 81.6 | 0.7738 | 0.2262 | 0.8819 | 0.1181 |
| (3) [BF$_4$]$^-$ as the anion | | | | | |
| 1 | 88.2 | 0.5021 | 0.4979 | 0.6876 | 0.3124 |
| (4) [PF$_6$]$^-$ as the anion but ionic liquid is in the aqueous solution | | | | | |
| 1 | 90.6 | 0.5021 | 0.4979 | 0.5756 | 0.4244 |

As shown in Table 8, among three kinds of ionic liquids, [bmim]$^+$[AlCl4]$^-$, [bmim]$^+$[BF$_4$]$^-$ and [bmim]$^+$[PF$_6$]$^-$, the ionic liquid of [bmim]$^+$[AlCl4]$^-$ is the best for separating cyclohexane and toluene because in the case of same liquid composition ($x_1$ =

0.5021, $x_2 = 0.4979$) the corresponding vapor composition of cyclohexane is the greatest. This means that in this case the relative volatility of cyclohexane to toluene is also the largest. Moreover, it can be seen that under the same liquid composition the relative volatility of cyclohexane to toluene, when the ionic liquid of [bmim]$^+$[AlCl4]$^-$ is used as the separating agent, is apparently larger than when no separating agent is used. Since cyclohexane and toluene can be regarded as the representatives of aromatics and non-aromatics, the ionic liquid of [bmim]$^+$[AlCl4]$^-$ should be a potential solvent for separating aromatics and non-aromatics.

### 3.4.4. The advantages and disadvantages of extractive distillation with ionic liquid

As mentioned above, extractive distillation with ionic liquid as the separating agent, like the combination of liquid solvent and solid salt, integrates the advantages of liquid solvent (easy operation) and solid salt (high separation ability). Whether it is the separation of polar or non-polar systems, extractive distillation with ionic liquids may be feasible. So it is concluded that extractive distillation with ionic liquid is another promising separation method besides extractive distillation with the combination of liquids solvent and solid salt.

Unfortunately, extractive distillation with ionic liquid as the separating agents has some disadvantages, that is, it often takes much time to prepare ionic liquid and the prices of the materials used for synthesizing ionic liquid are somewhat expensive. These disadvantages may hold back the wide application of this technique in industry, even if the advantages of this technique are very attractive.

## 4. EXPERIMENTAL TECHNIQUES OF EXTRACTIVE DISTILLATION

The expressions of selectivity and relative volatility are given in Eqs. (1) and (2), respectively. Since the activity coefficients depend on the phase compositions, and the role of the solvent tends to increase with an increase in its concentration, it is common practice to consider the situation at infinite dilution. Then from Eq. (2) the definition of selectivity at infinite dilution becomes:

$$S_{ij}^\infty = (\frac{\gamma_i^\infty}{\gamma_j^\infty}) \tag{6}$$

which also represents the maximum selectivity.

A relation analogous to Eq. (6) holds for the relativity volatility:

$$\alpha_{ij}^\infty = \frac{\gamma_i^\infty P_i^0}{\gamma_j^\infty P_j^0} \tag{7}$$

which also represents the maximum relativity volatility.

In order to evaluate the possible separating agents, the goal of the experimental technique is to obtain the value of selectivity and/or relativity volatility at infinite dilution, or activity coefficient at infinite dilution from which selectivity and relativity volatility can be deduced.

These values may be found in some certain data banks. But it is in most cases that the investigated systems aren't included. Thus, we have to rely on the experiments. There are four experimental methods to obtain selectivity, or relativity volatility, or activity coefficient at infinite dilution: direct method, gas-liquid chromatography method, ebulliometry method, and inert gas stripping and gas chromatography method.

## 4.1. Direct method

This is an older method, involving extrapolation of classical vapor-liquid equilibrium (VLE) data [73, 74]. By plotting experimental data in the form of $\lg\gamma$ versus liquid composition ($x$) and extrapolating the curve to $x = 0$, the activity coefficient at infinite dilution $\gamma^\infty$ is deduced. Therefore, it is required that all of the VLE at total concentration should be measured, which leads to much time consumption and getting inaccurate results.

There are many famous experimental instruments to measure the VLE at normal or vacuum pressure [44, 45, 75-77]. One kind of experimental instruments such as Rose still, Othmer still and their modifications can be bought in the market. The common characteristics of these instruments are that they belong to recycling vapor-liquid equilibrium still and the data under boiling, isobaric conditions are measured. Each experiment is done for each data.

Another kind of experimental instruments to measure the VLE at normal or vacuum pressure [10], such as Stage-Muller dynamic still and so on, are used to obtain the data under isothermal condition. Each experiment is done for each data.

In the recent years, one important experimental instrument, i.e. HSGC (Headspace-gas chromatography) method, to measure the VLE at normal or vacuum pressure has been developed to obtain the data under isothermal condition. Each experiment is done for many data.

In some references [78-80], this method is introduced. It combines a headspace sampler and a GC (gas chromatograph) in order to determine the composition of a vapor phase. Samples of different liquid (10 ml) are filled into vials (capacity 20 ml) and sealed with air-tight septa. To ensure thermodynamic equilibrium, the vials are mixed at equilibrium temperature for 24 h and then transferred into the headspace oven. Inside the oven of the headspace sampler, the system is mixed by an agitation mode and equilibrated again for another 5 h. A pneumatically-driven thermostated headspace sampler takes a sample of approximately 0.05 g out of the vials' vapor phase in equilibrium with the liquid phase. A specially designed sample interface delivers the sample to the GC column of the GC.

The HSGC method is the most promising direct method, and can finish the VLE measurement in a relatively short time. But it is a pity that the drawback of extrapolating the experimental data can't be overcome yet.

However, if the VLE at high pressure is required for the investigated system such as C3, C4 hydrocarbons, direct method may be very tedious and difficult to measure, and is better to be replaced by other methods such as gas-liquid chromatography method, and inert gas stripping and gas chromatography method.

One development of high pressure VLE technique is the electrobalance method and its corresponding modification [151-153] in which the vapor composition is determined by GC,

UV, et al., and the liquid composition is determined by weighting. In general, it is relatively difficult to analyze the liquid sample at high pressure, but for these methods no need to take the liquid sample out. The electrobalance method, which has many advantages such as a small amount of sample (short measurement time) and a high sensitivity, is also popular in the solubility measurement at near room temperature. In the electrobalance method, several researchers used a microbalance placed in a pressure vessel. For this case, however, measurements were limited to temperatures below 125 ℃ due to the microbalance operating conditions. New type gravimetric apparatus, in which temperatures of the balance and sample can be controlled independently, have been proposed for solubility measurements at high temperatures. Kleinrahm and Wagner [154] developed a unique balance, so-called a magnetic suspension balance (MSB), for accurate measurements of fluid densities.

Schematic diagram of the MSB (made in Rubotherm Prazisionsmesstechnik GmbH, Germany) for solubility measurement is shown in Fig. 22. An electronically controlled magnetic suspension coupling is used to transmit the measured force from the sample enclosed in a pressure vessel to a microbalance. The suspension magnet, which is used for transmitting the force, consists of a permanent magnet, a sensor core and a device for decoupling the measuring load. An electromagnet, which is attached at the underfloor weighing hook of a balance, maintains the freely suspended state of the suspension magnet via an electronic control unit. Using this magnetic suspension coupling, the measuring force is transmitted contactlessly from the measuring chamber to the microbalance, which is located outside the chamber under ambient atmospheric conditions.

The distinct merit of the MSB is that the microbalance can be tared and calibrated during measurements. This zero-point correction and calibration of sensitivity of the microbalance is important for measuring the gas solubility accurately, because long sorption time is generally needed. The MSB offers the possibility of lowering the suspension magnet in a controlled way to a second stationary position a few millimeters below the measuring position. At this position, a sample basket is set down on a support by the suspension magnet's descent, and the sample is decoupled from the balance. This so-called "zero-point position" allows a taring and calibration of the balance at any time, even while recording measurements.

In the calculation program, the Sanchez and Lacombe equation of state (S-L EOS) is commonly used for the liquid and/or vapor phases [155-157]:

$$\widetilde{P} = -\widetilde{\rho}^2 - \widetilde{T}\left[\ln(1-\widetilde{\rho}) + (1-\frac{1}{r})\widetilde{\rho}\right] \tag{7}$$

$$\widetilde{P} = \frac{P}{P^*}, \ \widetilde{\rho} = \frac{\rho}{\rho^*}, \ \widetilde{T} = \frac{T}{T^*}, \ r = \frac{MP^*}{RT^*\rho^*} \tag{8}$$

where characteristic parameters, $P^*$, $\rho^*$ and $T^*$ of the S-L EOS for pure component or mixture are evaluated with the following mixing rules.

$$P^* = \sum_i \sum_j \phi_i \phi_j P_{ij}^* \tag{9}$$

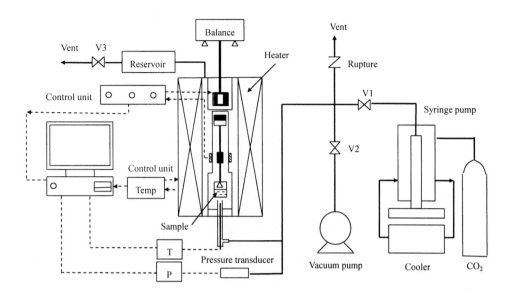

Fig. 22. Schematic diagram of measuring the VLE at high pressure.

$$P_{ij}^* = (1 - k_{ij})(P_i^* P_j^*)^{0.5} \tag{10}$$

$$T^* = P^* \sum_i \frac{\phi_i^0 T_i^*}{P_i^*} \tag{11}$$

$$\frac{1}{r} = \sum_i \frac{\phi_i^0}{r_i^0} \tag{12}$$

$$\phi_i^0 = \frac{(\phi_i P_i^* / T_i^*)}{\sum_j (\phi_j P_j^* / T_j^*)} \tag{13}$$

$$\phi_i = \frac{(w_i / \rho_i^*)}{\sum_j (w_j / \rho_j^*)} \tag{14}$$

$$\frac{1}{\rho^*} = \sum_i \frac{w_i}{\rho_i^*} \tag{15}$$

In Eqs. (9) - (15), $T_i^*$, $P_i^*$, $\rho_i^*$ and $r_i^0$ refer to the characteristic parameters of component $i$ in the pure state, $w_i$ is weight fraction and $k_{ij}$ is a binary interaction parameter determined by fitting the equation to the experimental data.

The swelling degree $S_w$ is solved by

$$S_w = \frac{v(P,T,S)}{v(P,T,0)} - 1 \tag{16}$$

and

$$v = 1/\rho \tag{17}$$

Consequently, the solubility is calculated in terms of the force balance,

$$W_g = W_F(P,T) - W_0(0,T) + \rho(P,T)[(V_P(P,T)(1 + S_w(P,T,S)) + V_B] \tag{18}$$

where $W_F$ and $W_0$ are readout of the balance at temperature $T$, pressure $P$ and zero

pressure, respectively. $W_g$ is the weight of the dissolved gases. $\rho$ is the gas density (g cm$^{-3}$)

and can be obtained from the website (NIST data bank): http:

//webbook.nist.gov/chemistry/form-ser.html. $V_P$ and $V_B$ (cm$^3$) are volumes of the sample

and a sample basket (including a sample holder and the measuring load decoupling device),

respectively. $V_B$ is determined by buoyancy measurements using the dissolved gases before

the experiment.

## 4.2. Gas-liquid chromatography method

This method is commonly used for determining experiments values of $\gamma^\infty$. It involves

measuring the retention behavior of a solute in an inert carrier gas stream passing through a

column containing a solvent-coated solid support [81-83].

The equation used in calculating $\gamma^\infty$ is:

$$\gamma_2^\infty = \frac{RT\phi_2 Z_m \exp(-V_2^\infty P/RT)n_1}{\phi_2^s P_2^s (V_R - V_m)} \tag{19}$$

where $Z_m$ is compressibility of he mixture, $\phi_2$ is vapor-phase fugacity coefficient of solute,

$V_R$ is retention volume of solute, $V_m$ is volume of the mobile phase, $V_2^\infty$ is molar volume

of solute at infinite dilution, and $n_1$ is moles of solvent on the column. This relation assumes

that equilibrium is achieved throughout the column, that the solute is sufficiently dilute to be

within the Henry's law region, and that the packing is inert relative to the solvent and the

solute.

There are several advantages in using the gas-liquid chromatography method. Firstly, it is the method that $\gamma^\infty$ may be found directly. Secondly, commercially available equipment can be used and the techniques are well established. Further, although solvent purity is crucial, solute purity isn't critical as separation is achieved by the chromatograph. Finally, the method is quick on the condition that the solvent is determined beforehand. Up to 30 data points may be measured in the course of a day's run. Conventional static technique in the direct method may take hours or days to obtain one or two limiting activity coefficients, and even the ebulliometric method to be discussed afterwards can measure only four values in a day's run with available equipment.

However, the most apparent defect of this method is that when one solute system has been given and different solvents are to be evaluated, the column packing should be made once for each solvent. We know that the preparation of the column packing is a very tedious work and the time consumed isn't endurable.

The minor defects of this method are encountered when solvents nearly as volatile as the solutes are used. The stripping of solvent from the column packing introduces impurities into the carrier gas and thus reduces the sensitivity of the detector. This means that some solutes can't be detected in certain volatile solvents.

### 4.3. Ebulliometric method

Ebulliometric method is a rapid and robust method of measuring pressure-temperature-liquid mole fraction ($P$-$T$-$x$) data [84, 85]. $P$-$T$-$x$ data are measured in an ebulliometer which is a one-stage total reflux boiler equipped with a vapor-lift pump to spray slugs of equilibrated liquid and vapor upon a thermometer well. The advantages of ebulliometric method are:

(1). Degassing is not required.

(2) Equipment is simple, inexpensive, and straightforward to use.

(3) Data can be measured rapidly.

(4) Either $P$-$x$ or $T$-$x$ data can be measured.

For a binary system, the procedure is to measure $\left(\dfrac{\partial T}{\partial x}\right)_P$ in a twin-ebulliometer (differential) arrangement where a pure component is charged to both ebulliometers and small amounts of the second component are added to one ebulliometer. Activity coefficient at infinite dilution $\gamma^\infty$ is derived by extrapolating $\left(\dfrac{\partial T}{\partial x}\right)_{x=0}$, which leads to obtaining inaccurate results. One formulation of solving $\gamma^\infty$ is:

$$\gamma_1^\infty = \frac{P}{P_1^0}\left[1-(\frac{\Delta H^V}{RT^2})(\frac{dT}{dx_1})_{x_1=0}\right] \qquad (20)$$

where

$$(\frac{dT}{dx_1})_{x_1=0} = T_1 - T_2 + (\frac{\Delta T}{x_1 x_2})_{x_1=0},$$

and $\Delta T = T - x_1 T_1 - x_2 T_2$.

$P$ is the total pressure, $P_1^0$ is the saturated pressure of solute 1, $T$ is the absolute temperature, $R$ is gas constant, and $\Delta H^V$ is the evaporation latent heat of solute 1.

When solute 1 is added into the solvent S, the concentration of solute 1 $x_1$ is known beforehand. Only if the boiling point different of the mixture is measured, then the activity coefficient can be calculated from Eq. (20). Moreover, if solute 1 is in the very dilution, then it is reasonable to think that the activity coefficient may be $\gamma_1^\infty$.

For extractive distillation, a ternary system consisting of solvent S, components A and B to be separated, are encountered while only one solvent S needs to be evaluated. In this case selectivity and/or relative volatility at infinite dilution in solvent S, can be obtained by carrying out two $\gamma^\infty$ ($\gamma_A^\infty$, $\gamma_B^\infty$) experiments. From this viewpoint, ebulliometric method is somewhat tedious.

### 4.4. Inert gas stripping and gas chromatography method

Leroi et al. [86] proposed a new method to measure $\gamma^\infty$, $S^\infty$ and $\alpha^\infty$, i.e. the combination of inert gas stripping and gas chromatography method, which has the merits of a high level of accuracy and reproducibility [87, 88]. One schematic diagram of the experimental apparatus is depicted in Fig. 23 [64]. The principle of the method is based on the variation of the vapor phase composition when the highly diluted components of the liquid mixture, controlled to be below 0.01 mol l⁻¹, are stripped from the solution by a constant flow of inert nitrogen gas with a flow rate 20 ml min⁻¹. In the stripping cell 9, the outlet gas flow is in equilibrium with the liquid phase, and gas is injected into the gas chromatograph 11 by means of a six-way valve 10 at periodic intervals. The peak areas of solutes are recorded by the integrating meter 12.

The most important part in this method is the equilibrium cell. One modification to the structure proposed by Lei et al. [64] involves enlarging the gas-liquid interface and increasing the contact time between gas and liquid phases by means of a spiral path. Two similar cells, a presaturation cell 8 and a stripping cell 9, are used in Fig. 23. The configuration of the modified equilibrium cell is illustrated in Fig. 24. The relative volatility can be measured by this method in a shorter time than by other methods.

According to this experimental principle, we can derive the expression of activity coefficient $\gamma_i^\infty$ at infinite dilution which is given by

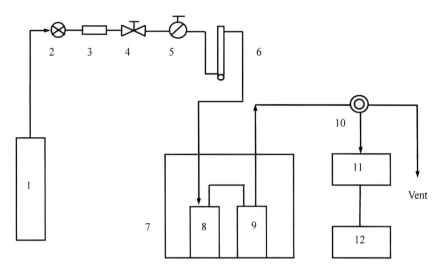

Fig. 23. Schematic diagram of experimental apparatus. (1) gas tank, (2) pressure-reducing valve, (3) gas purifier, (4) pressure-stabilizing valve, (5) current-stabilizing valve, (6) lather flowmeter, (7) thermostatic water bath, (8) presaturation cell, (9) stripping cell, (10) six-way valve, (11) gas chromatograph, (12) integrating meter.

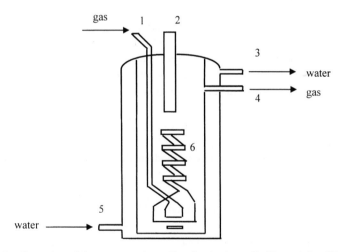

Fig. 24. Configuration of the presaturation cell and stripping cell. (1) gas inlet, (2) thermometer point, (3) water outlet, (4) gas outlet, (5) water inlet, (6) spiral gas path.

$$\gamma_i^{\infty} = \frac{NRT}{DP_i^0 t} \ln \frac{A_0}{A} \tag{21}$$

where $N$ is the total number of mole of solvent in the dilution cell at time $t$, and $A_0$ and $A$ are respectively the original peak areas of the solute and the peak area at intervals of the time $t$.

For extractive distillation, we usually take on the selectivity or relative volatility at infinite dilution as the standard of evaluating the solvent. The selectivity $S_{ij}^\infty$ and relative volatility $\alpha_{ij}^\infty$ at infinite dilution are respectively given by the following equations:

$$S_{ij}^\infty = \frac{\gamma_i^\infty}{\gamma_j^\infty} = \frac{\ln(A_{i0}/A_i)P_j^0}{\ln(A_{j0}/A_j)P_i^0} \tag{22}$$

$$\alpha_{ij}^\infty = \frac{\gamma_i^\infty P_i^0}{\gamma_j^\infty P_j^0} = \frac{\ln(A_{i0}/A_i)}{\ln(A_{j0}/A_j)} \tag{23}$$

It is evident that the selectivity $S_{ij}^\infty$ and relative volatility $\alpha_{ij}^\infty$ are consistent in evaluating the separation ability of the solvent. For the sake of simplicity, it is better to select $\alpha_{ij}^\infty$. Comparing with $\gamma_i^\infty$, the data of $\alpha_{ij}^\infty$ is easy to be obtained accurately because it needn't measure $N$ and $t$ which perhaps bring extra errors from the experiment. Thus if the peak area $A$ of the solute at different time is known, $\alpha_{ij}^\infty$ is able to be calculated according to Eq. (23) in which only the peak area $A$ is required. It has been verified by Lei et al. [64] that the results obtained by this method are reliable, and very suitable for the evaluation of different solvents.

The comparison of gas-liquid chromatography method (method 1) and inert gas stripping and gas chromatography method (method 2) for separating C4 mixture with ACN and DMF is shown in Table 9 [56, 64], where we use the subscript (1-5) for butane, 1-butene, 2-trans-butene, 2-cis-butene and 1,3-butadiene respectively. It can be seen that the difference between these two methods is below 15%.

Table 9
The comparison of gas-liquid chromatography method (method 1) and inert gas stripping and gas chromatography method (method 2)

| | ACN(20℃) | | | DMF(50℃) | | |
|---|---|---|---|---|---|---|
| | Method 1 | Method 2 | Deviation % | Method 1 | Method 2 | Deviation % |
| $\alpha_{15}^\infty$ | 3.48 | 3.82 | 9.44 | 3.82 | 3.32 | 13.09 |
| $\alpha_{25}^\infty$ | 2.13 | 2.11 | 0.01 | 2.40 | 2.12 | 11.67 |
| $\alpha_{35}^\infty$ | - | 1.81 | - | 1.94 | 1.79 | 7.73 |
| $\alpha_{45}^\infty$ | 1.47 | 1.63 | 10.88 | 1.65 | 1.61 | 2.42 |

It is concluded by comparison of these four methods that inert gas stripping and gas chromatography method should be firstly recommended, secondly the HSGC technique in the direct method and finally gas-liquid chromatography method. However, it is hard to find the references about ebulliometric method in the recent years, which may be a cool point.

## 5. CAMD OF EXTRACTIVE DISTILLATION

It is tiresome to choose the best solvent from thousands of different substances for a given system through experiments only. There are some methods to pre-select the possible solvents by calculation, such as Pierotti-Deal-Derr method, Parachor method, Weimer-Prausnitz method and CAMD, among which CAMD is both the recent development and the most important, and has completely replaced other calculation methods for screening solvents.

### 5.1. CAMD for screening solvents

CAMD (computer-aided molecular design) is developed in the 1980s, and widely used in such unit operations as gas absorption, liquid-liquid extraction, extractive distillation and so on [89-100]. CAMD breaks new way in selecting the possible solvents by largely reducing experiments. The application of CAMD in chemical engineering is mostly based on the UNIFAC group contribution. Incalculable feasible molecules will be formulated by UNIFAC groups in accordance with certain rules, but in terms of given target properties, the desired molecules are screened by computers. The groups of UNIFAC provide building blocks for assembling molecules. CAMD is essentially the inverse of property prediction by group contribution. Given a set of desirable properties, it is proposed to find a combination of structural groups, satisfying the property specifications. In most cases, more than one solution is produced. Thus, a screening is needed since only one of the alternatives must be chosen for the specified problem. At this point, factors such as corrosion, prices, source, etc. should be taken into consideration. Of course, it is a procedure after CAMD.

The solvents are divided into three parts, liquid solvent, solid salt and ionic liquid which can be designed respectively by CAMD.

### 5.1.1. CAMD of liquid solvents

(1) Procedure of CAMD for the liquid solvents

The UNIFAC groups are used as building blocks for the synthesis of liquid solvents. CAMD for liquid solvents is conducted in the following four steps.

(a) Group sorting and pre-selection

Molecular design makes full use of the group concept raised by Franklin, which is built on the UNIFAC groups [101-104]. The groups must be systematically ordered to facilitate their use.

There are different approaches to sorting the groups [105-108]. The approach we adopt is that a certain group is determined by the number of attachments present in a given group (or the valence number of the group) and the degree of difficulty that the group combines other groups (or the type of attachment). The UNIFAC groups are sorted and listed in Table 10. If a group belongs to several different classes, it means that this group takes on different types of characterization in different molecules.

Since the groups in Table 10 are based on the UNIFAC definition of functional group, they can be used to obtain the values of some properties related to UNIFAC-based methods. For the computation of other properties based on certain group contributions, but not related to UNIFAC-based methods, new parameter tables for the groups in Table 10 have been developed using information on the group contributions and their corresponding parameter tables.

Table 10

Attachment type characterization of UNIFAC groups

| Type of attachment | Groups | | | |
|---|---|---|---|---|
| (N,0) | $CH_3OH$ | $CH_3NO_2$ | $H_2O$ | $CH_3NH_2$ |
| | $CF_3$ | $CH_3SH$ | $CH_3CN$ | Furfural |
| | HCOOH | DMSO | $(CH_2OH)_2$ | $CH_2Cl_2$ |
| | $CHCl_3$ | ACRY | DMF | NMP |
| | $CH_3NH_2$ | $CS_2$ | $CCl_3F$ | $CHCl_2F$ |
| | $CHClF_2$ | $CClF_3$ | $CCl_2F_2$ | $C_4H_4S$ |
| | Morph | $AmHCH_3$ | $Am(CH_3)_2$ | |
| (M,1) | $(CH_3)$ | | | |
| (J,2) | $(CH_2)$ | | | |
| (L,1) | $(CH_2Cl)$ | | | |
| (L,2) | $(CHCl)$ | | | |
| (K,1) | $(CH_2=CH)$ | (OH) | $(CH_3CO)$ | (CHO) |
| | $(CH_3COO)$ | (HCOO) | $(CH_3O)$ | $(CH_2NH_2)$ |
| | $(CH_3NH)$ | $(C_5H_4N)$ | $(CH_2CN)$ | (COOH) |
| | $(CHCl_2)$ | $(CCl_3)$ | $(CH_2NO_2)$ | $(CH_2SH)$ |
| | (I) | (BR) | $(CH\equiv C)$ | $Cl-(C=C)$ |
| | $(SiH_3)$ | | | |
| (K,2) | $(CH=CH)$ | $(CH_2=C)$ | $(CH_3N)$ | $(C_5H_3N)$ |
| | $(CCl_2)$ | $(CHNO_2)$ | $(C\equiv C)$ | (DMF-2) |
| | (COO) | $(SiH_2)$ | $(SiH_2O)$ | |
| (K,1), (L,1) | $(CH_2CO)$ | $(CH_2COO)$ | $(CH_2O)$ | $(CHNH_2)$ |
| | $(CH_2NH)$ | $(FCH_2O)$ | | |
| (I,1) | (ACH) | (ACF) | | |
| (H,1) | $(ACCH_3)$ | (ACOH) | $(ACNH_2)$ | (ACCl) |
| | $(ACNO_2)$ | | | |
| (H,1), (M,1) | $(ACCH_2)$ | | | |
| (H,1), (J,2) | (ACCH) | | | |

The attachment types are indicated as ordered pairs $(i, j)$, where $i$ is the type of attachment and $j$, the number of $i$ attachment. Five types of attachments are put forward for non-aromatic groups.

N = Single molecules as a group having no attachment with other groups, such as $H_2O$

K = Severely restricted attachment, such as OH

L = Partially restricted attachment, such as $CH_2Cl$

M = unrestricted carbon attachment in linear dual valence or single valence groups, such as $CH_3$

J = unrestricted carbon attachment in radial dual valence groups, such as $CH_2$.

For aromatic molecules, two new attachments are introduced:

I = aromatic carbon ring attachment, such as ACH

H = substituted aromatic carbon ring attachment, such as ACCl

Types M and J attachments are extended to aromatic groups.

M = unrestricted attachment in a carbon linked to an aromatic carbon, such as $ACCH_2$

J = unrestricted attachment in a radial carbon linked to an aromatic carbon, such as ACCH.

The chemically feasible molecules will be generated from the above characterized UNIFAC groups in terms of the rules that must be met in the formation of a molecule and are to be mentioned afterwards. Not all of the groups in Table 10 need to be used in CAMD for a specific application. The groups may be screened in advance. The pre-selection of groups is conducted on two criteria:

The first: availability of interaction parameters. The prediction of solvent properties requires the availability of UNIFAC binary parameters. Therefore, the first criterion of the selection is the availability of binary parameters for the synthesis groups.

The second: elimination of unsteady compounds and corrosive, toxic materials. Sometimes for a specialized separation process, we need to bear in mind that a lot of some groups could be excluded in advance. For instance, some groups that may cause corrosion to the equipment are avoided for distillation processes.

By conforming to these criteria, the number of participating groups is reduced and the time the computer takes to fulfill the combination of groups is saved.

(b) Combination of groups

The most difficult part in molecular design is assembling groups into one molecule. The assembly must fit to the following rules:

The first: The chemical valence of a molecule must be zero.

The second: the neighborhood effect of groups must be avoided. Fortunately, many researchers have discussed the assembly rules from different standpoints and given different restrictive conditions. Their work is helpful for us to program CAMD.

The third: the group parameters and group interaction parameters must be known ahead of time. We have collected 109 groups and most of the UNIFAC binary parameters and group parameters are known to us. Many properties can be calculated from these parameters.

The fourth: in general, a molecule is composed of not more than eight groups and the polar groups cannot be over three. The groups in a molecule must accord with the following attachment criterion: $K = M + J/2 + 2$ for aliphatic compounds; $I + H = 6$, $H \leqslant 2$ or $H \leqslant 3$ for aromatic compounds; for aliphatic-aromatic compounds, these restriction must be satisfied simultaneously. Otherwise, this molecule may be unsteady under normal condition.

The number of the molecules combined by groups at random is often too large to be processed by a computer in a short time, which is the so-called combination explosion. In the molecular design, we should be cautious to avoid its occurrence.

Assuming that there are $n$ kinds of different groups and the molecule combined by them has only $m$ groups, then the number of ways to combine the molecules without other restrictions is: $\dfrac{(n+m-1)!}{(n-1)!m!}$.

When the number of groups is in the range of 2-$m$, the number of such molecules is:

$$\sum_{n=2}^{m} \frac{(n+m-1)!}{(n-1)!m!}.$$

If $n = 2$ and $m = 6$, the number of combination ways is very high, up to 18,551. Therefore, the assembled rules must be considered to effectively decrease the combination ways.

(c) Prediction of target properties

Different problems have different sets of properties as constrains. For CAMD for extractive distillation, such properties as relative volatility ($\alpha_{ij}^{\infty}$), selectivity ($S_{ij}^{\infty}$), solubility capacity (SP), molecular weight (MW) and boiling point ($T_b$) are important. Specification of the problem type identifies the corresponding target properties. As not all the target properties are computable, it is convenient to classify them as explicit target properties and implicit target properties. Prediction methods for explicit target properties are available and can be realized automatically by computer. Prediction methods for implicit properties aren't presently available and thus a combination of experience, information from the open references and experiments is needed to determine them. As a result, implicit properties are generally obtained manually. The prediction methods for explicit properties are listed in Table 11 [109, 110].

(d) Selection of additives

Although explicit properties can be computed according to the methods of Table 11, relative volatility among them is the key to extractive distillation. For this reason, compounds are ordered by the values of their relative volatility. But we know that evaluation of implicit properties is also required for the selection of additives. For CAMD for extractive distillation, implicit properties such as toxicity, cost, stability and material source are important. The compounds that don't satisfy the implicit properties are crossed out from the order. The remaining compounds ranked in front of the order are the objects of the best possible additives we seek. As a result, by CAMD the work to search for the best additives is reduced. The final additives will be tested with only a minimum number of experiments.

(2) Program of CAMD for the liquid solvents

Many CAMD algorithms have been proposed. These may be broadly classified as interactive, combinatorial, knowledge-based and mathematical programming methods. Recently, a new method, genetic algorithm, is put forward, which retains the efficiency of mathematical programming while still performing well in difficult search spaces.

In our work CAMD for extractive distillation has been programmed with Visual Basic for Windows. This program is a combinatorial method, and can provide much information necessary for the extractive distillation. CAMD is an easy-to-use software with user friendly interface. One with a little knowledge of extractive distillation can become familiar with it in a few hours. Fig. 25 shows a block diagram of CAMD for liquid solvents [111].

Table 11
Properties estimation for liquid solvents

| Property | Method |
|---|---|
| Relative volatility | $\alpha^{\infty}_{ij} = \dfrac{\gamma^{\infty}_i p^0_i}{\gamma^{\infty}_j p^0_j}$ |
| Selectivity | $S^{\infty}_{ij} = \dfrac{\gamma^{\infty}_i}{\gamma^{\infty}_j}$ |
| Solubility capacity | $SP_i = \dfrac{1}{\gamma^{\infty}_i} \dfrac{MW_i}{MW_j}$ |
| Molecular weight | Pure component data bank or by adding group parameters |
| Normal boiling point | Pure component data bank or by adding group parameters |
| Vapor Pressure | Pure component data bank or by Antoine equation |
| Azeotropic judgment | By drawing the curves of $x$-$y$ to judge |

(3) Application of CAMD for the liquid solvents
(a) Separation of C4 mixture with ACN

Extractive distillation is widely used in purification of 1,3-butadiene from C4 mixture. ACN is a frequently used basic solvent and is inconvenient to be completely replaced by other solvents to enhance the relative volatility of C4 mixture. Therefore, water is usually added to ACN as a cosolvent. But the mixed ACN has a disadvantage, that is, it tends to give rise to hydrolysis, which contributes to the loss of ACN and corrosion of equipment during production. So a more efficient additive is required to substitute water.

2-Butene and 1,3-butadiene are key components in C4 separation because their separation is the most difficult. So it is reasonable to only consider 2-butene and 1,3-butadiene as the representatives of C4 mixture for solvent evaluation. The restrictions of additives for the ACN/C4 system are listed as follows:

1. Preselected group type number: 10 ($CH_3OH$, $H_2O$, $CH_3$, $CH_2$, $CH_3COO$, $CH_3CO$, $COOH$, $OH$, $CH_2CN$, $CH_2NH_2$). The groups are selected on the basis of availability of interaction parameters and reduction of unsteady compounds and corrosive, toxic materials.
2. Expected group number: 2-6
3. Maximum molecular weight: 150
4. Minimum boiling point: 323.15 K
5. Maximum boiling point: 503.15 K

The conditions of selecting additives for the ACN/C4 system are given below:
1. Temperature: 303.15 K
2. Minimum relative volatility at infinite dilution: 1.35

3. Concentration of additive in ACN: 10 wt%

4. Key components: 2-butene and 1,3-butadiene, which are at infinite dilution

CAMD has been performed by authors on a PC (Pentium 2 286MHz) [111]. The calculation in this case needs about 10 minutes. The designed results of liquid solvents are obtained by means of CAMD, as shown in Table 12, where superscript $\infty$ stands for the condition at infinite dilution.

Table 12 shows the sorted list of possible molecules according to the values of relative volatility. Some have already been found in the references, such as $H_2O$ (No. 1). It indicates that the designed results are reliable in some degrees. After considering such implicit properties as toxic, boiling point, chemical stability and hydrolysis, the molecules of No. 1, 4, 5, 6 and 9 are regarded as the possible additives.

However, it is found from the reference [55] that C4/methanol and C4/ethanol can form azeotrope, which doesn't correspond with the CAMD results. It manifests that the acceptance of CAMD depends on the accuracy of the computation methods for the properties. Even though the prediction methods have been extensively used, they should be considered only as an approximation for the estimation of solvent properties. That is where the limitations of CAMD lie.

The additives generated by CAMD are tested by experiments, and the results are listed in Table 13 where the subscript (1-5) represents n-butane, 1-butene, trans-2-butene, cis-2-butene and 1,3-butadiene, respectively. It can be seen that water and ethylenediamine are good potential additives of ACN among all liquid solvents. If we consider the hydrolysis influence, ethylenediamine is the best substance as liquid solvent which is added to ACN.

Thus, by means of CAMD we can find the best additive, ethylenediamine, from a great number of liquid solvents with only a few experiments. It is shown that CAMD can effectively decrease the amounts of experiments and help researchers to explore the useful substances in a short time.

(b) Separation of aromatics and non-aromatics

In industry, the methods for separating aromatics and non-aromatics are liquid-liquid extraction and extractive distillation [112-115]. Liquid-liquid extraction as an old method is almost replaced by extractive distillation in most cases. It is assumed that cyclohexane and benzene are taken on as the representation of aromatics and non-aromatics respectively because they are the key components in the separation process. In order to find the potential solvents, CAMD is used. The restrictions for the solvents are listed as follows:

1. Maximum molecular weight: 240

2. Minimum boiling point: 383.15 K

3. Maximum boiling point: 553.15 K

4. Maximum melting point: 278.15 K

5. Minimum selectivity at infinite dilution: 3.0

6. Maximum activity coefficient of component i at infinite dilution in solvent ($\gamma_{i,s}^{\infty}$): 12

7. Maximum activity coefficient of solvent at infinite dilution in component i ($\gamma_{s,i}^{\infty}$): $\dfrac{P_i^0}{P_s^0}$.

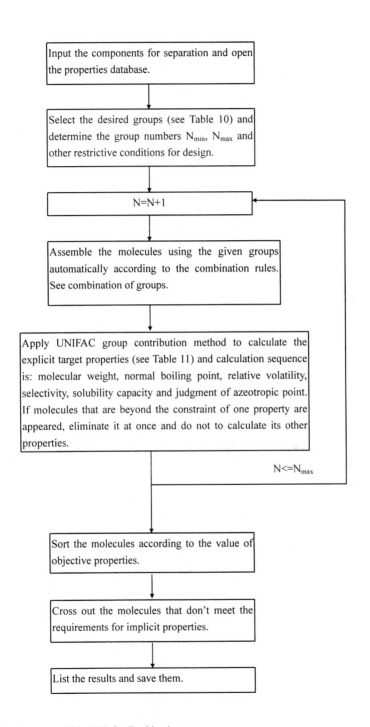

Fig. 25. Block diagram of CAMD for liquid solvents.

Table 12
Results of CAMD for liquid solvent for separating C4 mixture

| No. | Molecular structure | $MW$ | $T_b$ / K | $\alpha_{12}^{\infty}$ | $S_{12}^{\infty}$ | $SP$ | Azeotropic judgement | |
|---|---|---|---|---|---|---|---|---|
| | | | | | | | 1 and solvent | 2 and solvent |
| 1 | $H_2O$ | 18.0 | 373.2 | 1.66 | 2.16 | 0.38 | no | no |
| 2 | $2CH_2CN$ | 80.0 | 495.3 | 1.51 | 1.97 | 0.37 | no | no |
| 3 | $CH_2$ $CH_2CN$ $OH$ | 71.0 | 462.5 | 1.46 | 1.91 | 0.36 | no | no |
| 4 | $CH_3$ $3CH_2$ $CH_3COO$ | 116.0 | 399.2 | 1.46 | 1. 90 | 0.40 | no | no |
| 5 | $CH_3$ $CH_2$ $CH_3CO$ | 72.0 | 352.8 | 1.45 | 1.89 | 0.40 | no | no |
| 6 | $2CH_2NH_2$ | 60.0 | 390.4 | 1.44 | 1.88 | 0.38 | no | no |
| 7 | $CH_2$ $OH$ $CH_2NH_2$ | 61.0 | 443.5 | 1.42 | 1.85 | 0.36 | no | no |
| 8 | $CH_3OH$ | 32.0 | 337.8 | 1.42 | 1.86 | 0.38 | no | no |
| 9 | $CH_3$ $CH_2$ $OH$ | 46.0 | 351.5 | 1.39 | 1.81 | 0.39 | no | no |

Table 13
Experimental values of relative volatility of C4 mixture at different additives concentration

| Additives | Concentration / wt% | $T$ / ℃ | $\alpha_{15}^{\infty}$ | $\alpha_{25}^{\infty}$ | $\alpha_{35}^{\infty}$ | $\alpha_{45}^{\infty}$ |
|---|---|---|---|---|---|---|
| - | 0 | 28 | 3.61 | 2.08 | 1.77 | 1.58 |
| n-butanone | 10 | 28 | 3.55 | 2.08 | 1.75 | 1.57 |
| n-butanone | 20 | 29 | 3.37 | 2.06 | 1.71 | 1.56 |
| n-butyl acetate | 10 | 30 | 3.52 | 2.04 | 1.71 | 1.55 |
| n-butyl acetate | 20 | 30 | 3.30 | 2.01 | 1.69 | 1.53 |
| Ethyl alcohol | 10 | 30 | 3.58 | 2.09 | 1.77 | 1.57 |
| Ethyl alcohol | 20 | 30 | 3.54 | 2.11 | 1.77 | 1.59 |
| water | 10 | 29 | 4.06 | 2.16 | 1.84 | 1.62 |
| ethylenediamine | 10 | 30 | 3.77 | 2.11 | 1.79 | 1.62 |
| ethylenediamine | 20 | 30 | 3.81 | 2.15 | 1.83 | 1.60 |
| ethylenediamine | 100 | 30 | 3.62 | 2.13 | 1.89 | 1.58 |

By means of CAMD the potential solvents are found and listed in Table 14 [116] from which it can be seen that No. 1, NMP, and No. 2, DMF, are the potential solvents which have been reported as the solvents. So it means that the results of CAMD are reliable in some degrees.

It happened that Braam [117] also studied this system by using CAMD and provided the result (see Table 15). Aniline is suggested by Braam as a possible solvent, resulting in a relative volatility of 2.65 (UNIFAC) with cyclohexane in the distillate. The only solvent not to result in a higher relative volatility is acetonyl acetone. However, this solvent has a higher boiling point than aniline.

By comparison of Tables 14 and 15, it can be found that NMP is common for both cases in separating aromatics and non-aromatics. Unfortunately, DMF as a commonly used solvent is neglected in Table 15. The reason may be that DMF forms minimum boiling point azeotropes with non-aromatic hydrocarbons with 6-8 carbon atoms (e.g. cyclohexane and heptane) [118], which causes solvent losses with the distillate. In order to decrease the solvent losses, the addition of stream to the distillation column above the solvent feed has been recommended. Stream breaks the DMF-hydrocarbons azeotrope, and DMF entrained by the distillate can be recovered by washing it with water or ion exchange. But the post-treatment after extractive distillation is a little complicated, which contributes to no economics of the whole process.

Table 14

Results of CAMD for liquid solvents for separating aromatics and non-aromatics

| No. | Molecular structure | MW | $T_b$ / K | $T_f$ / K | $S_{ij}^{\infty}$ | Supply in the market? |
|---|---|---|---|---|---|---|
| 1 | NMP | 99.13 | 475.15 | 249.00 | 4.672 | Yes |
| 2 | DMF | 73.10 | 426.15 | 212.15 | 5.415 | Yes |
| 3 | 2CH$_2$, 1CH, 3CH$_3$COO | 218.23 | 529.70 | 273.56 | 5.124 | Yes |
| 4 | 4ACH, 2ACCOO, 2CH$_3$ | 194.20 | 539.86 | 271.07 | 3.963 | Yes |
| 5 | 5ACH, 1ACCH$_2$, 1CH$_3$COO | 150.19 | 475.25 | 253.04 | 3.418 | Yes |
| 6 | 5ACH, 1AC, 1CH$_3$CO | 120.16 | 467.24 | 258.05 | 3.604 | Yes |
| 7 | 1CH$_3$, 1CH$_2$COO, 1CH$_2$CN | 116.13 | 475.25 | 215.60 | 5.039 | No |
| 8 | 4ACH, 2AC, 1CH$_3$COO, 1CH$_3$O | 170.09 | 494.33 | 263.18 | 5.145 | No. |

Table 15

Generated solvents and predicted properties for cyclohexane (1) / benzene (2); adapted from the reference [117]

| Distillate: cyclohexane | | | | |
|---|---|---|---|---|
| No. | Solvent | $\alpha_{12}$ (UNIFAC) | $T_{boil}$ / K (Joback) | $T_f$ / K (Joback) |
| 1 | Acetonyl acetone | 2.58 | 444 | 257 |
| 2 | Dimethyl maleate | 2.79 | 457 | 236 |
| 3 | Dimethyl phthalate | 3.67 | 576 | 325 |
| 4 | NMP | 3.76 | 475 | 250 |
| Industrial solvent | | | | |
| 5 | aniline | 2.65 | 436 | 267 |

(c) Separation of ethanol and water

This system forms a minimum-boiling azeotrope, and extractive distillation may be used as an alternative to azeotropic and pressure-swing distillation. The solvents generated by CAMD for this system and the experimental results are given in Table 16 [117].

The solvent given in the reference [64] for this system is ethylene glycol. This solvent will allow the recovery of ethanol in the distillate with a predicted relative volatility of 2.54 (UNIFAC). In contrast to this, the first generated solvent, hexachlorobutadiene, is predicted to cause a relative volatility more than three times this value. This solvent was tested and performed very well (see Table 16).

Solvents were also generated to reverse the relative volatility of the system to facilitate the recovery of water in the distillate, which are listed in Table 16. No solvents could be found before for this interesting separation.

(d) Other systems

The systems of acetone/methanol, ethanol/ethyl acetate and methanol/methyl acetate, have been studied by Braam [117] by using CAMD. For separating acetone and methanol, dimethyl sulfoxide (DMSO) generated by CAMD is found to be the best alternative to water which is commonly used in industry; for separating ethanol and ethyl acetate, diethylene glycol and DMSO generated by CAMD were tested by experiments to be the potential alternatives; for separating methanol and methyl acetate, tetrachloroethylene generated by CAMD was tested and found to perform very well. Therefore, it is suggested that tetrachloroethylene should replace 2-methoxy ethanol which is currently used as the industrial solvent.

*5.1.2. CAMD of solid salts*

(1) Procedure of CAMD for the solid salts

There is only one reference [111] with regard to the work of CAMD for solid salts despite the frequent use of solid salts as separating agents in chemical engineering, say nothing of ionic liquids.

Table 16

The solvents generated by CAMD and the experimental results; adapted from the reference [117]

| Distillate: ethanol | | | |
| --- | --- | --- | --- |
| Solvents | $\alpha_{12}$ (UNIFAC) | $T_b$ / K (Joback) | $T_f$ / K (Joback) |
| hexachlorobutadiene | 8.41 | 523 | 248 |
| dimethyl sulfoxide | 3.44 | 462 | 292 |
| industrial solvent ethylene glycol | 2.54 | 471 | 258 |
| **Distillate: water** | | | |
| Solvents | $\alpha_{21}$ (UNIFAC) | $T_b$ / K (Joback) | $T_f$ / K (Joback) |
| dodecane | 4.72 | 474 | 224 |
| tridecane | 4.83 | 497 | 236 |
| tetrdecane | 4.95 | 520 | 247 |
| diethylbenzene | 5.85 | 455 | 228 |
| industrial solvent none available | N/A | N/A | N/A |
| **Experimental results** | | | |
| components | liquid mole fraction | $\alpha_{12}$ | |
| ethanol (1) | 0.163 | | |
| water (2) | 0.181 | | |
| hexachlorobutadiene | | 7.26 | |

With the development of extractive distillation, the extractive distillation with the combination of liquid solvent and solid salt is becoming increasingly applied in industry. Therefore, it is interesting to consider the salting effect in regard to CAMD.

As a simplified presumption, a salt is thought to be composed of one positive ion and one negative ion which are regarded as the groups of salts. It is consequently easy to assemble the ions into molecules in the same way as liquid solvents. These ions are collected and listed in Table 17. In the procedure of CAMD the combination rule is simply that the chemical valence of a salt must be zero, which is impossible to lead to combination explosion because most molecules are composed of just two groups: one positive ion and the other negative ion. Therefore, both liquid solvents and solid salts can be designed in one CAMD program, and this goes a further step in the application of CAMD.

The key of CAMD for solid salts is the selection of an appropriate thermodynamic method to obtain the target properties. The prediction methods for explicit properties are given in Table 18, in which relative volatility and selectivity are closely related with salting coefficients. Salting coefficients are calculated by the scaled particle theory (see chapter 1).

Table 17
Ions of CAMD for solid salts

| Ions | Groups | | | |
|------|--------|---|---|---|
| cation | $Li^+$ | $Na^+$ | $K^+$ | $Rb^+$ |
| | $Cs^+$ | $Ag^+$ | $NH_4^+$ | $Cu^+$ |
| | $Mg^{2+}$ | $Ca^{2+}$ | $Sr^{2+}$ | $Ba^{2+}$ |
| | $Be^{2+}$ | $Fe^{2+}$ | $Zn^{2+}$ | $Cu^{2+}$ |
| | $Al^{3+}$ | $Fe^{3+}$ | | |
| anion | $OH^-$ | $F^-$ | $Cl^-$ | $Br^-$ |
| | $I^-$ | $SCN^-$ | $NO_3^-$ | $NO_2^-$ |
| | $CH_3COO^-$ | $ClO_4^-$ | $BF_4^-$ | |

Table 18
Properties estimation for solid salts

| Property | Method |
|----------|--------|
| Salting coefficients $k_s$ | Calculated by Scaled Particle Theory |
| Activity coefficient | Calculated by UNIFAC method |
| Relative volatility without salt | $\alpha_{ij}^\infty = \dfrac{\gamma_i^\infty p_i^0}{\gamma_j^\infty p_j^0}$ |
| Selectivity without salt | $S_{ij}^\infty = \dfrac{\gamma_i^\infty}{\gamma_j^\infty}$ |
| Relative volatility with salt | $\alpha_s^\infty = \alpha^\infty \exp(k_s c_s)$ |
| Selectivity with salt | $S_s^\infty = S^\infty \exp(k_s c_s)$ |
| Molecular weight | Pure component data bank or by adding group parameters [158] |
| Vapor Pressure | Pure component data bank or by Antoine equation [110] |

(2) Application of CAMD for the solid salts

There is only one example reported in the reference [111], i.e. the system of ACN/C4. The additive constraints for the system of ACN/C4 are listed as follows.

1. Pre-selected group type number: 8 ($Li^+$, $Na^+$, $K^+$, $Br^-$, $Cl^-$, $I^-$, $SCN^-$, $NO_3^-$). The groups are selected because the salts formed by them are often met and easy to be obtained in the industry.

2. Expected group number: 2

3. Temperature: 303.15

4. Minimum infinite dilute relative volatility: 1.40

5. Concentration of additive in ACN: 10 wt%

6. Key components: 2-butylene and 1,3-butadiene that are at infinite dilution

CAMD has been performed by authors on a PC (Pentium 2 286 MHz). The calculation in this case needs a few minutes. But it is known that the salts must meet the requirement of implicit properties, e.g. chemical stability, oxidation, material source and solubility in ACN, etc. These factors taken in, the outputs of the selected salts are listed in Table 19.

If the price factor is further considered, the solid salts NaSCN and KSCN are more attractive than NaI and KI. On this basis, some experiments are performed to investigate the effect of solid salts. Table 20 shows the experimental values at 30°C with salt concentration 10 wt%. The subscripts (1-5) represent n-butane, 1-butene, trans-2-butene, cis-2-butene and 1,3-butadiene, respectively.

A comparison of the results of Table 13 and Table 20 demonstrates that the addition of salt is more efficient than the addition of liquid solvents (including water) for improving the separation ability of ACN. Because NaSCN and KSCN are more economic than NaI and KI, they are the desired additives. Simulation of extractive distillation process for ACN and ACN/NaSCN with a 10 wt% salt concentration is carried out. The results show that the required solvent flowrate decreases 11.8% and much energy is saved in the extractive distillation column when solid salt is added. Therefore, the mixture of ACN and solid salt performs more satisfactorily than a single ACN solvent. Apart from this, the mixture has no problem about dissolution, reuse and transport of salt, which is often brought on by using a single salt.

Table 19
Results of CAMD for solid salts

| No. | Molecular structure | $MW$ | $\alpha_{45}^{\infty}$ | $S_{45}^{\infty}$ |
|-----|------|------|------|------|
| 1 | $Na^{+}SCN^{-1}$ | 81.0 | 1.57 | 2.00 |
| 2 | $Na^{+}I^{-1}$ | 150.0 | 1.56 | 1.98 |
| 3 | $K^{+}SCN^{-1}$ | 97.0 | 1.55 | 1.98 |
| 4 | $K^{+}I^{-1}$ | 166.0 | 1.55 | 1.97 |

Table 20
Experimental values of relative volatility of C4 mixture at infinite dilution

| Solid salt | $\alpha_{15}^{\infty}$ | $\alpha_{25}^{\infty}$ | $\alpha_{35}^{\infty}$ | $\alpha_{45}^{\infty}$ |
|-----|------|------|------|------|
| NaSCN | 4.03 | 2.19 | 1.85 | 1.64 |
| KSCN | 4.00 | 2.18 | 1.85 | 1.64 |

*5.1.3. CAMD of ionic liquids*

It is a pity that up to data, there is no report of CAMD for ionic liquids. But we know that ionic liquids may become potential excellent separating agents in extractive distillation. So studying CAMD for ionic liquids is a very interesting thing.

Since ionic liquid is still a kind of salt, the program of CAMD for ionic liquids should be similar to that for solid salts. Likewise, an ionic liquid can be assumed to be composed of one positive ion and one negative ion which are regarded as the groups of salts. Consequently, it is easy to assemble the ions into molecules. In the procedure of CAMD the combination rule for ionic liquids is also same to that for solid salts. So CAMD for ionic liquids, solid salts and liquid solvents can be placed in a same program, which facilitates the exploitation and use of CAMD.

These ions are collected and listed in Table 21, in which $R$ is alkyl [70].

Table 21
Typical cation/anion combinations in ionic liquids

| Cations | Anions | Coordination ability of anions |
|---|---|---|
| Alkylammonium, alkylphosophonium, N-alkylpyridinium, N, N-dialkylimidazolium $PR_4^+$, $SR_4^+$, $NR_4^+$ | $[BF_4]^-$, $[PF_6]^-$, $[SbF_6]^-$, $[CF_3SO_3]^-$, $[CuCl_2]^-$, $[AlCl_4]^-$, $[AlBr_4]^-$, $[AlI_4]^-$, $[AlCl_3Et]^-$, $[NO_3]^-$, $[NO_2]^-$, $[SO_4]^{2-}$ | Weak (neutral) |
| | $[Cu_2Cl_3]^-$, $[Cu_3Cl_4]^-$, $[Al_2Cl_7]^-$, $[Al_3Cl_{10}]^-$ | None (acidic) |

The key of CAMD for ionic liquids is the selection of an appropriate thermodynamic method to obtain the target properties. For extractive distillation, the most importance of all in the explicit properties is activity coefficient which can be used to calculate relative volatility or selectivity and thus evaluate the possible ionic liquids.

Today, there are two methods commonly used for predicting activity coefficient, i.e. UNIFAC (UNIQUAC functional group activity coefficient) model and MOSCED (modified separation of cohesive energy density) model. The UNIFAC model is the most widely used, and based on the contributions of the constituent groups in a molecule. Since the first version of UNIFAC, several revisions and extensions have been proposed to improve the prediction of activity coefficient. The MOSCED model is an extension of regular solution theory to mixtures that contain polar and hydrogen bonding components. The cohesive energy density is separated into dispersion forces, dipole forced, and hydrogen bonding with small corrections made for asymmetry. However, group parameters of the selected ionic liquid can't be found in the present limited parameter table of UNIFAC and MOSCED models. In this case, COSMO-RS (Conductor-like Screening Model for Real Solvents) model is selected and used as an alternative to UNIFAC and MOSCED models [119, 120].

COSMO-RS model is a novel and efficient method for the priori prediction of thermophysical data of liquids, and has been developed since 1994 [121-123]. It is based on unimolecular quantum chemical calculations that provide the necessary information for the evaluation of molecular interactions in liquids. It can be applied to nearly any system for which no group parameters are available in the UNIFAC and MOSCED models. Moreover, COSMO-RS model is open for a large number of qualitative improvements and functional extensions. A correction for misfit charge interactions will improve the accuracy of the electrostatic part and enable calculations for ions.

In COSMO-RS model, the activity coefficient of component $i$ is related with chemical potential and given as follows:

$$\gamma_i = \exp\left(\frac{\mu_i - \mu_i^0}{RT}\right) \qquad (24)$$

where $\mu_i$ is the chemical potential of component $i$ in the mixture, $\mu_i^0$ the chemical potential in the pure liquid substance, $T$ the absolute temperature and $R$ gas constant. The chemical potential can be solved by using the exact equation resulting from statistical thermodynamics.

Therefore, it is possible to select the potential ionic liquid by means of CAMD. But it is believed that the ionic liquids must meet the requirement of implicit properties that can't be obtained by calculation. In this case it is better to set up a data bank of commonly used ionic liquids. Tables 22a and 22b show some implicit properties of commonly used ionic liquids [70], which can be input into the CAMD program for screening ionic liquids. By combination of explicit and implicit properties, the best ionic liquid can be found by CAMD.

Table 22a
Some implicit properties of commonly used ionic liquids; adapted from the reference [70]

| Ionic liquid[a] | $x^b$ | Color (with impurities) | Density (g ml$^{-1}$) | Liquid temperature (°C) | |
|---|---|---|---|---|---|
| | | | | lowest | highest |
| [bmim]BF$_4$ | | Light yellow | 1.320 | -48.96 | 399.20 |
| [bmim]PF$_6$ | | Light yellow | 1.510 | 13.50 | 388.34 |
| [bmim]Cl / AlCl$_3$ | 0.50 | Light brown | 1.421 | -88.69 | 263.10 |
| | 0.55 | Light brown | 1.456 | -94.44 | 286.59 |
| | 0.60 | Light brown | 1.481 | -95.87 | 316.34 |
| [emim]Br / AlCl$_3$ | 0.50 | Purplish black | 1.575 | 13.61 | 272.51 |
| | 0.55 | Brownish black | 1.656 | 6.45 | 294.02 |
| | 0.60 | Brownish black | 1.995 | -19.08 | 345.34 |
| [emim]PF$_6$ | | yellow | 1.426 | 2.71 | 304.65 |
| N-butylpyridine / AlCl$_3$ | 0.50 | yellow | 1.412 | 18.80 | 240.00 |
| | 0.55 | Brownish yellow | 1.430 | 33.73 | 245.39 |
| | 0.60 | Brownish yellow | 1.497 | 18.11 | 260.2 |
| (CH$_3$)$_3$NHCl / 2AlCl$_3$ | 0.66 | Brownish yellow | 1.621 | -67.90 | 80.25 |

Table 22b

Some implicit properties of commonly used ionic liquids; adapted from the reference [70]

| Ionic liquid[a] | $x^b$ | Solubility in common solvents[c] | | | | | | | |
|---|---|---|---|---|---|---|---|---|---|
| | | water | methanol | acetone | chloroform | Petroleum ether | hexane | Acetic anhydride | toluene |
| [bmim]BF$_4$ | | s | s | s | s | i | i | i | i |
| [bmim]PF$_6$ | | i | s | s | s | i | i | s | i |
| [bmim]Cl / AlCl$_3$ | 0.50 | r | r | s | s | i | i | s | s |
| | 0.55 | r | r | s | s | i | i | s | s |
| | 0.60 | r | r | s | s | i | i | s | s |
| [emim]Br / AlCl$_3$ | 0.50 | r | r | s | i | i | i | s | i |
| | 0.55 | r | r | s | i | i | i | s | i |
| | 0.60 | r | r | s | i | i | i | s | i |
| [emim]PF$_6$ | | i | i | s | s | i | i | s | i |
| N-butyl-pyridine / AlCl$_3$ | 0.50 | r | r | s | i | i | i | s | s |
| | 0.55 | r | r | s | i | i | i | s | s |
| | 0.60 | r | r | s | i | i | i | s | s |
| (CH$_3$)$_3$NHCl / 2AlCl$_3$ | 0.66 | r | r | s | i | i | i | s | s |

[a][bmim]=1-butyl-3-methylimidazolium, [emim]=1-ethyl-3-methylimidazolium.
[b]Apparent mole fraction of AlCl$_3$
[c]s: soluble; i: insoluble; r: may react with each other.

## 5.2. Other methods for screening solvents

### 5.2.1. Pierotti-Deal-Derr method

Apart from CAMD, the potential solvents may be evaluated by other prediction methods. Pierrotti-Deal-Derr method is one of those methods [5] that can be used to predict the activity coefficient at infinite dilution and thus deduce relative volatility at infinite dilution, the most important explicit property in extractive distillation.

Activity coefficients at infinite dilution are correlated to the number of carbon atoms of the solute and solvent ($n_1$ and $n_2$). For the members of homologous series H(CH$_2$)$_{n1}$X$_1$ (solute) in the members of the homologous series H(CH$_2$)$_{n2}$Y$_2$:

$$\lg \gamma_1^\infty = A_{12} + \frac{F_2}{n_2} + B_2 \frac{n_1}{n_2} + \frac{C_1}{n_1} + D_0 (n_1 - n_2)^2 \qquad (25)$$

where the constants are functions of temperature, $B_2$ and $F_2$ are functions of the solvent

series, $C_1$ is a function of the solute series, $A_{12}$ is a function of both, and $D_0$ is independent of both.

For zero members of a series, e.g. water for alcohol, no infinite value for $\gamma^\infty$ is obtained. Instead, by convention, any terms containing an $n$ for the zero members are incorporated in the corresponding coefficient. So for n-alcohols in water:

$$\lg \gamma_1^\infty = K + B_2 n_1 + C_1 / n_1 \tag{26}$$

Notice that the term $D_0(n_1 - n_2)^2$ is incorporated into the $K$ constant because $D_0$ is smaller than the other coefficients by a factor of 1000. Therefore, this term is insignificant. In Eq. (26) only $K$ is a function of the solute and solvent. $B_2$ is always the same when water is the solvent and $C_1$ is the same for n-alcohol solutes. This is shown better from the following homologous series in water at 100°C.

$$\text{n-Alcohols: } \lg \gamma_1^\infty = -0.420 + 0.517 n_1 + \frac{0.230}{n_1} \tag{27}$$

$$\text{n-Aldehydes: } \lg \gamma_1^\infty = -0.650 + 0.517 n_1 + \frac{0.320}{n_1} \tag{28}$$

where the coefficient $B$ is the same in both cases.

### 5.2.2. Parachor method

Activity coefficients at infinite dilution are obtained from the following relationship:

$$\lg \gamma_i^\infty = \frac{1}{2.303 RT} \left[ U_1^{1/2} - C U_2^{1/2} \right]^2 \tag{29}$$

$$U_i = \Delta H_i^V - RT \tag{30}$$

where $U_i$ is potential energy of component i, $\Delta H_i^V$ is enthalpy of evaporation, $C$ is a constant, a function of temperature, the parachor ratio of the two components, and the number of carbon atoms in the solute and solvent molecules, $T$ is the absolute temperature, and $R$ is gas constant. The same variety of systems covered in the Pierrotti-Deal-Derr method is also included in this approach.

### 5.2.3. Weimer-Prausnitz method

Starting with the Hildebrand-Schatchard model for non-polar mixtures, Weimer and Prausnitz developed an expression for evaluating values of hydrocarbons in polar solvents:

$$RT \ln \gamma_2^\infty = V_2[(\lambda_1 - \lambda_2)^2 + \tau_1^2 - 2\psi_{12}] + RT[\ln \frac{V_2}{V_1} + 1 - \frac{V_2}{V_1}] \tag{31}$$

where $V_i$ is the molar volume of pure component i, $\lambda_i$ is the non-polar solubility parameter, $\tau_i$ is the polar solubility parameter, $T$ is the absolute temperature, and $R$ is gas constant. The subscript 1 represents the polar solvent and subscript 2 is the hydrocarbon solute with

$$\psi_{12} = k\tau_1^2 \tag{32}$$

Later Helpinstill and Van Winkle suggested that Eq. (31) is improved by considering the small polar solubility parameter of the hydrocarbon (olefins and aromatics):

$$RT\ln\gamma_2^\infty = V_2[(\lambda_1 - \lambda_2)^2 + (\tau_1 - \tau_2)^2 - 2\psi_{12}] + RT[\ln\frac{V_2}{V_1} + 1 - \frac{V_2}{V_1}] \tag{33}$$

$$\psi_{12} = k(\tau_1 - \tau_2)^2 \tag{34}$$

For saturated hydrocarbons,

$$\psi_{12} = 0.399(\tau_1 - \tau_2)^2 \tag{35}$$

For unsaturated hydrocarbons,

$$\psi_{12} = 0.388(\tau_1 - \tau_2)^2 \tag{36}$$

For aromatics,

$$\psi_{12} = 0.447(\tau_1 - \tau_2)^2 \tag{37}$$

The term $\psi_{12}$ corresponds to the induction energy between the polar and non-polar components. Since no chemical effects are included, the correlation should not be used for solvents showing strong hydrogen bonding.

Unfortunately, the parameters of these three methods are limited, which leads to their very narrow application in extractive distillation. So these methods are rarely reported in the recent references.

However, CAMD based on group contribution methods is very desirable. By means of CAMD, the experiment working is greatly decreased in a search for the best solvents. CAMD as a useful tool plays an important role in finding the solvent and shortening the search time. It is believed that with the development of CAMD, it would be possible to be extended and applied in many more fields.

## 6. THEORY OF EXTRACTIVE DISTILLATION

Until now, there are two theories related to extractive distillation, i.e. Prausnitz and Anderson theory which gives semi-quantitative explanations, and scaled particle theory which gives complete quantitative results. The two theories provide a molecular basis for explaining why some solvents can increase the relative volatility of the components to be separated and why more selective than others.

## 6.1. Prausnitz and Anderson theory

Separation of hydrocarbon mixtures by extractive distillation has been practiced industrially for many years, even though there has been only limited understanding of the fundamental phase equilibria which forms the thermodynamic basis of this operation. In general, it is known that the addition of polar solvents to hydrocarbon mixture results in increased volatilities of paraffins relative to naphthenes, olefins, diolefins and alkynes, and in increased volatilities of naphthenes relative to aromatics. Therefore, the addition of a polar solvent enables facile separation by distillation of certain mixtures which otherwise can only be separated with difficulty. Prausnitz and Anderson theory [124] tries to explain the solvent selectivity in extractive distillation of hydrocarbons from the viewpoint of molecular thermodynamics and intermolecular forces. The interaction forces between the solvent and the component are broadly divided into two types, i.e. physical force and chemical force. The true state in the solution is undoubtedly a hybrid of these two forces.

### 6.1.1. Physical force

The selectivity is related to the various energy terms leading to the desired nonideality of solution which is the basis of extractive distillation, and can be expressed in such a manner.

$$RT \ln S_{23} = \left[ \delta_{1p}^2 (V_2 - V_3) \right] + \left[ V_2 (\delta_{1n} - \delta_2)^2 - V_3 (\delta_{1n} - \delta_3)^2 \right] + \left[ 2V_3 \xi_{13} - 2V_2 \xi_{12} \right] \tag{38}$$

where subscripts (1-3) represent solvent, the light component and the heavy component to be separated by extractive distillation, respectively, and $V_i$ is the molar volume of component $i$.

The three bracketed terms in Eq. (38) show, respectively, the separate contributions of physical force to the selectivity, i.e. the polar effect, the dispersion effect and the inductive effect of the solvent. It is convenient to rewrite Eq. (38) as

$$RT \ln S_{23} = P + D + I \tag{39}$$

where

$$P = \delta_{1p}^2 (V_2 - V_3)$$

$$D = V_2 (\delta_{1n} - \delta_2)^2 - V_3 (\delta_{1n} - \delta_3)^2$$

$$I = 2V_3 \xi_{13} - 2V_2 \xi_{12}$$

It is found that the polar term $P$ is considerably larger than the sum of $D$ and $I$, and frequently very much larger. Thus, Eq. (39) becomes

$$RT \ln S_{23} = \delta_{1p}^2 (V_2 - V_3) \tag{40}$$

Of course, in the special case where components 2 and 3 are identical in size, the polar term vanishes. This means that the physical force can't play a role in separating hydrocarbon mixture. In this case, the chemical force is dominant, and can be used to explain the separation phenomena, as is discussed in the following text.

Eq. (40) not only shows the effect of molecular size but also predicts that when one separates hydrocarbons of different molar volumes, the selectivity is sensitive to the polar solubility parameter. It indicates that the effectiveness of a solvent depends on its polarity, which should be large, and on its molar volume, which should be small.

One example of separating C4 mixture is given to illustrate the physical force. It is known that the order of the molar volume of C4 mixture is as follows:

Butane > butene > butadiene > butyne

According to Eq. (40), the order of volatilities of C4 mixture is in the same order:

Butane > butene > butadiene > butyne

However, in order to have a much higher selectivity, the polar solubility parameter of the solvents should be as great as possible. That is why such solvents as ACN, DMF and NMP with high polarity and small molar volume, are used for this separation.

Another example is concerned with the separation of ethane (1) and ethylene (2) by extractive distillation [125]. The experimental results are listed in Table 23, in which the solvent polarity is characterized by dipole moment. It can be seen from Table 23 that with the increase of dipole moment, relative volatility and selectivity at infinite dilution also approximately rise, which is consistent with the Prausnitz and Anderson theory.

Table 23
The solvents for separating ethane and ethylene

| Solvents | Dipole moment $\mu$ (debyes) | Relative volatility $\alpha_{12}^{\infty}$ | Selectivity $S_{12}^{\infty}$ |
|---|---|---|---|
| toluene | 1.23 | 0.84 | 0.83 |
| xylene | o-1.47 | 0.90 | 0.89 |
| | m-1.13 | | |
| | p-0 | | |
| Tetrahydrofuran | 5.70 | 0.97 | 0.96 |
| Butyl acetate | 6.14 | 0.95 | 0.94 |
| Ethyl acetate | 6.27 | 1.01 | 1.00 |
| Pyridine | 7.44 | 1.02 | 1.01 |
| Acetone | 8.97 | 1.08 | 1.07 |
| ACN | 11.47 | 1.24 | 1.23 |
| DMF | 12.88 | 1.20 | 1.19 |
| Dimethyl sulfoxide | 13.34 | 1.19 | 1.18 |
| NMP | 13.64 | 1.14 | 1.13 |

*6.1.2. Chemical force*

For components of identical size, solvent polarity isn't useful and selectivity on the basis of a physical effect isn't promising. In such cases selectivity must be based on chemical force which will selectively increase the interaction between the solvent and the components. However, examples of separating components of identical size are rare.

The chemical viewpoint of solutions considers that nonideality in solution arises because of association and solvation. In accordance with this concept, the true species in solution are loosely bonded aggregates consisting of two (or more) molecules of the same species (association) or of different species (solvation). That is to say, the solvent and the component can form complexes. Such complexes are believed to be the result of acid-base interactions following the Lewis definitions that a base is an electron donor and that an acid is an electron accepter.

For example, for the separation of C4 hydrocarbons, the fluidity of electron cloud is different for the group C-C, C=C, C=C-C=C, C ≡ C and in the following order:

C-C < C=C < C=C-C=C < C ≡ C

The greater the fluidity is, the easier the group is to be polarized. Accordingly, the base of C4 hydrocarbons is in the same order:

C-C < C=C < C=C-C=C < C ≡ C

which means that the chemical force between solvent and butyne is the greatest, while it is the smallest between solvent and butane.

So the volatilities of C4 hydrocarbons are in the following order:

Butane > butane > butadiene > butyne

which is consistent with the conclusion resulting from experiment.

A schematic diagram of complex formation between DMF and cis-2-butene is given in Fig. 26.

In fact, in the solution physical and chemical forces exist at the same time. Just in some cases one is predominant, and the other is minor.

The limitation of this theory is that it is only suitable for the separation of non-polar systems. But in extractive distillation polar-polar and non-polar – polar systems are often met, and at this time the molecular interaction may be more complicated. But the idea of physical and chemical forces is valuable and may be adopted. For instance, for the separation of acetic acid and water with tributylamine as the separating agent by extractive distillation, the chemical force between acetic acid and tributylamine is very strong and has been verified by IR and GC-MS techniques.

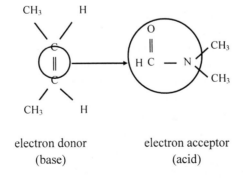

electron donor                    electron acceptor
(base)                              (acid)

Fig. 26. Schematic diagram of chemical interaction (Arrow shows donation of electrons).

## 6.2. Scaled particle theory

*6.2.1. The relationship of salt effect and relative volatility at infinite dilution*

Due to the unique advantages, extractive distillation with the combination of liquid solvent and solid salt as an advanced technique has been caught on more attention. However, selection of solid salt needs theoretical guidance. In this case, scaled particle theory is recommended to apply for extractive distillation, and tries to interpret the effect of solid salt on extractive distillation. Moreover, it provides a theoretical support for optimizing the solvents if needed.

Nowadays there are many theories [126-129] about salt effect such as the electrostatic theory of Debye-McAulay in dilute electrolyte solutions, internal pressure theory of McDevit-Long, salt effect nature of Huangziqing, electrolyte solution theory of Pitzer and scaled particle theory. But only scaled particle theory is the most promising.

An important state of extractive distillation to select and evaluate solvents is infinite dilution of non-electrolyte. It is assumed that the system of extractive distillation with the combination of liquid solvent and solid salt is composed of solvent, salt, components A and B both of which are non-electrolyte. But when components A and B are in the case of infinite dilution, it is known that component B has no influence on the system composed of solvent, salt and component A. On the other hand, component A has also no influence on the system composed of solvent, salt and component B. Therefore, even though in the most cases scaled particle theory applies for ternary system, it can extend to the systems containing many components while all non-electrolytes are at finite dilution. So it is convenient to deal with the problem on the extractive distillation with the combination of liquid solvent and solid salt.

According to the procedure of establishing fundamental equations in scaled particle theory, the vapor partial pressures of non-electrolytes in the solutions are always constant under the same temperature in the case of salt and no salt.

Assume that the system of extractive distillation is composed of solvent, components A and B with low concentrations. The liquid molar fractions of components A and B are respectively $x_{01}$, $x_{02}$, corresponding concentration $c_{01}$, $c_{02}$ (mol l$^{-1}$). When the system attains the state of vapor-liquid equilibrium, the vapor molar fractions of component A and component B are respectively $y_{01}$, $y_{02}$, corresponding vapor pressure $P_{01}$, $P_{02}$. The total pressure is $P_0$ and component A is more volatile than component B in the solvent. It is required to keep the temperature and solute vapor partial pressure $P_{01}$, $P_{02}$ constant when adding salt to the system. In the case of adding salt, it is presumed that the liquid molar fractions of component A and component B are respectively $x_1$, $x_2$ corresponding concentration $c_1$, $c_2$ (mol l$^{-1}$) and vapor partial pressure $P_1 = P_{01}$, $P_2 = P_{02}$. The total pressure of the system with salt is $P$. Therefore, we can write

$$\frac{c_{01}/c_1}{c_{02}/c_2} = \frac{x_{01}/x_1}{x_{02}/x_2} = \frac{y_{02}x_{01}}{y_{01}x_{02}} \frac{y_1/x_1}{y_2/x_2} \frac{y_{01}y_2}{y_{02}y_1} = \frac{\alpha_s}{\alpha_0} \frac{P_{01}P_2}{P_{02}P_1} = \frac{\alpha_s}{\alpha_0} \tag{41}$$

where $\alpha_s$ and $\alpha_0$ are relative volatilities of components A and B at finite dilution with salt and no salt. From Eq. (3), Eqs. (42) and (43) are obtained:

$$\log \frac{c_{01}}{c_1} = k_{s1} c_s \tag{42}$$

$$\log \frac{c_{02}}{c_2} = k_{s2} c_s \tag{43}$$

Regarding the definition of relative volatility and combining Eqs. (41) - (43), it is derived:

$$\frac{a_s}{a_0} = 10^{(k_{s1}-k_{s2})c_s} \tag{44}$$

When component A and B are at infinite dilution, it becomes:

$$\frac{a_s^\infty}{a_0^\infty} = 10^{(k_{s1}-k_{s2})c_s} \tag{45}$$

where $a_s^\infty$ and $a_0^\infty$ represent relative volatilities at infinite dilution with salt and no salt.

Eq. (45) presents the relationship of salting coefficients and relative volatilities at infinite dilution, and constructs a bridge between microscale and macroscale. Even if the calculated salting coefficients aren't accurate due to the current limitation of scaled particle theory, it isn't difficult to judge the magnitude of $k_{s1}$ and $k_{s2}$ by the conventional thermodynamics knowledge and decide whether it is advantageous to improve the relative volatilities with salt. From Eq. (45) it is concluded that if $k_{s1} > k_{s2}$ with low salt concentration, the relative volatilities of components to be separated will be increased by adding salt; the more great the difference between $k_{s1}$ and $k_{s2}$, the more apparent the effect of improving relative volatility.

The value of $\alpha_0^\infty$ can be derived from experiment or calculation using a vapor pressure equation and liquid activity coefficient equation. According to Eq. (45), it is convenient to obtain the values of $\alpha_s^\infty$ just by calculating salting coefficients according to scaled particle theory.

Differentiating Eqs. (42) and (43) with respect to $c$, we can write in an ordinary expression:

$$\lim_{c_s \to 0} \log \frac{c_0}{c} = k_s c_s \tag{46}$$

In terms of scaled particle theory, we obtain

$$-\left( \frac{\partial \log c}{\partial c_s} \right) = k_s = \left[ \frac{\partial(\bar{g}_1^h / 2.3kT)}{\partial c_s} \right]_{c_s \to 0} + \left[ \frac{\partial(\bar{g}_1^s / 2.3kT)}{\partial c_s} \right]_{c_s \to 0} + \left[ \partial(\ln \sum_{j=1}^{m} \rho_j)/\partial c_s \right]_{c_s \to 0}$$

$$= k_\alpha + k_\beta + k_\gamma \tag{47}$$

How to derive the expressions of $k_\alpha$, $k_\beta$ and $k_\gamma$ in terms of characteristic parameters of nonelectrolyte and the ions of the salt has been clarified in chapter 1.

### 6.2.2. Case study
(1) The system of DMF / C4

The method of separating C4 by extractive distillation with DMF is widely used in industry. Herein, the investigated system is composed of DMF, salt NaSCN with weight fraction 10% in DMF, and C4 hydrocarbons. We use the number 1, 2, 3, 4, 5 and s to represent n-butane, butene-1, trans-2-butene, cis-2-butene, 1,3-butadiene and salt respectively [130].

According to scaled particle theory, the three terms $k_\gamma$, $k_\beta$, $k_\alpha$ are calculated, and the sequence of salting coefficients is $k_{s1} > k_{s2} > k_{s3} > k_{s4}$, which is reasonable because the fluidity of electron cloud of C4 hydrocarbons is different and thus the interaction between salt and C4 hydrocarbons is in the following order: Butane < butene < butadiene < butyne. Furthermore, $k_{s1}, k_{s2}, k_{s3}$ and $k_{s4}$ are larger than zero, which means that the salt effect of C4 hydrocarbons is salting out.

The relative volatilities at infinite dilution of C4 hydrocarbons with salt are calculated by using Eq. (45). To evaluate the accuracy, the calculated values are compared with the experimental values, as shown in Table 24 from which it can be seen that the calculated values are in good agreement with the experimental values. It indicates that scaled particle theory is successful for this system.

Table 24
Comparison of relative volatilities at infinite dilution between calculated and experimental values at $T_1 = 303.15K$ and $T_2 = 323.15K$

|  | $a_{s15}^\infty$ | | $a_{s25}^\infty$ | | $a_{s35}^\infty$ | | $a_{s45}^\infty$ | |
|---|---|---|---|---|---|---|---|---|
|  | $T_1$ | $T_2$ | $T_1$ | $T_2$ | $T_1$ | $T_2$ | $T_1$ | $T_2$ |
| Calculated value | 4.43 | 3.75 | 2.50 | 2.26 | 2.05 | 1.87 | 1.74 | 1.66 |
| Experimental Value | 4.53 | 3.73 | 2.55 | 2.28 | 2.11 | 1.94 | 1.85 | 1.76 |
| Relative error % | 2.21 | 0.54 | 1.96 | 0.88 | 2.84 | 3.61 | 5.95 | 5.68 |

(2) The system of ACN / C3

The method of separating C3 by extractive distillation with ACN was reported in the reference [131]. The investigated system is composed of ACN, salt NaSCN with weight fraction 10% in ACN, and C3 (propane and propylene). The expressions of $k_\alpha$, $k_\beta$, $k_\gamma$ are deduced in the same way as that of the system DMF / C4.

The salt coefficients, $k_{s1}$ and $k_{s2}$, and relative volatilities at infinite dilution at different temperatures are listed in Table 25 in which we use the number 1, 2 and s to represent propane, propylene and salt respectively.

Table 25
Comparison of relative volatilities at infinite dilution between calculated and experimental values at different temperatures

| $T$ (K) | $k_{s1}$ | $k_{s2}$ | $\alpha_{012}^\infty$ | $\alpha_{s12}^\infty$ | | |
|---|---|---|---|---|---|---|
| | | | | Calculated value | Experimental value | Relative error (%) |
| 289.65 | 0.6053 | 0.5483 | 1.69 | 1.92 | 2.01 | 4.47 |
| 298.15 | 0.6034 | 0.5462 | 1.67 | 1.90 | 1.96 | 3.06 |
| 303.15 | 0.6000 | 0.5450 | 1.65 | 1.87 | 1.90 | 1.58 |
| 312.15 | 0.6000 | 0.5428 | 1.64 | 1.86 | 1.84 | 1.09 |
| 324.15 | 0.5977 | 0.5399 | 1.62 | 1.84 | 1.79 | 2.79 |

Table 25 tells us that scaled particle theory is reliable in predicting the system of ACN/C3 with relative error less than 5%. In general the solvents ACN and DMF are optimized by adding a little of water to improve relative volatilities of the nonelectrolytes. But we know that ACN and DMF are prone to hydrolyze, which limits their use in industry. The salts haven't this problem and can substitute for water to increase the separation ability of solvents.

(3) The system of ethylene glycol / ethanol / water

In the separation of aqueous ethanol by extractive distillation with ethylene glycol added salt, the concerned system is composed of ethylene glycol, salt potassium acetate (KAc) with weight fraction 10% in ethylene glycol, ethanol (1) and water (2). In many plants the mixture of ethanol and water are separated by extractive distillation with ethylene glycol added KAC. In terms of scaled particle theory, the reason could be explored. Because the interaction between water / KAc are stronger than that between alcohol / KAc, it is evident that $k_{s1} > k_{s2}$.

Therefore it is judged from Eq. (45) that KAc should enhance the separation ability of ethylene glycol. But formerly the interpretation of the phenomena was very vague [46-48].

However, at present the three terms of salt coefficients in scaled particle theory aren't related to hydrogen bond force between solutes and solvents. Therefore it is difficult to calculate salt coefficients precisely in terms of scaled particle theory. But we can qualitatively predict the salt effect according to Eq. (45).

(4) The system of ethylene glycol / acetone / methanol

In general, acetone and methanol are separated by extractive distillation [132, 133]. Herein, the investigated system is composed of ethylene glycol, salt potassium acetate (KAc) with weight fraction 10% in ethylene glycol, acetone (1) and methanol (2). Like the system of ethylene glycol / ethanol / water, we can qualitatively predict that KAc is bound to increase relative volatilities of acetone to methanol because $k_{s1} > k_{s2}$, even though $k_{s1}$ and $k_{s2}$ are unknown. In order to accurately calculate $k_{s1}$ and $k_{s2}$ for the polar solutes systems, it requires the scaled particle theory to be improved in the future.

Therefore, the relationship between microscale salt coefficients and macroscale relative volatilities at infinite dilution can be set up in terms of scaled particle theory. For separating non-polar system, e.g. DMF / C4 and ACN / C3, the relative volatilities at infinite dilution with salt correspond well with experimental values. The reason may be that C4 and C3 are non-polar components and the sizes of them aren't large, which lead to the accurate results. It is of interest to compare the relative contribution of the three terms $k_\alpha$, $k_\beta$ and $k_\gamma$ to the salting coefficient $k_s$, and it is found that $k_\gamma$ is small. Therefore, the salting coefficients $k_s$ mainly depends upon the relative magnitudes of $k_\alpha$ and $k_\beta$.

However, for separating polar solutes systems salt coefficients aren't easy to be accurately calculated in terms of scaled particle theory. But this doesn't influence our analysis on whether it is advantageous to add salt to a system because in most cases it isn't difficult for us to qualitatively judge the relative values of salt coefficient of each component.

Although the application of the scaled particle theory to the calculation of salt effect has the great advantages that the required molecular parameters are readily available, it is limited in some degrees. For polar solutes, the scaled particle theory can only provide qualitative analysis according to Eq. (45). The reason may be that the hydrogen bonding between polar solutes and polar solvent is very complicated and greater than van der Waals bonding. By now, only van der Waals bonding is considered in the scaled particle theory. Consequently, quantitative calculation for polar solutes is very difficult and inaccurate. We know any theory has its own deficiency, but we believe that with the development of scaled particle theory the problem will be solved one day.

Anyway, scaled particle theory is extended to solve the problem of extractive distillation with the combination of solvent and salt, and will promote the development of extractive distillation which is always an important separation method.

## 7. MATHEMATIC MODELS OF EXTRACTIVE DISTILLATION

For design of distillation process, two types of modeling approaches have been developed: the equilibrium (EQ) stage model and the non-equilibrium (NEQ) stage model. The most difference between these two models is that the mass and heat transfer rates should be considered in every tray in the NEQ stage model. For extractive distillation, the process can be simulated either by EQ stage model or by NEQ stage model.

## 7.1. EQ stage model

### 7.1.1. Model equations

The schematic diagrams of a tray column for EQ stage model are shown in Fig. 27. In general, the assumptions adopted are as follows [134-136]:

1. Operation reaches steady state;
2. System reaches mechanical equilibrium in every tray;
3. The vapor and liquid bulks are mixed perfectly and assumed to be at thermodynamic equilibrium;
4. Heat of mixing can be neglected;
5. Reactions take place in the liquid phase; if reactions take place in the vapor phase, the actual vapor compositions are replaced by transformed superficial compositions.
6. The condenser and reboiler are considered as an equilibrium tray.

The equations that model EQ stages are known as the MESHR equations. MESHR is an acronym referring to the different types of equation. The M equations are the material balance equations. The total material balance takes the form

$$\frac{dM_j}{dt} = V_{j+1} + L_{j-1} + F_j - (1+r_j^V)V_j - (1+r_j^l)L_j + \sum_{k=1}^{r}\sum_{i=1}^{c} v_{i,k} R_{k,j} \varepsilon_j. \tag{48}$$

where $F$, $V$, $L$ are feed, vapor and liquid flowrates respectively, $t$ is time, $R_{k,j}$ is reaction rate, $v_{i,k}$ is the stoichiometric coefficient of component $i$ in reaction $k$, $\varepsilon$ is reaction volume, and $M_j$ is the hold-up on stage $j$.

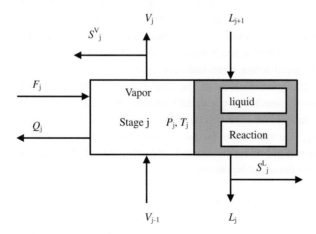

Fig. 27. Schematic representation of an EQ stage.

With very few exceptions, $M_j$ is considered to be the hold-up only of the liquid phase. It is more important to include the hold-up of the vapor phase at higher pressures. The component material balance (neglecting the vapor hold-up) is:

$$\frac{dM_j x_{i,j}}{dt} = V_{j+1} y_{i,j+1} + L_{j-1} x_{i,j-1} + F_j z_{i,j} - (1+r_j^V) V_j y_{i,j} - (1+r_j^L) L_j x_{i,j} + \sum_{k=1}^{r} v_{i,k} R_{k,j} \varepsilon_j \quad (49)$$

where $x$, $y$ are mole fractions in the liquid and vapor phase, respectively. In the material balance equations given above $r_j$ is the ratio of sidestream flow to interstage flow:

$$r_j^V = S_j^V / V_j , \quad r_j^L = S_j^L / L_j \quad (50)$$

The E equations are the phase equilibrium relations:

$$y_{i,j} = K_{i,j} x_{i,j} \quad (51)$$

where $K_{i,j}$ is chemical equilibrium constant.

The S equations are the summation equations:

$$\sum_{i=1}^{c} x_{i,j} = 1 , \quad \sum_{i=1}^{c} y_{i,j} = 1 \quad (52)$$

The enthalpy balance is given by

$$\frac{dM_j H_j}{dt} = V_{j+1} H_{j+1}^V + L_{j-1} H_{j-1}^L + F_j H_j^F - (1+r_j^V) V_j H_j^V - (1+r_j^L) L_j H_j^L - Q_j \quad (53)$$

where $H$ is molar enthalpy, and $Q$ is heat duty.

There is no need to take separate account in Eq. (53) of the heat generated due to chemical reaction since the computed enthalpies include the heats of formation.

The R equations are the reaction rate equations. For the reactive extractive distillation it is known that chemical reaction is reversible, and the reaction rate is assumed to be zero. That is to say, the chemical equilibrium is reached in every tray.

Under steady-state conditions all of the time derivatives in the MESH equations are equal to zero. Newton's method (or a variant thereof) for solving all of the independent equations simultaneously is an approach nowadays widely used. But other methods also are frequently used, e.g. the relaxation method where the MESH equations are written in unsteady-state form and are integrated numerically until the steady-state solution has been found, is used to solve the above equations.

### 7.1.2. Case studies

EQ stage model is used for process simulation of extractive distillation in order to obtain the necessary information for various purposes. In what follows we discuss the steady state of extractive distillation and extractive distillation with chemical reaction (i.e. reactive extractive distillation) in the EQ stage model.

(1) Steady state analysis

Analysis of operation state is still a topic of considerable interest in the distillation community. Especially in the catalytic and azeotropic distillation processes the non-linearity phenomena are very prominent, and multiple steady state is easy to appear (see chapter 4). In this chapter only steady state is analyzed.

For the separation of C4 mixture by extractive distillation with DMF, the operation state is analyzed by changing the feeding location of solvent, which is a sensitive parameter [159]. The feeding location of solvent into the extractive distillation column for recovering butane is variable, but other operating conditions are kept constant. The relation of the top molar composition of butane with the feeding stage of solvent is shown in Fig. 28 (the stage is numbered from the top to the bottom). When the solvent is fed in the vicinity of No. 20 stage, the composition change is very abrupt and two operation states (OS) are found. However, only No. 1 operation state (OS) is desirable because in this case the top molar composition of butane is much higher.

Similarly, only the feeding location of solvent into the extractive distillation column for recovering 1,3-butadiene is variable, but other operating conditions are kept constant. In this case the top molar composition of 1,3-butadiene is stable because the amount of vinylacetylene (VAC) is so small that the 1,3-butadiene composition is not affected. The function of this column is to remove VAC from 1,3-butadiene, so the change of the bottom flowrate of VAC should be obvious. The relation of the bottom flowrate of VAC (kmol h$^{-1}$) with the feeding stage of solvent S is shown in Fig. 29 (the stage is numbered from the top to the bottom). It can be seen from Fig. 29 that when solvent S is fed in the vicinity of No. 30 stage, two operation states are also found. However, only No. 1 operation state is desirable because in this case the amount of VAC removed from 1,3-butadiene is greater. Therefore, by using EQ stage model, we can find which operation state is the best or the worst.

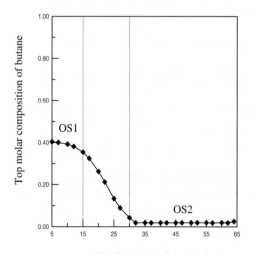

Feeding stage of solvent S

Fig. 28. The relation of the top composition of butane with the feeding stage of solvent S in the extractive distillation column.

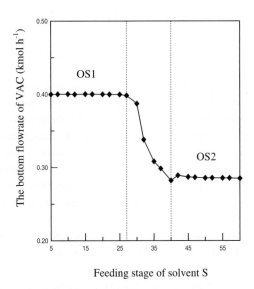

Fig. 29. The relation of the bottom flowrate of VAC with the feeding stage of solvent in the extractive distillation column.

(2) Reactive extractive distillation

In the EQ stage model, if there exists no chemical reaction, the independent equations are simplified, only solving MESH equations. However, for reactive extractive distillation, EQ stage model is somewhat complicated, especially when there exist more than one chemical reactions. Separation of acetic acid and water with tributylamine as the separating agent is just the case.

As mentioned before, the following reversible chemical reaction may take place:

$$HAc \quad + \quad R_3N \quad \rightleftharpoons \quad R_3NH^+ \cdot {}^-OOCCH_3$$

where HAc, $R_3N$ and $R_3NH^+ \cdot {}^-OOCCH_3$ represent acetic acid, tributylamine and the salt formed by the reaction, respectively.

In terms of the chemical equilibrium constant $K_p$ at 25℃, the relation of chemical equilibrium constant with temperature can be deduced by Eq. (4).

On the other hand, it is known that aggregation of acetic acid molecules in the vapor phases occurs. If only aggregation of two molecules is considered, the following reversible chemical reaction may take place:

$$2A_1 \quad \rightleftharpoons \quad A_2$$

where $A_1$ is the monomer of acetic acid, and $A_2$ is the dimer of acetic acid.

In this case, the chemical equilibrium constant $K_{A_2}$ is written as

$$K_{A_2} = \frac{\eta_{A_2}^V}{P(\eta_{A_1}^V)^2} \tag{54}$$

In addition, $K_{A_2}$, the function of temperature, can also be calculated by the following empirical equation [137, 138],

$$\lg K_{A_2} = \varepsilon_{A_2} + \omega_{A_2} / T \tag{55}$$

where $\varepsilon_{A_2} = -10.4205$, $\omega_{A_2} = 3166$.

The key to the simulation of extractive distillation process is the selection of an accurate VLE model to solve the EQ stage model of the column. In general, the Wilson, NRTL and UNIQUAC equations, which are suitable for systems composed of many components and can deduce the multi-components system from binary systems, are used.

Although the VLE data can be obtained by experiments and then correlated by the iterative solution by using the maximum likelihood regression, the interaction parameters of the VLE model may be various under different assumptions. Table 26 shows the interaction parameters of Wilson, NRTL and UNIQUAC equations for the system of water (1) and acetic acid (2) under two cases: considering two-molecule aggregation and not, respectively; Table 27 shows the interaction parameters of Wilson, NRTL and UNIQUAC equations for the system of water (1) and tributylamine (2); Table 28 shows the interaction parameters of Wilson, NRTL and UNIQUAC equations for the system of acetic acid (1) and tributylamine (2) under three cases: not considering two-molecule aggregation and reversible chemical reaction; only considering reversible chemical reaction, and considering two-molecule aggregation and reversible chemical reaction at the same time, respectively.

In these tables an average deviation $\Delta y_1$ is calculated by

$$\text{Average deviation} = \frac{\sum_{i=1}^{N} |\Delta y|}{N}, \text{ where } N \text{ is the times of experiments.}$$

It can be seen from Table 26 that the difference of interaction parameters between two cases is somewhat obvious, but the average deviation $\Delta y_1$ in both cases is small enough.

The reason may be that the activity coefficients of water and acetic acid aren't changed apparently at the total concentration, and the interaction parameters of the VLE models can be correlated well. Even so, it manifests that the influence of two-molecule aggregation on the interaction parameters can't be ignored.

Table 26
Interaction parameters of Wilson, NRTL and UNIQUAC equations for the system of water (1) and acetic acid (2) (unit: J mol$^{-1}$)

| Equation | $A_{12}$ | $A_{21}$ | $\Delta y_1$ |
|---|---|---|---|
| 1. Not considering two-molecule aggregation | | | |
| Wilson | 3766.34 | -3301.3 | 0.0204 |
| NRTL ($\alpha = 0.47$) | 87.86 | 363.04 | 0.0205 |
| UNIQUAC | -2882.96 | 4149.57 | 0.0241 |
| 2. Considering two-molecule aggregation | | | |
| Wilson | 3566.01 | -818.99 | 0.0164 |
| NRTL ($\alpha = 0.47$) | 2861.58 | -120.11 | 0.0164 |
| UNIQUAC | -1553.11 | 2321.62 | 0.0185 |

Table 27
Interaction parameters of Wilson, NRTL and UNIQUAC equations for the system of water (1) and tributylamine (2) (unit: J mol$^{-1}$)

| Equation | $A_{12}$ | $A_{21}$ | $\Delta y_1$ |
|---|---|---|---|
| Wilson | 16425.00 | 22378.10 | 0.0213 |
| NRTL ($\alpha = 0.3$) | 9245.76 | 2940.54 | 0.0202 |
| UNIQUAC | 791.39 | 302.74 | 0.0191 |

Table 28
Interaction parameters of Wilson, NRTL and UNIQUAC equations for the system of acetic acid (1) and tributylamine (2) (unit: J mol$^{-1}$)

| Equation | $A_{12}$ | $A_{21}$ | $\Delta y_1$ |
|---|---|---|---|
| 1. Not considering two-molecule aggregation and reversible chemical reaction | | | |
| Wilson | 7373.29 | -3558.84 | 0.1158 |
| NRTL ($\alpha = 0.2$) | -1553.97 | 5263.68 | 0.1160 |
| UNIQUAC | -1707.19 | 4830.12 | 0.1156 |
| 2. Only considering reversible chemical reaction | | | |
| Wilson | 7550.79 | -3478.02 | 0.1182 |
| NRTL ($\alpha = 0.2$) | 346.69 | 3413.18 | 0.1181 |
| UNIQUAC | -1551.01 | 4697.44 | 0.1200 |
| 3. Considering two-molecule aggregation and reversible chemical reaction simultaneously | | | |
| Wilson | 8731.21 | -2265.00 | 0.0580 |
| NRTL ($\alpha = 0.2$) | 1496.1 | 4118.9 | 0.0596 |
| UNIQUAC | -1505.47 | 5463.82 | 0.0552 |

In Table 28, the difference of interaction parameters among three cases is apparent.

Moreover, the average deviation $\Delta y_1$ in the third case is very small. The reason may be that this case is approximate to the actual situation, and the VLE model can accurately describe the real activity coefficients. Besides, it manifests that the influence of two-molecule aggregation and reversible chemical reaction on the interaction parameters can't be ignored.

By using EQ stage model, the extractive distillation process of separating water and acetic acid is simulated. In the simulation three cases are concerned, i.e. 1. not considering two-molecule aggregation and reversible chemical reaction (EQ1 model); 2. only considering reversible chemical reaction (EQ2 model); 3. considering two-molecule aggregation and reversible chemical reaction simultaneously (EQ3 model).

The experimental data, as well as the calculated results from the EQ stage models (EQ1, EQ2 and EQ3 models), are listed in Table 29. It is shown that the values of EQ3 model are closer to experimental results than those of EQ1 and EQ2 models. That means that EQ3 model reflects the real state of the system of water / acetic acid / tributylamine more accurately.

In order to go a further step to investigate the difference among EQ1, EQ2 and EQ3 models, the composition and temperature distributions along the extractive distillation column under the same operation condition are given in Tables 30 and 31, where subscripts 1, 2, 3 and 4 represent water, acetic acid, tributylamine and salt produced by the reaction, respectively. The tray is numbered from the bottom to the top. It can be seen from Table 30 that the calculated results of EQ1 model correspond well with those of EQ2 model on a large part of trays. However, in the vicinity of the tray feeding the mixture of acetic acid and water, i.e. from No. 9 to No.15, the difference of liquid composition distribution is enlarged, which may be due to the relatively great influence of the feed mixture on the chemical reaction between acetic acid and tributylamine. But the difference of the composition and temperature distributions between EQ3 model and EQ1 (or EQ2) model is apparent. This indicates that the influence of two-molecule aggregation is more obvious than that of reversible chemical reaction. In summary, EQ3 model is more suitable for the design and optimization of extractive distillation process than EQ1 and EQ2 models.

On the other hand, the effect of the solvent, tributylamine, on the separation of water and acetic acid can be verified by simulation. It can be seen from Table 29 that water composition at the bottom of the extractive distillation column is nearly equal to zero. That is to say, the mixture of water and acetic acid can be effectively separated by extractive distillation with tributylamine as the separating agent.

## 7.2. NEQ stage model

The NEQ stage model for extractive distillation should follow the philosophy of rate-based models for conventional distillation. Unfortunately, it has been rarely reported on the NEQ stage model for extractive distillation. The reason may be that is building an NEQ stage model for extractive distillation process isn't as straightforward as it is for the EQ stage model where, at most, if the chemical reaction is involved, we need to simply add an equation to take account of the effect of chemical reaction on a tray or section of packing.

Table 29
The comparison of the experimental and calculated values

| No. | $T_{top}$ / °C | | | | Water concentration at the top, wt% | | | |
|---|---|---|---|---|---|---|---|---|
| | Exp. | EQ1 | EQ2 | EQ3 | Exp. | EQ1 | EQ2 | EQ3 |
| 1 | 99.9 | 99.7 | 99.5 | 99.7 | 99.74 | 96.20 | 96.30 | 96.46 |
| 2 | 100.2 | 99.8 | 99.9 | 99.6 | 99.81 | 94.09 | 92.10 | 97.51 |
| 3 | 99.6 | 100.0 | 99.5 | 99.7 | 99.77 | 98.78 | 99.30 | 96.08 |
| 4 | 99.9 | 100.5 | 99.8 | 99.5 | 99.86 | 94.73 | 94.77 | 98.55 |

Table 30
The composition distribution along the extractive distillation column for the EQ1, EQ2 and EQ3 models

| No. | $x_1$ | | | $x_2$ | | | $x_3$ | | | $x_4$ | | |
|---|---|---|---|---|---|---|---|---|---|---|---|---|
| | EQ1 | EQ2 | EQ3 | EQ1 | EQ2 | EQ3 | EQ1 | EQ2 | EQ3 | EQ1 | EQ2 | EQ3 |
| 1 | 0.0004 | 0.0002 | 0.0 | 0.4970 | 0.4762 | 0.4906 | 0.5029 | 0.4904 | 0.4769 | 0.0 | 0.0331 | 0.0326 |
| 3 | 0.0068 | 0.0036 | 0.0 | 0.9350 | 0.9266 | 0.8419 | 0.0582 | 0.0552 | 0.1318 | 0.0 | 0.0145 | 0.0263 |
| 5 | 0.0301 | 0.0164 | 0.0 | 0.9168 | 0.9142 | 0.8577 | 0.0531 | 0.0549 | 0.1177 | 0.0 | 0.0145 | 0.0245 |
| 7 | 0.1200 | 0.0680 | 0.0 | 0.8270 | 0.8580 | 0.8584 | 0.0530 | 0.0590 | 0.1169 | 0.0 | 0.0149 | 0.0246 |
| 9 | 0.3430 | 0.2268 | 0.0012 | 0.6015 | 0.7046 | 0.8569 | 0.0553 | 0.0556 | 0.1178 | 0.0 | 0.0130 | 0.0242 |
| 11 | 0.6184 | 0.4690 | 0.0474 | 0.3255 | 0.4562 | 0.8133 | 0.0561 | 0.0638 | 0.1152 | 0.0 | 0.0108 | 0.0241 |
| 13 | 0.4862 | 0.5194 | 0.1237 | 0.3006 | 0.3512 | 0.5910 | 0.2133 | 0.1164 | 0.2411 | 0.0 | 0.0129 | 0.0441 |
| 15 | 0.4885 | 0.6162 | 0.1617 | 0.3061 | 0.2759 | 0.5971 | 0.2054 | 0.0981 | 0.2149 | 0.0 | 0.0098 | 0.0263 |
| 17 | 0.6334 | 0.6576 | 0.1619 | 0.2318 | 0.2096 | 0.5430 | 0.1348 | 0.1224 | 0.2607 | 0.0 | 0.0104 | 0.0344 |
| 19 | 0.6612 | 0.6618 | 0.1486 | 0.2029 | 0.2091 | 0.5256 | 0.1359 | 0.1205 | 0.3008 | 0.0 | 0.0086 | 0.0250 |
| 21 | 0.7236 | 0.7523 | 0.2890 | 0.1386 | 0.1274 | 0.3986 | 0.1378 | 0.1145 | 0.2927 | 0.0 | 0.0058 | 0.0196 |
| 23 | 0.7939 | 0.7414 | 0.8163 | 0.0722 | 0.0688 | 0.0925 | 0.1340 | 0.1870 | 0.0874 | 0.0 | 0.0028 | 0.0038 |
| 25 | 0.9730 | 0.9721 | 0.9680 | 0.0115 | 0.0112 | 0.0106 | 0.0155 | 0.0166 | 0.0213 | 0.0 | 0.0001 | 0.0 |

Table 31
The temperature distribution along the extractive distillation column for the EQ1, EQ2 and EQ3 models

| No. | $T / K$ | | |
|---|---|---|---|
| | EQ1 | EQ2 | EQ3 |
| 1 | 404.83 | 404.90 | 424.45 |
| 3 | 392.51 | 392.37 | 419.40 |
| 5 | 391.41 | 391.26 | 419.19 |
| 7 | 388.34 | 388.14 | 419.18 |
| 9 | 382.65 | 382.43 | 418.97 |
| 11 | 377.76 | 377.65 | 411.03 |
| 13 | 378.06 | 376.89 | 395.89 |
| 15 | 378.73 | 376.62 | 394.63 |
| 17 | 376.89 | 375.60 | 392.60 |
| 19 | 376.44 | 375.62 | 393.41 |
| 21 | 376.44 | 374.16 | 384.81 |
| 23 | 374.12 | 373.22 | 374.53 |
| 24 | 373.30 | 372.88 | 373.36 |
| 25 | 372.84 | 372.69 | 372.83 |

As we know, the NEQ stage model is more complicated than the EQ stage model. In the NEQ stage model, the design information on the column configuration must be specified so that mass transfer coefficients, interfacial areas, liquid hold-ups, etc. can be calculated. Therefore, for any new invented configuration of the column, many experiments have to be done in advance to obtain the necessary model parameters. Evidently, it is too tedious and much time will be spent on the design of extractive distillation process. Fortunately, as pointed out by Lee and Dudukovic [139], a close agreement between the predictions of the EQ and NEQ stage models can be found if the tray efficiency or HETP (height equivalent of a theoretical plate) is known. So the EQ stage model is widely used in extractive distillation.

The model equations of NEQ stage for extractive distillation are similar to those for reaction distillation [140-144]. The schematic diagrams of a tray column for NEQ stage model are shown in Fig. 30. In general, the assumptions adopted are as follows:

1. Operation reaches steady state;
2. System reaches mechanical equilibrium in every tray;
3. The vapor and liquid bulks are mixed perfectly and assumed to be at thermodynamic equilibrium;
4. Heat of mixing can be neglected;
5. There is no accumulation of mass and heat at the interface;
6. The condenser and reboiler are considered as an equilibrium tray;
7. Reactions take place in the liquid bulk within the interface ignored; if reactions take place in the vapor phase, the actual vapor compositions are replaced by transformed superficial compositions;

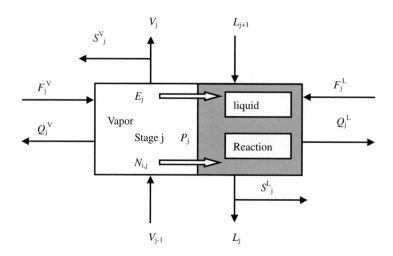

Fig. 30. Schematic representation of an NEQ stage.

8. The heat generated due to chemical reaction is neglected because heat effect is often not apparent for the reaction concerned.

In addition, in the NEQ stage model it is assumed that the resistance to mass and energy transfer is located in thin film adjacent to the vapor-liquid interface according to the two-film theory; See Fig. 31.

The time rate of change of the number of moles of component $i$ in the vapor($M_i^V$) and

liquid($M_i^L$) phases on stage $j$ are given by the following balance relations:

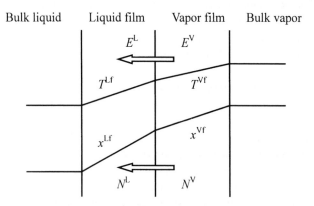

Fig. 31. Two-film model of the NEQ stage.

$$\frac{dM_{i,j}^V}{dt} = V_{j+1} y_{i,j+1} - V_j y_{i,j} + z_{i,j}^V F_{i,j}^V - N_{i,j}^V \tag{56}$$

$$\frac{dM_{i,j}^L}{dt} = L_{j-1} x_{i,j-1} - L_j x_{i,j} + z_{i,j}^L F_{i,j}^L + N_{i,j}^L + \sum_{k=1}^{r} v_{i,k} R_{k,j} \varepsilon_j^L \tag{57}$$

where $N_{i,j}$ is the interfacial mass transfer rate.

The overall molar balances are obtained by summing Eqs. (56) and (57) over the total number of components $c$ in the mixture:

$$\frac{dM_j^V}{dt} = V_{j+1} - V_j + F_j^V - \sum_{k=1}^{c} N_{k,j}^V \tag{58}$$

$$\frac{dM_j^L}{dt} = L_{j-1} - L_j + F_j^L + \sum_{k=1}^{c} N_{k,j}^L + \sum_{i=1}^{c} \sum_{k=1}^{r} v_{i,k} R_{k,j} \varepsilon_j^L \tag{59}$$

The mole fractions of the vapor and liquid phases are calculated from the respective phase molar hold-ups:

$$y_{i,j} = \frac{M_{i,j}^V}{M_j^V}, \quad x_{i,j} = \frac{M_{i,j}^L}{M_j^L} \tag{60}$$

Only $c-1$ of these mole fractions are independent because the phase mole fractions sum to unity:

$$\sum_{k=1}^{c} y_{k,j} = 1, \quad \sum_{k=1}^{c} x_{k,j} = 1 \tag{61}$$

The energy balance for the vapor and liquid phases are written as follows:

$$\frac{dE_j^V}{dt} = V_{j+1} H_{j+1}^V - V_j H_j^V + F_j^V H_j^{VF} - E_j^V - Q_j^V \tag{62}$$

$$\frac{dE_j^L}{dt} = L_{j-1} H_{j-1}^L - L_j H_j^L + F_j^L H_j^{LF} + E_j^L - Q_j^L \tag{63}$$

where $H_j^V$ and $H_j^L$ represent the molar enthalpy of vapor and liquid phases, respectively.

There is no need to take separate account in Eqs. (62) and (63) of the heat generated due to chemical reaction since the computed enthalpies include the heats of formation.

As mentioned above, the importance to model the NEQ is to set up the relation of interfacial mass and energy transfer rates. The molar transfer rate $N_i^L$ in the liquid phase is related to the chemical potential gradients by the Maxwell-Stefan equation:

$$\frac{x_i^l}{RT^l} \frac{\partial \mu_i^l}{\partial \eta} = \sum_{k=1}^{c} \frac{x_i^l N_k^l - x_k^l N_i^l}{C_i^l \kappa_{i,k}^l A} \tag{64}$$

$\kappa_{i,k}^l$ represents the mass transfer coefficient of the $i-k$ pair in the liquid phase; this coefficient is estimated from information on the corresponding Maxwell-Stefan diffusivity $D_{i,k}^l$ [145]. How to solve the Maxwell-Stefan equation has been clarified in detail in chapter 1. The summation equations at the interface are:

$$\sum_{k=1}^{c} y_{i,j}^{Vf} = 1, \quad \sum_{k=1}^{c} x_{i,j}^{lf} = 1 \tag{65}$$

The interphase energy transfer rates $E^{lf}$ have conductive and convective contributions

$$E^{lf} = -h^{lf} A \frac{\partial T^{lf}}{\partial \eta} + \sum_{i=1}^{c} N_i^{lf} H_i^{lf} \tag{66}$$

where $A$ is the interfacial area and $h^{lf}$ is the heat transfer coefficient in the liquid phase. A relation analogous to Eq. (66) holds for the vapor phase.

At the vapor-liquid interface we assume phase equilibrium

$$y_{i,j} \mid_1 = K_{i,j} x_{i,j} \mid_1 \tag{67}$$

We also have continuity of mass and energy

$$N_i^{Vf} \mid_1 = N_i^{lf} \mid_1, \quad E^{Vf} \mid_1 = E^{lf} \mid_1 \tag{68}$$

Under steady-state conditions all of the time derivatives in the above equations are equal to zero. Similar to EQ stage model, the model equations can be solved using Newton's method (or a variant thereof). Other methods such as the combination of modified relaxation and Newton-Raphson methods where the equations are written in unsteady-state form and are integrated numerically until the steady-state solution has been found, can also be used to solve the above equations.

But both in the EQ stage model and in the NEQ stage model, the thermodynamic and physical properties are required. Moreover in the NEQ stage model the mass and energy transfer models are also necessary. A list of the thermodynamic, physical properties and mass and energy transfer correlations [110, 146-148] available in our program is provided in Table 32.

Though sophisticated NEQ stage model is available readily, detailed information on the hydrodynamics and mass transfer parameters for the various hardware configurations is woefully lacking in the open reference. Moreover, such information may have vital consequences for the calculated results of the distillation column. There is a crying need for research in this area. It is perhaps worth noting here that modern tools of computational fluid dynamics could be invaluable in developing better insights into hydrodynamics and mass transfer in the distillation column. But the computing time may be greatly prolonged.

Table 32
Thermodynamic, physical properties and mass and energy transfer models used in the EQ and NEQ stage models

| $K$-value models | $$k_i = \frac{\gamma_i P_i^0}{P}$$ |
|---|---|
| Equations of state | Virial equation |
| Molar volume | Data from the references |
| Enthalpy | The modified Peng-Robinson (MPR) equation for the vapor enthalpy |
|  | The liquid phase enthalpy deduced from vapor phase enthalpy and evaporation heat |
| Activity coefficient | UNIFAC equation |
| Enthalpy | The modified Peng-Robinson (MPR) equation for the vapor enthalpy |
|  | The liquid phase enthalpy deduced from vapor phase enthalpy and evaporation heat |
| Activity coefficient | UNIFAC equation |
| Vapor pressure | Antoine equation |
| Viscosity | Orrick-Erbar equation for the liquid phase |
|  | Wilke equation for the vapor phase |
| Surface tension | Data from the references |
| Thermal conductivity | Data from the reference |
| Binary mass transfer coefficient | AIChE method |
| Multi-component mass transfer coefficient | The generalized Maxwell-Stefan equation |
| Binary diffusion coefficient | Fuller-Schettler-Giddings equation for the vapor phase |
|  | Lusis-Ratcliff equation for the liquid phase at infinite dilution; Vignes method for the liquid phase at finite dilution |
| Heat transfer coefficient | Calculated from the Chilton-Colburn $j$-factor |

In the field of reaction distillation (RD), recent NEQ modeling works have exposed the limitations of EQ stage model for final design and for the development of control strategies. NEQ stage model has been used for commercial RD plant design and simulation. Thus, it is conjecturable that NEQ stage model would be widely applied in extractive distillation, and EQ stage model only has its place for preliminary design. The research on NEQ stage model for extractive distillation should be strengthened in the near future.

In the end, it should be pointed out that, some contents are adapted from the authors' reference [160].

## REFERENCES

[1] N. Hilal, G. Yousef and M.Z. Anabtawi, Sep. Sci. Technol., 37 (2002) 3291-3303.

[2] J.E. Ewanchyna and C. Ambridge, Cana. J. Chem. Eng., No. 2 (1958) 19-36.

[3] E. R. Hafslund, Chem. Eng. Prog., 65 (1969) 58-64.

[4] L. Berg, AIChE J., 29 (1983) 961-966.

[5] Z.T. Duan, Petrochemical Technology (China), 7 (1978) 177-186.

[6] I. Sucksmith, Chem. Eng., 28 (1982) 91-95.

[7] S.O. Momoh, Sep. Sci. Technol., 26 (1991) 729-742.

[8] L. Berg, Chem. Eng. Prog., 65 (1969) 52-57.

[9] S.K. Min, N. Sangyoup, C. Jungho and K. Hwayong, Korean J. Chem. Eng., 19 (2002) 996-1000.

[10] H.S. Wu, L.C. Wilson and G.M. Wilson, Fluid Phase Equilib., 43 (1988) 77-89.

[11] G.S. Shealy, D. Hagewiesche and S.I. Sandler, J. Chem. Eng. Data, 32 (1987) 366-369.

[12] B.H. Chen, Z.G. Lei and J.W. Li, J. Chem. Eng. Japan, 36 (2003) 20-24.

[13] Z.G. Lei, J.C. Zhang and B.H. Chen, J. Chem. Tech. Biotechnol., 77 (2002) 1251-1254.

[14] W.W. Coogler, Oil Gas J., 22 (1967) 99-101.

[15] R.R. Bannister and E. Buck, Chem. Eng. Prog., 65 (1969) 65-68.

[16] S. Takao, Hydro. Proc., 45 (1966) 151-154.

[17] S. Takao, Oil Gas J., 77 (1979) 81-82.

[18] H. Klein and H.M. Weitz, Hydro. Proc., 47 (1968) 135-138.

[19] R.W. King and H. Mondria, High Purity Alkadienes by Extractive Distillation, US Patent No. 3 317 627 (1967).

[20] D. Barba, C. Dagostini and A. Pasquinelli, Process for the Separation of Butadiene by Plural Stage Extractive Distillation, US Patent No. 4 128 457 (1978).

[21] Z.G. Lei, R.Q. Zhou and Z.T. Duan, Chem. Eng. J., 85 (2002) 379-386.

[22] R.B. Eldridge, Ind. Eng. Chem. Res., 32 (1993) 2208-2212.

[23] D.J. Safarik and R.B. Eldridge, Ind. Eng. Chem. Res., 37 (1998) 2571-2581.

[24] W.H. Wang, W.Z. Ren and W.Y. Xu, Guangxi Huagong (China), 28 (1999) 20-23.

[25] W.Z. Ren, S.A. Xu and L.X. Wang, Qilu Petrochemical Engineering (China), 26 (1998) 95-99.

[26] Z.G. Lei, R.Q. Zhou, Z.T. Duan and H.Y. Wang, Computers and Applied Chemistry (China), 16 (1999) 265-267.

[27] J.Y. Wang, L.S. Tian, W.C. Tang and S.X. Gui, Petroleum Processing and Petrochemicals (China), 33 (2002) 19-22.

[28] L.S. Tian, Y.M. Zhang, M. Zhao and W.C. Tang, Petroleum Processing and Petrochemicals (China), 332 (2001) 6-9.

[29] Department of Chemical Engineering, Chem. Eng. (China), No. 1 (1975) 79-88.

[30] Research group on trays, J. Tsinghua Univ., No. 1 (1977) 112-122.

[31] Z.T. Duan, J.J. Peng and R.Q. Zhou, Multi-overflow Compound Slant-hole Tray, Chinese Patent No. 97100551.6 (1997).

[32] J.J. Peng, Z.T. Duan and B.J. Li, Computers and Applied Chemistry (China), 13 (1996) 49-53.

[33] Q.S. Li, Z.T. Zhang, B.S. Zhao, Y.Q. Han and E.J. Wang, Modern Chemical Industry (China), 21 (2001) 39-41.

[34] Q.S. Li and Z.T. Zhang, Chemical Industry and Engineering Progress (China), 20 (2001) 32-35.

[35] C.Y. Song, Q.S. Li and Z.T. Zhang, Chemical Industry and Engineering Progress (China), 20 (2001) 39-42.

[36] P. Bai, H. Wu and S.Q. Zhu, Petrochemical Technology (China), 30 (2001) 563-566.

[37] X.B. Cui, Z.C. Yang, Y.R. Zhai and Y.J. Pan, Chinese J. Chem. Eng., 10 (2002) 529-534.

[38] P. Lang, H. Yatim, P. Moszkowicz and M. Otterbein, Comput. Chem. Eng., 18 (1994) 1057-1069.

[39] S.M. Milani, Trans IchemE, 77 (1999) 469-470.

[40] I.M. Mujtaba, Trans IchemE, 77(1999) 588-596.

[41] B.T. Safrit, A.W. Westerberg, U. Diwekar and O.M. Wahnschafft, Ind. Eng. Chem. Res., 34 (1995) 3257-3264.

[42] A.I. Johnson and W.F. Furter, Cana. J. Chem. Eng., 38 (1960) 78-87.

[43] D. Barba, V. Brandani and G.D. Giacomo, Chem. Eng. Sci., 40 (1985) 2287-2292.

[44] W.F. Furter, Chem. Eng. Comm., 116 (1992) 35-40.

[45] W.F. Furter, Sep. Purif. Methods, 22 (1993) 1-21.

[46] Z.T. Duan, L.H. Lei and R.Q. Zhou, Petrochemical Technology (China), 9 (1980) 350-353.

[47] L.H. Lei, Z.T. Duan and Y.F. Xu, Petrochemical Technology (China), 11 (1982) 404-409.

[48] Q.K. Zhang, W.C. Qian and W.J. Jian, Petrochemical Technology (China), 13 (1984) 1-9.

[49] L. Berg, Dehydration of Acetic Acid by Extractive Distillation, US Patent No. 5 167 774 (1992).

[50] L. Berg, Dehydration of Acetic Acid by Extractive Distillation, US Patent No. 4 729 818 (1987).

[51] R.W. Helsel, Chem.Eng. Prog., No. 5 (1977) 55-59.

[52] J. Golob, V. Grilc and B. Zadnik, Ind. Eng. Chem. Process Des. Dev., 20 (1981) 435-440.

[53] B. Dil, Y.Y. Yang and Y.Y. Dai, J. Chem. Eng. Chinese Univ., 7 (1993) 174-179.

[54] J.H. Ma and S.X. Sun, Chinese J. Appl. Chem., 14 (1997) 70-73.

[55] L.L. Chen (ed.), Solvent Handbook, Chemical Industry Press, Beijing, 1997.

[56] Chemical Engineering Department of Zhejiang University, Petrochemical Technology (China), 2 (1973) 527-533.

[57] H.Z. Lu, J. Chem. Eng. (China), (3) (1977) 1-15.

[58] F.M. Lee and R.E. Brown, Extractive Distillation of Hydrocarbons Employing Solvent Mixture, US Patent No. 4 921 581 (1990).

[59] R.E. Brown and F.M. Lee, Extractive Distillation of Hydrocarbon Feeds Employing Mixed Solvent, US Patent No. 4 954 224 (1990).

[60] X.B. Cui, Z.C. Yang and T.Y. Feng, Chemical Industry and Engineering (China), 18 (2001) 215-220.

[61] S.H. Zhu, W.C. Tang and L.S. Tian, Petrochemical Technology (China), 31 (2002) 24-27.

[62] L.S. Tian, W.C. Tang and S.J. Wu, Petroleum Processing and Petrochemicals (China), 32 (2001) 5-8.

[63] Z.G. Lei, R.Q. Zhou and Z.T. Duan, J. Chem. Eng. Japan, 35 (2002) 211-216.

[64] Z.G. Lei, H.Y. Wang, R.Q. Zhou and Z.T. Duan, Chem. Eng. J., 87 (2002) 149-156.

[65] J. Gmehling and U. Onken (eds.), Vapor-Liquid Equilibrium Data Collection, Dechema, Frankfurt, 1977.

[66] T. Welton, Chem. Rev., 99 (1999) 2071-2083.

[67] C.M. Gordon, Appl. Catal. A: Gen., 222 (2001) 101-117.

[68] H. Olivier, J. Mol. Catal., 146 (1999) 285-289.

[69] R. Hagiwara and Y. Ito, J. Fluorine Chem., 105 (2000) 221-227.

[70] D.B. Zhao, M. Wu, Y. Kou and E.Z. Min, Catal. Today, 74 (2002) 157-189.

[71] S.O. Gregory and M.A.O. Mahdi, J. Mol. Catal. A: Chem., 187 (2002) 215-225.

[72] M. Arlt, M. Seiler, C. Jork and T. Schneider, DE Patent No. 10114734 (2001).

[73] S.R.M. Ellis and D.A. Jonah, Chem. Eng. Sci., 17 (1962) 971-976.

[74] A.K. Hilmi, S.R.M. Ellis and P.E. Barker, Brit. Chem. Eng., 15 (1970) 1321-1324.

[75] M.R. Jose, G. Cristina, A.B. Miguel and R. Aitor, Fluid Phase Equilib., 156 (1999) 89-99.

[76] M.R. Jose, G. Cristina, O.L. Salomer, L. Juan and F. Marian, Fluid Phase Equilib., 182 (2001) 171-187.

[77] B. Beatriz, T.S. Maria, B. Sagrario, L.C. Jose and C. Jose, Fluid Phase Equilib., 175 (2000) 117-124.

[78] M. Seiler, W. Arlt, H. Kautz and H. Frey, Fluid Phase Equilib., 201 (2002) 359-379.

[79] M. Seiler, D. Kohler and W. Arlt, Sep. Purif. Technol., 29 (2002) 245-263.

[80] A.B. Fawzi, A.A. Fahmi and S. Jana, Sep. Purif. Technol., 18 (2000) 111-118.

[81] E.R. Thomas, B.A. Newman and T.C. Long, J. Chem. Eng. Data, 27 (1982) 399-405.

[82] M.D. Donohue, D.M. Shah and K.G. Conally, Ind. Eng. Chem. Fundam., 24 (1985) 241-246.

[83] A. Vega, F. Diez and R. Esteban, Ind. Eng. Chem. Res., 36 (1997) 803-807.

[84] W.C. Qian, Z.T. Duan and L.H. Lei, Petrochemical Technology (China), 10 (1981) 241-246.

[85] J.D. Olson, Fluid Phase Equilib., 52 (1989) 209-218.

[86] J.C. Leroi, J.C. Masson and H. Renon, Ind. Eng. Chem. Process Des. Dev., 16 (1977) 139-144.

[87] P. Duhem and J. Vidal, Fluid Phase Equilib., 2 (1978) 231-235.

[88] Z.Q. Chen, F.T. Tian and S.Y. Jin, Chemical Industry and Engineering (China), 2 (1988) 1-9.

[89] V. Venkatasubramanian, K. Chan and J.M. Caruthers, ACS Symp. Ser., 589 (1995) 396-414.

[90] K.G. Joback and G. Stephanopoulos, Adv. Chem. Eng., 21 (1995) 257-311.

[91] N. Churi and L.E.K. Achenie, Ind. Eng. Chem. Res., 53 (1996) 3788-3794.

[92] E.C. Marcoulaki and A.C. Kokossis, Comput. Chem. Eng., 22 (1998) S11-S18.

[93] M.L. Mavrovouniotis, Comput. Chem. Eng., 22 (1998) 713-715.

[94] A.H. Meniai, D.M.T. Newsham and B. Khalfaoui, Trans. IChemE, 76 (A11) (1998) 942-950.

[95] J.E. Ourique and A.S. Telles, Comput. Chem. Eng., 22 (1998) S615-S618.

[96] E.N. Pistikopoulos and S.K. Stefanis, Comput. Chem. Eng., 22 (1998) 717-733.

[97] V.S. Raman and C.D. Maranas, Comput. Chem. Eng., 22 (1998) 747-763.

[98] M. Sinha, L.E.K. Achenie and G.M. Ostrovsky, Comput. Chem. Eng., 23 (1999) 1381-1394.

[99] P.M. Harper, R. Gani, T. Ishikawa and P. Kolar, Fluid Phase Equilib., 158-160 (1999) 337-347.

[100] M. Hostrup, P.M. Harper and R. Gani, Comput. Chem. Eng., 23 (1999) 1395-1414.

[101] J.L. Franklin, Ind. Eng. Chem. Res., 41 (1949) 1070.

[102] S.S. Jorgensen, B. Kolbe, J. Gmehling and P. Rasmussen, Ind. Eng. Chem. Proc. Des. Dev., 18 (1979) 714-722.

[103] J. Gmehling, P. Rasmussen and A. Fredenslund, Ind. Eng. Chem. Proc. Des. Dev., 21 (1982) 118-127.

[104] E.A. Macedo, U. Weidlich, J. Gmehling and P. Rasmussen, Ind. Eng. Chem. Proc. Des. Dev., 22 (1983) 676-678.

[105] R. Gani and E.A. Brignole, Fluid Phase Equilib., 13 (1983) 331-340.

[106] R. Gani and A.A. Fredenslund, Fluid Phase Equilib., 82 (1993) 39-46.

[107] R. Gani, B. Nielsen and A. Fredenslund, AIChE J., 37 (1991) 1318-1332.

[108] E.J. Pretel, P.A. Lopez, S.B. Bottini and E.A. Brignole, AIChE J., 40 (1994) 1349-1360.

[109] K.G. Joback and R.C. Reid, Chem. Eng. Comm., 57 (1987) 233-243.

[110] R.C. Reid, J.M. Prausnitz and B.E. Poling (eds.), The Properties of Gases and Liquids, McGraw-Hill, New York, 1987.

[111] Z.G. Lei, H.Y. Wang, R.Q. Zhou and Z.T. Duan, Comput. Chem. Eng., 26 (2002) 1213-1221.

[112] B.S. Rawat and I.B. Gulati, J. Chem. Tech. Biotechnol., 31 (1981) 25-32.

[113] W. Wardencki and A.H.H. Tameesh, J. Chem. Tech. Biotechnol., 31 (1981) 86-92.

[114] Y. Zheng, R. Cai and W. Zhao, Fuel and Chemical Process (China), 28 (1997) 90-92.

[115] B. Liu, Y. Zheng and W. Zhao, J. Chem. Eng. Chinese Univ., 9 (1995) 259-262.

[116] Z.S. Yang, C.L. Li and J.Y. Wu, Computers and Applied Chemistry (China), 18 (2001) 553-557.

[117] V.D. Braam and N. Izak, Ind. Eng. Chem. Res., 39 (2000) 1423-1429.

[118] R. Cesar, C. Jose, V. Aurelio and V.D. Fernando, Ind. Eng. Chem. Res., 36 (1997) 4934-4939.

[119] J. Gmehling, Fluid Phase Equilib., 144 (1998) 37-47.

[120] E.R. Thomas and L.R. Eckert, Ind. Eng. Chem. Process Des. Dev., 23 (1984) 194-209.

[121] A. Klamt, J. Phys. Chem., 99 (1995) 2224-2235.

[122] A. Klamt, V. Jonas, T. Burger and J.C.W. Lohrenz, J. Phys. Chem. A 101 (1998) 5074-5085.

[123] A. Klamt and F. Eckert, Fluid Phase Equilib., 172 (2000) 43-72.

[124] J.M. Prausnitz and R. Anderson, AIChE J., 7 (1961) 96-101.

[125] B. Yi, Z. Xu, Z.G. Lei, Y.X. Liu, R.Q. Zhou and Z.T. Duan, J. Chem. Ind. Eng. (China), 52 (2001) 549-552.

[126] Y. Li (ed.), Thermodynamics of Metal Extraction, Tsinghua University Press, Beijing, 1988.

[127] S.K. Shoor and K.E. Gubbins, J. Phys. Chem., 73 (1969) 498-505.

[128] W.L. Masterton and T.P. Lee, J. Phys. Chem., 74 (1970) 1776-1782.

[129] R.A. Pierotti, Chem. Rev., 76 (1976) 717-726.

[130] Z.G. Lei, R.Q. Zhou and Z.T. Duan, Fluid Phase Equilib., 200 (2002) 187-201.

[131] B. Liao, Z.G. Lei, Z. Xu, R.Q. Zhou and Z.T. Duan, Chem. Eng. J., 84 (2001) 581-586.

[132] C.P. Zhao, C.M. Zhao and M.X. Chen, Petrochemical Technology (China), 18 (1989) 236-239.

[133] C. Zhao, M.X. Chen and C.M. Zhao, Chem. Eng. (China), 20 (1992) 58-60.

[134] H. Komatsu, J. Chem. Eng. Japan, 10 (1977) 200-205.

[135] H. Komatsu and C.D. Holland, J. Chem. Eng. Japan, 10 (1977) 292-297.

[136] J. Jelinek and V. Hlavacek, Chem. Eng. Comm., 2 (1976) 79-85.

[137] X.L. Shi, H.L. Pan, Q.X. Zhang, W.Z. Li and Y. Li, Journal of East China University of Science and Technology, 25 (1999) 325-329.

[138] X.L. Shi, C.L. Yang, Q.X. Zhang and Y.A. Li, Petrochemical Technology (China), 28 (1999) 549-553.

[139] J.H. Lee and M.P. Dudukovic, Comput. Chem. Eng., 23 (1998) 159-172.

[140] R. Taylor and R. Krishna, Chem. Eng. Sci., 55 (2000) 5183-5229.

[141] R. Baur, A.P. Higler, R. Taylor and R. Krishna, Chem. Eng. J., 76 (2000) 33-47.

[142] R. Baur, R. Taylor and R. Krishna, Chem. Eng. Sci., 56 (2001) 2085-2102.

[143] A.P. Higler, R. Taylor and R. Krishna, Chem. Eng. Sci., 54 (1999) 2873-2881.

[144] A.P. Higler, R. Taylor and R. Krishna, Chem. Eng. Sci., 54 (1999) 1389-1395.

[145] Z.G. Lei, C.Y. Li, B.H. Chen, E.Q. Wang and J.C. Zhang, Chem. Eng. J., 93 (2003) 191-200.

[146] R. Krishna and J.A. Wesselingh, Chem. Eng. Sci., 52 (1997) 861-911.

[147] J.S. Tong (ed.), The Fluid Thermodynamics Properties, Petroleum Technology Press, Beijing, 1996.

[148] H.A. Kooijman and R. Taylor, Ind. Eng. Chem. Res., 30 (1991) 1217-1222.

[149] Z.G. Lei, C.Y. Li and B.H. Chen, Chinese J. Chem. Eng., 11 (2003) 515-519.

[150] Z.G. Lei, C.Y. Li and B.H. Chen, Sep. Purif. Technol., 36 (2004) 131-138.

[151] Y. Sato, K. Fujiwara, T. Takikawa, S. Takishima and H. Masuoka, Fluid Phase Equilib., 162 (1999) 261-276.

[152] Y. Sato, T. Takikawa, S. Takishima and H. Masuoka, J. Supercrit. Fluids, 19 (2001) 187-198.

[153] Y. Sato, T. Takikawa, M. Yamane, S. Takishima and H. Masuoka, Fluid Phase Equilib., 194-197 (2002) 847-858.

[154] R. Kleinrahm and W. Wagner, J. Chem. Thermodyn., 18 (1986) 739-760.

[155] I.C. Sanchez and R.H. Lacombe, Macromolecules, 11 (1978) 1145-1156.

[156] I.C. Sanchez and R.H. Lacombe, J. Phys. Chem., 80 (1976) 2352-2362.

[157] P.A. Rodgers, J. Appl. Polym. Sci., 48 (1993) 1061-1080.

[158] F.J. Millero, Chem. Rev., 71 (1971) 147-176.

[159] Z.G. Lei, B.H. Chen and J.W. Li, Chinese J. Chem. Eng., 11 (2003) 297-301.

[160] Z.G. Lei, C.Y. Li and B.H. Chen, Sep. Purif. Rev., 32 (2003) 121-213.

# Chapter 3. Azeotropic distillation

Azeotropic distillation as an early and important special distillation process is commonly used in laboratory and industry. Since 1950s, along with extractive distillation, it has gained a wide attention. Like extractive distillation, the entrainer, i.e. the third component added to the system, is also the core of azeotropic distillation. In the process design and synthesis, the graphical method (in most cases refer to as triangular diagram) is often employed. In general, the operation process is very sensitive to some parameters in the case of more than one azeotrope formed, and thus the phenomenon of multiple steady-state (MSS) tends to appear.

## 1. INTRODUCTION

Azeotropic distillation is accomplished by adding to the liquid phase, a volatile third component, which changes the volatility of one of the two components more than the other so that the components are separated by distillation. The two components to be separated often are close boiling point components which do or do not azeotrope in the binary mixture, but sometimes they are components which do azeotrope although they aren't close boiling point components [1-37]. It is likely that in some cases, one system can be separated either by azeotropic distillation or by extractive distillation, for instance, alcohol/water, acid/water, etc.

The added third component, sometimes called the entrainer, may form a ternary azeotrope with the two components being separated. However, it must be sufficiently volatile from the solution so that it is taken overhead with one of the two components in the azeotropic distillation column. If the entrainer and the component taken overhead separate into two liquid phases when the vapor overhead is condensed, the entrainer phase is refluxed back to the column. The other phase can be fractionated to remover the dissolved entrainer and the residual amount of the other component before it is discarded in the solvent (entrainer) recovery column). Alternatively, this second liquid phase is recycled to some appropriate place in the main process scheme.

Azeotropic distillation is usually divided into two types: homogeneous azeotropic distillation and heterogeneous azeotropic distillation. They are illustrated in Figs. 1 and 2, respectively. In homogeneous azeotropic distillation, phase split doesn't appear in the liquid along the whole column. Whereas, in heterogeneous azeotropic distillation, two liquid phases exist in some regions of the composition space. Heterogeneous azeotropic distillation is more widely used for separating the close boiling point components or azeotropes than homogeneous azeotropic distillation. In particular, the case of heterogeneous mixtures without decanter at the top of azeotropic distillation column can be looked upon as the one of homogeneous mixtures, and at the same time the liquid composition on a tray or a section of packing is replaced by the overall liquid composition. The operation, controlling and optimization of the azeotropic distillation column have been studied in [56-59].

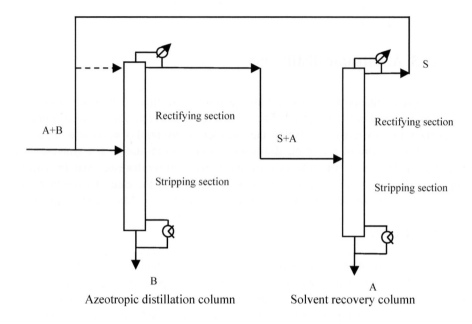

Fig. 1. The double-column process for homogeneous azeotropic distillation.

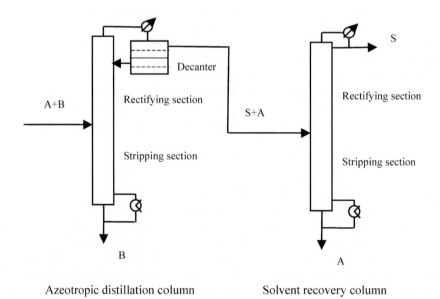

Fig. 2. The double column process for heterogeneous azeotropic distillation.

147

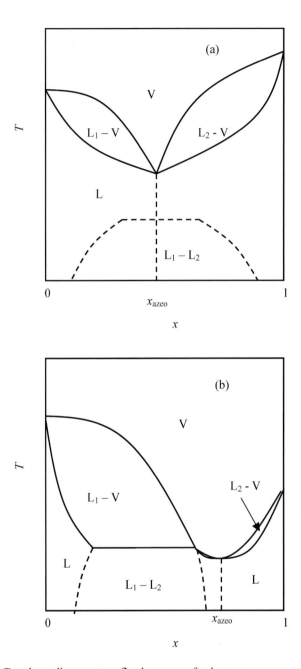

Fig. 3. Binary *T-x* phase diagram at a fixed pressure for homogeneous azeotrope; (a) azeotropic boiling point above but not far from the upper critical solution temperature; adapted from the reference: S. Widagdo and W.D. Seider [29]; (b) azeotropic boiling point below the upper critical solution temperature (usual behavior).

For the binary azeotrope to be separated, the typical binary $T$-$x$ phase diagrams at a fixed pressure are illustrated in Fig. 3 for homogeneous azeotrope and in Fig. 4 for heterogeneous azeotrope. At an azeotropic point, the relative volatility of binary azeotrope is unity because the overall liquid phase composition $x_i = y_i$. In Fig. 3 (a) the L - L two phases region may not exist; or even if exist, the operating temperature in the distillation column is over the highest temperature of L-L two phases region; In Fig. 3 (b) the azeotropic boiling point is below the upper critical solution temperature (usual behavior). In Fig. 4 two vapor-liquid envelops ($L_1 - V$ and $L_2 - V$) overlap with liquid-liquid envelop ($L_1$ - $L_2$) at one line which is also the tie line of liquid-liquid equilibria. Provided overlapping only at one point (i.e. azeotropic point) which is also the critical point of liquid-liquid equilibria, then heterogeneous azeotrope becomes homogeneous azeotrope because in this case $x_1 = x_2 = x_{azeo}$.

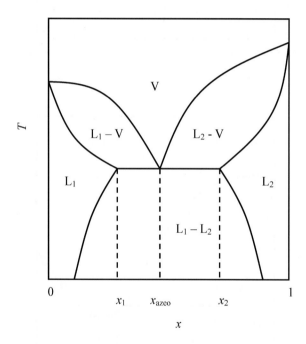

Fig. 4. Binary $T$-$x$ phase diagram at a fixed pressure for heterogeneous azeotrope; adapted from the reference: S. Widagdo and W.D. Seider [29].

Note that forming homogeneous azeotrope doesn't mean that the separation method is homogeneous azeotropic distillation, and forming heterogeneous azeotrope also doesn't mean that the separation method is heterogeneous azeotropic distillation, while depending on physical property of the entrainer used. Homogeneous and heterogeneous azeotropic distillation correspond to the real state of the mixture, consisting of the components to be separated and the entrainer, in the liquid phase on a tray or a section of packing in a distillation column.

In addition, either for homogeneous or for heterogeneous azeotropic distillation, the entrainer must be vaporized into the column top and thus much energy is consumed compared with extractive distillation. For this reason, the cases of azeotropic distillation are used less than those of extractive distillation.

## 2. ENTRAINER SELECTION

In principle, any substance can be an entrainer but not every substance can "break" the azeotrope and/or improve relative volatility of the close boiling point mixture. Separation factor (i.e. relative volatility) is an important physical quantity in azeotropic distillation. It is desirable to assess the influence that the entrainer will have on the vapor-liquid equilibrium whenever any column is designed and simulated [31, 32].

In some cases the entrainer and the two components of $i$ and $j$ being separated can produce three-phase (vapor-liquid-liquid) equilibrium. Two liquid phases may be in equilibrium with a vapor phase. For the three-phase equilibrium the solubilities of components $i$ and $j$ in the upper liquid phase are denoted respectively by $x_i^I$ and $x_j^I$, the solubilities of components A and B in the lower liquid phase respectively by $x_i^{II}$ and $x_j^{II}$, and the corresponding activity coefficients by $\gamma$ in the upper liquid phase and $\Gamma$ in the lower liquid phase, respectively. The relative volatility for components $i$ and $j$ is related to the overall composition $X_i$ according to

$$\alpha_{ij} = \frac{\gamma_i \Gamma_i P_i^0 \theta_i}{\gamma_j \Gamma_j P_j^0 \theta_j} \frac{(x_i^{II} - X_i)\Gamma_j + (X_i - x_i^I)\gamma_j}{(x_i^{II} - X_i)\Gamma_i + (X_i - x_i^I)\gamma_i} \tag{1}$$

where $\theta_i$ and $\theta_j$ are correction factor for high pressure. At low or even middle pressure, it can be approximately regarded as $\theta_i = \theta_j \approx 1$.

Over the two-liquid phase region Eq. (1) gives relative volatilities for three-phase equilibrium. In the composition range where only an upper liquid phase and a vapor phase exist, Eq. (1) reduces to

$$\alpha_{ij} = \frac{\gamma_i P_i^0 \theta_j}{\gamma_j P_j^0 \theta_i} \tag{2}$$

When only a lower liquid phase is in equilibrium with a vapor phase, Eq. (1) becomes

$$\alpha_{ij} = \frac{\Gamma_i P_i^0 \theta_j}{\Gamma_j P_j^0 \theta_i} \tag{3}$$

Similar relations have already been obtained in chapter 1 (Eqs. (63) and (64)), where for the low pressure, $\theta_i = \theta_j \approx 1$; and for the middle pressure, $\theta_i = \exp\left[\dfrac{(V_{iL} - B_{ii})(P - P_i^0)}{RT}\right]$,

$\theta_j = \exp\left[\dfrac{(V_{jL} - B_{ii})(P - P_j^0)}{RT}\right]$. The notation meaning refers to chapter 1 (Eqs. (63) and (64)).

*Example:* Deduce Eq. (1), the expression of relative volatility in the case of vapor-liquid-liquid three-phase equilibrium (VLLE).

*Solution:* Eqs. (2) and (3) are only suitable for vapor-liquid two-phase equilibrium (VLE), in which the liquid mixture is miscible and homogeneous. However, Eq. (1) is a generalization.

A schematic representation of VLLE is diagrammed in Fig. (5). In a closed system, at a fixed temperature, we obtain:

$$y_i = \frac{\gamma_i P_i^0 \theta_i}{P} x_i^I \ , \ y_j = \frac{\gamma_j P_j^0 \theta_j}{P} x_j^I \quad \text{(according to VLE)} \tag{4}$$

$$\gamma_i x_i^I = \Gamma_i x_i^{II} \ , \ \gamma_j x_j^I = \Gamma_j x_j^{II} \quad \text{(according to LLE and the equivalence of activities)} \tag{5}$$

$$L = L_1 + L_2 , \ LX_i = L_1 x_i^I + L_2 x_i^{II} \ , \ LX_j = L_1 x_j^I + L_2 x_j^{II} \quad \text{(according to material balance)} \tag{6}$$

$$X_i = \frac{L_1 x_i^I + L_2 x_i^{II}}{L_1 + L_2} , \ X_j = \frac{L_1 x_j^I + L_2 x_j^{II}}{L_1 + L_2} \quad \text{(according to the lever-arm rule)} \tag{7}$$

$$\frac{L_1}{L_2} = \frac{x_i^{II} - X_i}{X_i - x_i^I} = \frac{x_j^{II} - X_j}{X_j - x_j^I} \quad \text{(according to the lever-arm rule)} \tag{8}$$

where $L_1$ and $L_2$ are the mass in the upper and lower liquid phase (unit: mole), respectively.

In terms of the definition of relative volatility, it can be derived:

$$\alpha_{ij} = \frac{y_i / X_i}{y_j / X_j} = \frac{\gamma_i P_i^0 \theta_i x_i^I}{\gamma_j P_j^0 \theta_j x_j^I} \frac{X_j}{X_i} = \frac{\gamma_i P_i^0 \theta_i x_i^I}{\gamma_j P_j^0 \theta_j x_j^I} \frac{(L_1 x_j^I + L_2 x_j^{II})}{(L_1 x_i^I + L_2 x_i^{II})}$$

$$= \frac{\gamma_i P_i^0 \theta_i x_i^I}{\gamma_j P_j^0 \theta_j x_j^I} \frac{\dfrac{x_i^{II} - X_i}{X_i - x_i^I} x_j^I + x_j^{II}}{\dfrac{x_i^{II} - X_i}{X_i - x_i^I} x_i^I + x_i^{II}} = \frac{\gamma_i P_i^0 \theta_i}{\gamma_j P_j^0 \theta_j} \frac{\dfrac{x_i^{II} - X_i}{X_i - x_i^I} + \dfrac{x_j^{II}}{x_j^I}}{\dfrac{x_i^{II} - X_i}{X_i - x_i^I} + \dfrac{x_i^{II}}{x_i^I}}$$

$$= \frac{\gamma_i P_i^0 \theta_i}{\gamma_j P_j^0 \theta_j} \frac{\dfrac{x_i^{II} - X_i}{X_i - x_i^I} + \dfrac{\gamma_j}{\Gamma_j}}{\dfrac{x_i^{II} - X_i}{X_i - x_i^I} + \dfrac{\gamma_i}{\Gamma_i}} = \frac{\gamma_i \Gamma_i P_i^0 \theta_i}{\gamma_j \Gamma_j P_j^0 \theta_j} \frac{(x_i^{II} - X_i)\Gamma_j + (X_i - x_i^I)\gamma_j}{(x_i^{II} - X_i)\Gamma_i + (X_i - x_i^I)\gamma_i} \tag{9}$$

which gives rise to Eq. (1).

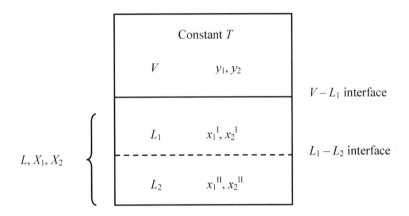

Fig. 5. Schematic representation of VLLE (vapor-liquid-liquid three-phase equilibrium) in a closed system.

*Example:* Use Eqs. (30) – (32) in chapter 1, and deduce the expression of $\theta_i$.

*Solution:* By Eq. (32) in chapter 1,

$$K_i = \frac{y_i}{x_i} = \frac{\gamma_i P_i^0 \phi_i^s (PF)_i}{P \overline{\phi}_i^V} \tag{10}$$

In comparison with Eq. (4), for any pressure,

$$\theta_i = \frac{\phi_i^s (PF)_i}{P \overline{\phi}_i^V} \tag{11}$$

The notation meaning in Eq. (11) refers to chapter 1 (Eqs. (30) – (32)). It is evident that under lower and middle pressures Eq. (11) can be further simplified and the simplification procedure has already been made in chapter 1 (1.1. The equilibrium ratio).

Since relative volatility is an important index for evaluating the possible entrainers, it can be seen from Eq. (1) that activity coefficient models of both liquid-liquid equilibrium and vapor-liquid equilibrium should be known beforehand in order to predict relative volatility. The liquid composition and activity coefficient in the isothermal liquid-liquid equilibrium are solved by numerical iteration from the equations (Eqs. (5) – (8)) in order to look for the compositions for which the activities are equal in the two phases for each individual component (Eq. (5)). On this basis, the vapor composition in the isothermal vapor-liquid equilibrium is calculated by using Eq. (4). However, one problem arises, that is, both one activity coefficient model suitable for VLE and another activity coefficient model suitable for LLE should be utilized in determining the same $\gamma_i$ and $\gamma_j$. In general, one model can't solve this problem, due to accuracy limitation of activity coefficient models under different conditions. Thus, it adds to some difficulty in calculation. But this contributes to one way to select the potential entrainers, herein, called calculation method.

Of course, experiment method is the most reliable in selecting the entrainers, but apparently much time and money will be spent. Only until a limited few of possible entrainers are found by means of calculation method, should the experiments be made.

Besides, simple rules using maps of distillation lines and residue curves (see what follows) have been suggested for screening many possible entrainers. Since drawing distillation lines and residue curves may be a very tedious thing, it is advisable to straightforwardly apply the conclusions generalized by the researchers from many maps. Stichlmair and Herguijuela [38] enumerate some experience rules for entrainer selection, which are given in Table 1.

Table 1
Experience rules for entrainer selection

| Mixtures with a minimum-boiling azeotrope: |
| --- |
| • Low boiler (lower than the original azeotrope). |
| • Medium boiler that forms a minimum-boiling azeotrope with the low boiling species. |
| • High boiler that forms minimum-boiling azeotropes with both species. At least one of the new minimum-boiling azeotropes has a lower boiling temperature than the original azeotrope. |
| Mixtures with a maximum-boiling azeotrope: |
| • Low boiler (lower than the original azeotrope). |
| • Medium boiler that forms a minimum-boiling azeotrope with the low boiling species. |
| • High boiler that forms minimum-boiling azeotropes with both species. At least one of the new maximum-boiling azeotropes has a higher boiling temperature than the original azeotrope. |

Some cases using azeotropic distillation as the separation method are listed in Table 2. The interested readers can compare the entrainers used in practice with the rules of selecting entrainers.

Based on computer-aided design, an advanced knowledge integrating system for the selection of solvents for azeotropic distillation is developed by Thomas and Hans [34]. The mass separating agents are searched in the system SOLPERT (Solvent Selecting Expert System) by means of combining databases, heuristic and numerical methods. Several inherent limitations of solvent screening methods, e.g. database search, empirical methods and group contribution methods, are circumvented and minimized to gain an maximum of usable information for solvent selection. The solvent selection of SOLPERT consists of four steps, as listed in Table 3.

Table 2
Some azeotropic distillation cases

| No. | Components to be separated | Entrainers |
|---|---|---|
| 1 | Ethanol / water; isopropanol / water, tert-butanol / water | Benzene, toluene, hexane, cyclohexane, methanol, diethylether, methyl-ethyl-ketone (MEK) [2, 4, 5, 13, 16, 21, 23] |
| 2 | Acetone / n-heptane | Toluene [28] |
| 3 | Acetic acid / water | n-butyl acetate [17] |
| 4 | Isopropanol / toluene | Acetone [20] |

Table 3
The steps in SOLPERT for selecting entrainer; adapted from the reference [34].

| Steps | Criterions | Applied methods |
|---|---|---|
| 1. Selection of classes | | |
| | Hydrogen bonding | Heuristic rules |
| | Polarity | Empirical approaches |
| | | Databases |
| 2. Selection of chemical similar groups | | |
| | Hildebrand's solubility parameter | Heuristic rules |
| | Snyder's classes | Generalization to structural groups |
| | Solvatochromic parameters | Databases |
| 3. Proposal of suitable solvents | | |
| | Lower and upper limit of boiling point | Heuristic rules |
| | Miscibility | Empirical approaches |
| | azeotropy | Numerical methods |
| | | Group contribution methods |
| | | Databases |
| 4. Ranking of proposed solvents | | |
| | Selectivity | Heuristic rules |
| | Miscibility | Numerical methods |
| | Relative volatility | Group contribution methods |
| | Performance of entrainers for azeotropic distillation | Databases |

It is evident that the solvent selection of SOLPERT also holds for extractive distillation.

Anyway, data bank search, for its convenience and reliability, is no less a good method [60-64]. A successful work has been done in [60, 61], where a computerized bank of azeotropic data is now available. Data from the references have been tested before storage. Newly measured azeotropic and azeotropic data are also included. This data bank now contains approximately 36000 entries (information) on azeotropic or azeotropic behavior or approximately 19000 non-electrolyte systems involving approximately 1700 compounds. It can be used reliably in process synthesis computations, e.g. for design calculations of distillation columns, selection of the best solvents for azeotropic distillation and also for further development of group contribution methods or for fitting reliable $g^E$-model parameters.

## 3. MATHEMATICAL MODELS

To describe the azeotropic distillation process, especially for the azeotropic distillation column, it is necessary to set up the reliable mathematical models. Once mathematical model is determined, it is possible to extend to the synthesis of distillation column sequence. Formerly, graphical method was popular for the analysis of azeotropic distillation column. Recently, the equilibrium (EQ) stage and non-equilibrium (NEQ) stage models, similar as those in extractive and catalytic distillation, have received more attention.

### 3.1. Graphical method

The principal concern in the graphical method is how to construct the maps of residual curve, distillation line and their boundaries so that the feasible composition space can be examined. After these are finished, the actual operating line with a certain reflux ratio between minimum and total reflux is plotted in the composition space. Consequently, the feasibility of a given separation process can be evaluated and the possible multiple steady-state may be found.

#### 3.1.1. Residual curves

Residual-curve maps have been widely used to characterize azeotropic mixture, establish feasible splits by distillation at total reflux and for the synthesis and design of column sequences that separate azeotropic mixture. Schreinemakers [29, 39, 40] defined a residual curve as the locus of the liquid composition during a simple distillation process. Residue curves are conceived for $n$-component systems, but can be plotted only for ternary or, with more powerful graphical tools and some imagination, quaternary systems.

Since the residual curve is the locus of the liquid composition remaining from a differential vaporization process, we write by a stepwise procedure for a simple distillation still with only one theoretical stage:

$$L_0 = L + V \tag{12}$$

$$L_0 x_{i,0} = Lx_i + Vy_i \tag{13}$$

where $L_0$ is the amount of liquid in still at start of vaporization increment; $L$ is the amount

of liquid in still at end of vaporization increment; $y_i$ is mean equilibrium vapor composition

over increment. A schematic representation of a simple distillation is diagrammed in Fig. 6.

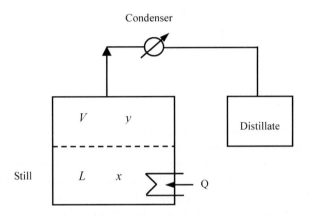

Fig. 6. Schematic representation of a simple distillation still.

In a difference time $dt$, we have

$$-dL = dV \tag{14}$$

$$-d(Lx_i) = y_i dV \tag{15}$$

or $-Ldx_i - x_i dL = y_i dV \tag{16}$

which arranges to

$$-Ldx_i = (y_i - x_i)dV \tag{17}$$

or $Ldx_i = (y_i - x_i)dL \tag{18}$

That is,

$$\frac{dx_i}{dL/L} = y_i - x_i = (K_i - 1)x_i \tag{19}$$

and for any two components $i$ and $j$,

$$\frac{dx_i}{dx_j} = \frac{y_i - x_i}{y_j - x_j} \tag{20}$$

Introduce a dimensionless time $dt = dL/L$, Eq. (19) becomes

$$\frac{dx_i}{dt} = y_i - x_i = (K_i - 1)x_i \tag{21}$$

which is the residual curve equation. Note that it is set for a fixed pressure.

In fact, Eq. (21) represents a group of ordinary (constant-coefficient) difference equations, and thus can be solved by such mathematical tools as Gear, Runge-Kutta (L-K) methods, etc. During the calculation, bubble point subroutine, as mentioned in chapter 1 (1.1. the equilibrium ratio), should be called as a sub-function in every $dt$ advance so as to solve the temperature because the equilibrium ratio $K_i$ is the function of temperature and liquid composition.

Map representation of residual curves in composition space for three-component system may be drawn in orthogonal or equilateral cartesian coordinates; or transformations can be made in the coordinate system as long as the linear appearance of straight lines remains invariant. The two types of coordinate systems will be used interchangeably throughout.

As examples, Figs. 7 and 8 respectively show the residual curves for the methanol / ethanol / water system at 101.3 kPa in the homogeneous mixture and for the ethanol / water / benzene at 101.3 kPa in the heterogeneous mixture. It can be seen that arrows may be assigned. The arrows are in the direction of time increasing (or temperature increasing) because the concentration of heavy components in the simple distillation still will become higher and higher as time goes on, which results in temperature increasing.

### 3.1.2. Operating lines

Similar as the $x$-$y$ diagram for two-component system, if composition space is only used to perform the calculations, it is necessary to assume constant molar overflow. As shown in Fig. 9, both $x_D$ (the distillate composition) and $x_B$ (the bottom composition) are predetermined as a given separation task, and the line DFB represents a limiting overall material balance. The operating lines in Fig. 9 are composed of real lines with arrows representing VLE and dashed lines without arrows representing material balance. Real lines and dashed lines are arranged alternately. The number of arrows is equivalent to the separation stages including reboiler, but excluding condenser. This means that there are six separation stages in Fig. 9.

Fig. 10 illustrates how to draw operating lines in the homogeneous azeotropic distillation column. The operating lines are also identical to those in the heterogeneous azeotropic distillation column, but without decanter at the top.

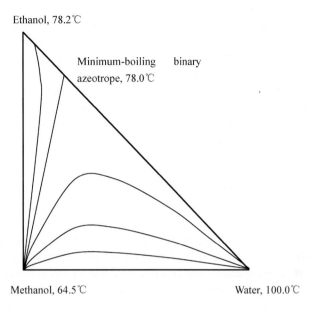

Fig. 7. The residual curves for the methanol / ethanol / water system at 101.3 kPa in the homogeneous mixture; adapted from the reference [21].

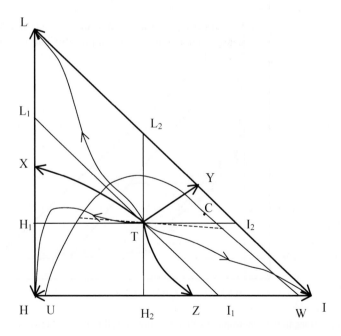

Fig. 8. The residual curves for the ethanol (L) / water (H) / benzene (I) system at 101.3 kPa in the heterogeneous mixture; adapted from the reference [26] with a little modification.

As an example, there are six separation stages ($n = 6$) in Fig. 10. The stages are counted downward along the column. At the bottom the liquid composition vector $\underline{x_6}$ is in equilibrium with the vapor composition vector, which is denoted by an arrow with a starting point B($\underline{x_6}$) to an ending point $V_6(\underline{y_6})$. Due to the limiting material balance among B, $V_6$ and $L_5$, point $L_5(\underline{x_5})$ lies in the line connecting B($\underline{x_6}$)- $V_6(\underline{y_6})$. Then, an equilibrium line with arrow is drawn from point $L_5(\underline{x_5})$ to point $V_5(\underline{y_5})$. Similarly, due to the limiting material balance among B, $V_5$ and $L_4$, point $L_4(\underline{x_4})$ lies in the dashed line connecting B($\underline{x_6}$)- $V_5(\underline{y_5})$. In the same way, other operating lines can be drawn. Therefore, it is easy to obtain that at the feed tray,

$$J = J' + F \tag{22}$$

$$V_3 + L_3 = L_2 + V_4 + F \tag{23}$$

On the other hand, the above material balance also can be derived from Fig. 11 by constraining the region at plane A1 - plane A2.

Operating lines in the heterogeneous azeotropic distillation column are illustrated in Fig. 12, where the vector $\underline{x}$ is the overall liquid composition, except $\underline{x_{D1}}$ and $\underline{x_{D2}}$. Note that although the foregoing operating lines are discussed just for tray column, it can be extended to packing column by transformation with the concept of HETP (height equivalent of a theoretical plate).

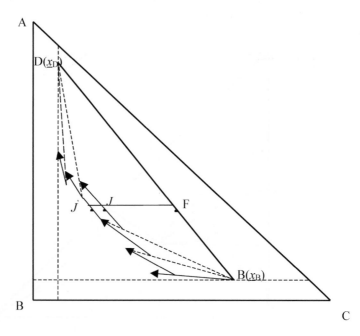

Fig. 9. Operating lines in the azeotropic distillation column; reflux ratio is between minimum and total reflux ratios.

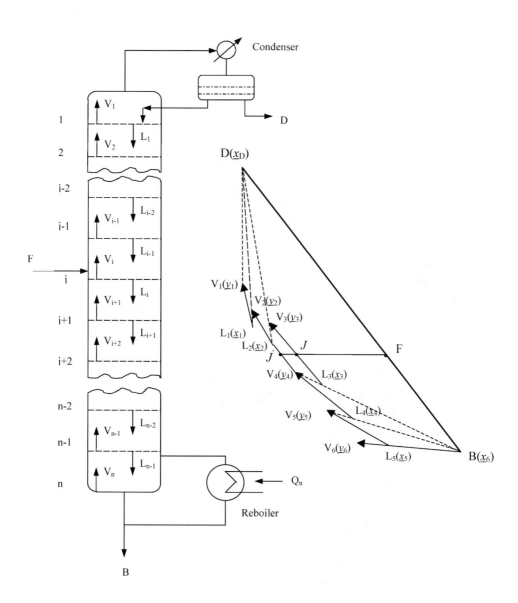

Fig. 10. Operating lines in the homogeneous azeotropic distillation column; reflux ratio is between minimum and total reflux ratios.

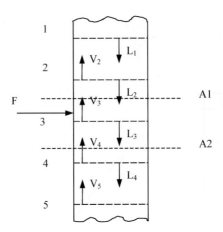

Fig. 11. Material balance at the feed tray.

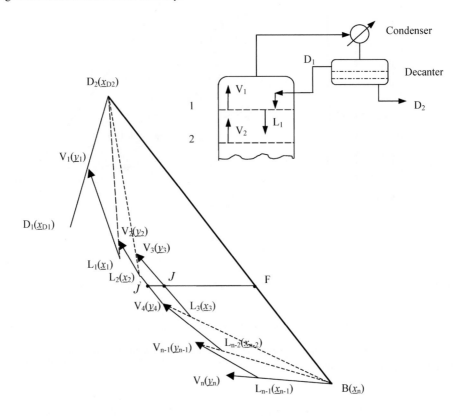

Fig. 12. Operating lines in the heterogeneous azeotropic distillation column; reflux ratio is between minimum and total reflux ratios.

### 3.1.3. Distillation lines

Like the two-component system, in the multi-component azeotropic distillation, there still exits a minimum reflux ratio (corresponding to infinite number of stages) and total reflux (corresponding to minimum stages). It can be imaged that in the composition space, as shown in Fig. 9, while reflux ratio is becoming larger, the operating lines are far away from the line DFB. This implies that at minimum reflux ratio at least one operating line lies in the line DFB and thus it is difficult to further draw other operating lines. For example, Fig. 13 shows one case at minimum reflux ratio.

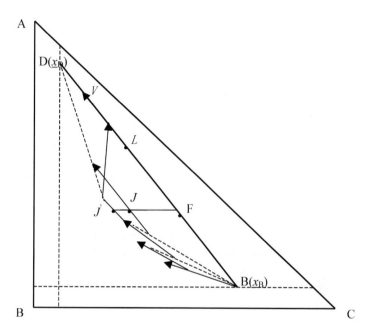

Fig. 13. Operating lines at minimum reflux ratio.

However, at total reflux where there is no feed and no products, operating lines are composed of consecutive real lines with arrows because in this case $y_{n+1} = x_n$. The operating lines at total reflux are called distillation lines. However, we often like to take on the smooth curves fitting the tie points connecting real lines with arrows as distillation lines (as shown in Fig. 14).

Let us study the difference between residual curves and distillation lines for the tray and packing columns:

(1) For the packing column, Eqs. (14), (18) and (21) used for plotting residual curves are also strictly satisfied because its composition change along the whole column is continuous, not abrupt as the tray column.

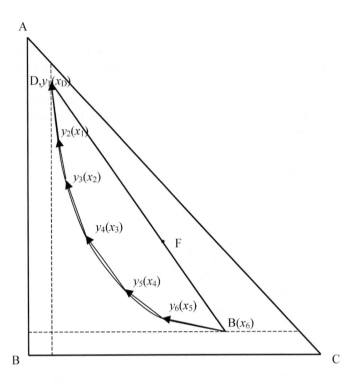

Fig. 14. Operating lines at total reflux (i.e. distillation lines).

As shown in Fig. 15, for a $dh$ differential section of packing column, the total material balance is:

$$-dL = dV \tag{24}$$

The component material balance is:

$$Ldx_i = (y_i - x_i)dL \tag{25}$$

and

$$\frac{dx_i}{dt} = y_i - x_i = (K_i - 1)x_i \tag{26}$$

which correspond with Eqs. (14), (18) and (21) in the form, respectively.

The subtle difference from simple distillation is that $L$ and $V$ represent the flowrate of liquid and vapor phases, respectively. Consequently, distillation lines are exactly the residual curves for the packing column. Note that Eq. (24) implies that the packing column is at total reflux in order to ensure the equality $-dL = dV$ because if having any feed, $-dL \neq dV$ in the vicinity of the feed position. Moreover, here the assumption of constant molar flow for the vapor and liquid phases isn't imposed because the equation of $-dL = dV$ already indicates that there is the change of the flowrate of $L$ and $V$ along the column.

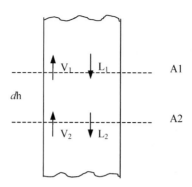

Fig. 15. A $dh$ differential section of packing column.

(2) For the tray column, distillation lines are slightly different from residual curves. The main reason is that its composition change along the whole column is stepwise, not continuous as the packing column.

Suppose that at total reflux the mathematical differential element is enlarged enough to be in the vicinity of one tray, as shown in Fig. 11 (A1 plane - A2 plane but no feed), and the following equations analogous to Eqs. (14), (18) and (21) are satisfied:

$$-\Delta L = \Delta V \tag{27}$$

$$L\Delta x_i = (y_i - x_i)\Delta L \tag{28}$$

$$\frac{\Delta x_i}{\Delta L / L} = y_i - x_i = (K_i - 1)x_i \tag{29}$$

For this reason, the residual curves are a good approximation to the operating lines at total reflux. Then, what arises at a limited reflux ratio (and having feed)? In this case, the difference equation is identical to that at total reflux except on the feed tray. That is why the operating lines at different reflux ratio have the similar shape.

Since the mathematical differential element for the tray column is enlarged not to be a limit, real lines with arrows are the chords of distillation lines, e.g. $y$ - $x$ (the tie line BM$_1$ vector), as shown in Fig. 16. But in accordance with Eq. (21), $y - x$ should be tangent to residual curves at point B. Note that when the residual curve is linear (e.g. for binary mixtures), the chords and residual curves are collinear (i.e. distillation lines overlap with residual curves).

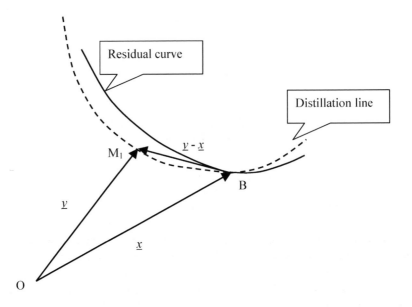

Fig. 16. Difference between residual curve and distillation line for the tray column.

### 3.1.4. Simple distillation and distillation line boundaries

In composition space, one or more residue curves (or distillation lines) divide the whole into regions with distinct pairs of starting and terminal points. Hence, the lines separating two or more distillation regions are called simple distillation boundaries (correspond to residue curves) or distillation lines boundaries (correspond to distillation lines). For homogeneous mixture, the simple distillation and distillation line boundaries can't be crossed; but for heterogeneous mixture, these boundaries may be crossed because in this case phase split occurs in LLE regions and thus two liquid phases may fall into different distillation regions.

As illustrated in Fig. 8 where it is supposed there is no decanter at the top of distillation column and LLE curve isn't utilized, the system forms three binary azeotropes representing with points X, Y and Z, and one ternary azeotropes representing with point T. Points X, Y, Z, as well as the vertices L, H and I, are called the singular (fixed) points in which the driving force for change in the liquid composition is zero; that is, $dx_i / dt = 0$ or $y_i = x_i$. Lines TX, TY, TZ divide the whole into three regions (LXTYL, HZTXH, IYTZI) with distinct pairs of starting and terminal points; for instance, in region LXTYL the starting point is T and the terminal point is L. So lines TX, TY and TZ are simple distillation boundaries or distillation lines boundaries. But if one point lies in LLE region (UCWU), the boundaries may be crossed due to phase split.

It is believed that the residual curves are a good approximation to the operating lines at total reflux. Rather, as discussed above, the distillation lines are equivalent to the operating lines at total reflux. For the packing column, simple distillation boundaries are exactly the distillation line boundaries; for the tray column, the appropriate boundaries are the separatrices on the maps of distillation lines [49]. In the case of infinite reflux and infinite number of trays (or infinitely high columns), the product compositions can reside on the simple distillation or distillation line boundaries.

## 3.2. EQ and NEQ stage models

The EQ stage model is widely applied to analyze and design the columns for either homogeneous or heterogeneous azeotropic distillation. However, as pointed out by Krishnamurthy and Taylor [1, 41, 42], concentration profiles predicted by the NEQ (or rate-based model) stage model may be different from those by the EQ stage model even with revisions to tray efficiency. The difference is caused by the difference in mass transfer resistances and the diffusional interaction effects especially for heterogeneous azeotropic distillation. Sometimes the mass transfer resistance in the liquid phase can't be neglected in the two-liquid region for heterogeneous azeotropic distillation. Therefore, the EQ stage model which essentially excludes the liquid phase mass transfer, may be an improper model for predicting the exact profiles of concentration and flowrate in the azeotropic distillation column. Alternatively, the NEQ stage model has recently been applied to analyze and design the azeotropic distillation column [1, 5]. However, for homogeneous and heterogeneous azeotropic distillation, the equation forms are different when using the EQ or NEQ stage model.

### 3.2.1. Homogeneous azeotropic distillation

Either for the EQ stage model or for the NEQ stage model, the equation forms of homogeneous azeotropic distillation are the same as those as in chapter 2 (7. Mathematical model of extractive distillation). But it is known that the amount of entrainer plays an important in the simulation of separation performance of homogeneous azeotropic distillation column. It can be obtained from the graphical method, which is shown in Fig. 17 where point $N$ represents the feed composition consisting of the components $A$ and $B$ to be separated and point $M$ represents the azeotropic composition with minimum boiling point.

In terms of lever-arm rule, the relation of the flowrates of entrainer $S$ and feed $F$ is written as

$$\frac{S}{F} = \frac{AN \bullet BM}{CM} = \frac{EN}{CM} = \frac{ON}{OC} \tag{30}$$

The value of the flowrate of entrainer $S$ can be used as an initial guess in the rigorous EQ and/or NEQ stage models.

### 3.2.2. Heterogeneous azeotropic distillation

In this case it is certain that phase split will occur in some regions of the tray column. The number of liquid phase can be determined by the modified plane phase stability test. If the fluid is split into two liquid phases, the concentration and flowrate of each phase are calculated by LLE calculation.

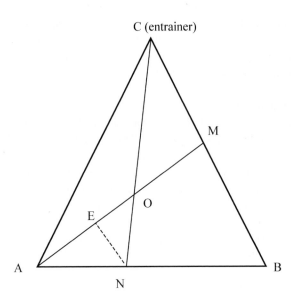

Fig. 17. Graphical method for calculating the amount of entrainer for homogeneous azeotropic distillation with one minimum boiling point azeotrope.

Note that for the packing column, due to the mixing effect of packing on the liquid phase, the concentration and flowrate of two liquid phases can be regarded as a whole.

As discussed in chapter 1 (1.1. The equilibrium ratio), a fundamental thermodynamic criteria for equilibrium between phases is the equality of chemical potentials for each component in all phases. That is, for component $i$ in two phases, $\text{I}$ and $\text{II}$,

$$\mu_i^I = \mu_i^{II} \tag{31}$$

This condition is able to reduce to the well-known iso-activity criterion for liquid-liquid equilibrium, which states

$$a_i^I = a_i^{II} \tag{32}$$

The activity of component $i$ ($a_i$) can be expressed in terms of an activity coefficient ($\gamma_i$), which is defined by the relationship:

$$a_i = \gamma_i x_i \tag{33}$$

Eq. (31), therefore, becomes

$$\gamma_i^I x_i^I = \gamma_i^{II} x_i^{II} \tag{34}$$

where $x_i^I$ represents the mole fraction and $\gamma_i^I$ the activity coefficient of component $i$ in phase I. The activity coefficient is, in turn, a function of composition and temperature:

$$\gamma_i = \gamma_i(x_1, x_2, \bullet \bullet \bullet, x_n, T) \tag{35}$$

which can be calculated by using an appropriate model. Some commonly used models have been listed in chapter 1 (1.1. The equilibrium ratio).

The liquid-liquid flash problem that is required to be solved for three-phase distillation is to calculate the equilibrium phase compositions and phase fractions for a given overall composition ($z_i$) and system temperature ($T$). If two equilibrium phases exist under these conditions, then the problem involves $2n + 1$ equilibrium and mass balance equations:

$$x_i^I \gamma_i^I - x_i^{II} \gamma_i^{II} = 0 \tag{36}$$

$$sx_i^I + (1-s)x_i^{II} = z_i \tag{37}$$

$$\sum_{i=1}^{n} x_i^I = 1 \tag{38}$$

The unknowns are the phase compositions ($x_i^I$ and $x_i^{II}$) and the fraction of the overall mixture occupied by phase I ($s$). The summation of phase I mole fraction has been included in Eq. (38), and then excludes the corresponding sum for phase II as it is implicit in the equation group since the feed mole fractions ($z_i$) are constrained to sum to one.

In principle, it is possible to find the values of $x_i^I$, $x_i^{II}$ and $s$ that satisfy Eqs. (36) - (38) for the systems at liquid-liquid two-phase equilibrium. However, the calculation is complicated by the fact that there are multiple solutions to this equation group [43-45].

Classical thermodynamic analysis states that the necessary and sufficient condition of a single stable phase with an initial overall mole fraction, $z_1$, is:

$$\left( \frac{\partial^2 G_m}{\partial z_1^2} \right)_{T,P} > 0 \tag{39}$$

where $G_m$ is the Gibbs free energy per mole of mixture (J mol$^{-1}$). That is to say, $G_m$ is a concave function of composition. Phase split will occur when

$$\left( \frac{\partial^2 G_m}{\partial z_1^2} \right)_{T,P} < 0 \tag{40}$$

and the boundary condition of phase stability is:

$$\left( \frac{\partial^2 G_m}{\partial z_1^2} \right)_{T,P} = 0 \tag{41}$$

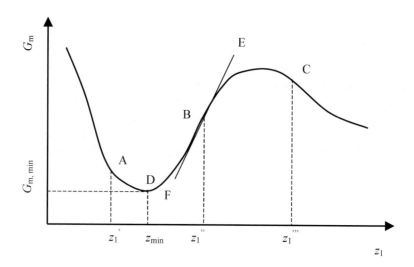

Fig. 18. Schematic representation of phase stability; line EF is the tangent of inflexion point B.

The schematic representation of judging phase stability is shown in Fig. 18, where at point A the single phase is stable and at point C unstable. The boundary is at point B which is an inflexion point. Apparently, point D is at equilibrium where the Gibbs free energy per mole of mixture for all phases $G_{m,\min}$ is a minimum.

However, it isn't an effective way to judge phase stability by applying Eqs. (39) - (41) because in general the molar Gibbs free energy is very difficult to be gotten directly. In practice, one method of tangent plane stability analysis is often adopted. Besides judging function, the key advantage is to produce excellent initial estimates for the phase split calculation. The interested reader can refer to the references [43-48], which presents the extensive algorithms.

The method used for solving Eqs. (34) - (38) simultaneously is by using the Newton-Raphson method. With respect to the stability test, the equations are written as deviation functions:

$$D_i = x_i^I \gamma_i^I - x_i^{II} \gamma_i^{II} \tag{42}$$

$$D_{i+n} = z_i - s x_i^I - (1-s) x_i^{II} \tag{43}$$

$$D_{2n+1} = 1.0 - \sum_{i=1}^{n} x_i^I \tag{44}$$

The variable vector is given as

$$x^T = (x_1^I, \bullet \bullet \bullet, x_n^I, x_1^{II}, \bullet \bullet \bullet, x_n^{II}, s) \tag{45}$$

which is revised according to

$$x^{(t+1)} = x^{(t)} + \Delta x \quad (t = \text{iteration number}) \tag{46}$$

The correction vector $\Delta x$ is calculated by

$$J\Delta x = -D \tag{47}$$

where $J$ is the Jacobian matrix housing the partial derivatives of Eqs. (42) – (44). The equation group in Eq. (47) can be solved by standard Gaussian elimination. The iterations are continued until the mean square error condition for the residuals ($D$) satisfies a certain tolerance.

Experience has shown that the step length corrections ($\Delta x$) calculated by Eq. (47) may be too large, thus forcing the iterations to diverge or converge on the trivial solution. To avoid this, the maximum corrections are limited to xmax for the mole fraction and smax for the phase fraction:

$$x\max = 0.10 \tag{48}$$
$$s\max = 0.15 \tag{49}$$

Using this scheme and the initial conditions of the stability test usually allow the liquid-phase split to be calculated in less than six-time iterations.

Here the concept of binodal curve and spinodal curve is briefly mentioned. Binodal curve is made up of the points satisfying Eq. (34), while spinodal curve satisfying Eq. (41). Both separate one-phase and two-phase regions. In the composition space, binodal curve and spinodal curve are close but don't overlap. The interval between binodal curve and spinodal curve is the partially stable region where a little strong external perturbation can lead to phase split. But the region enclosed by spinodal curve is thermodynamically unstable. A schematic representation of binodal curve and spinodal curve [65-67] is illustrated in Fig. 19.

*Example:* As shown in Fig. 8, please plot the binodal curve at liquid-liquid two-phase equilibrium for the system of ethanol (L) / water (H) / benzene (I) system at 101.3 kPa.

*Solution:* This problem is better to subdivide six parts, each of which can be finished by one or two students as one group. As shown in Fig. 20, the work of subprograms 1, 2, 3, 4 and plotting the binodal curve could be done each by one group. The sixth group has the obligation to connect subprograms 1, 2, 3 and 4 as a whole, and cooperate with the group in charge of plotting the binodal curve.

For the NEQ stage model, in the heterogeneous liquid (usually two-liquid phase) region, the effect of liquid phase mass and heat transfer resistances near the liquid-liquid interface is considered. However, this effect doesn't exist in the homogeneous region. Fig. 21 illustrates the vapor-liquid-liquid three-phase mass and heat transfers on the tray based on two-film model.

Consequently, at steady-state, we have continuity of mass and energy at the interfaces I and II:

$$N_i^{Vf}\big|_{\mathrm{I}} = N_i^{Lf}\big|_{\mathrm{I}}, \quad E^{Vf}\big|_{\mathrm{I}} = E^{Lf}\big|_{\mathrm{I}} \tag{50}$$

$$N_i^{L1f}\big|_{\mathrm{II}} = N_i^{L2f}\big|_{\mathrm{II}}, \quad E^{L1f}\big|_{\mathrm{II}} = E^{L2f}\big|_{\mathrm{II}} \tag{51}$$

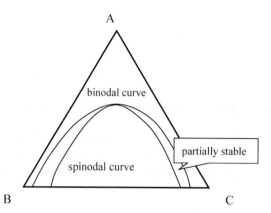

Fig. 19. A schematic representation of binodal curve and spinodal curve.

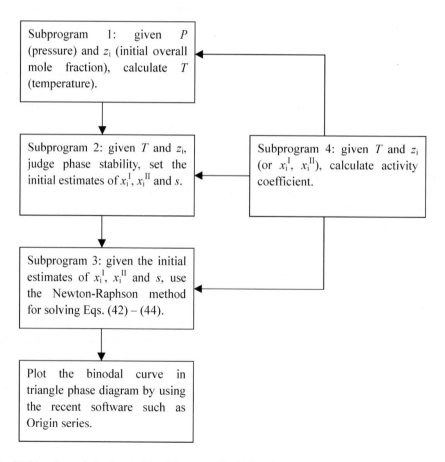

Subprogram 1: given $P$ (pressure) and $z_i$ (initial overall mole fraction), calculate $T$ (temperature).

Subprogram 2: given $T$ and $z_i$, judge phase stability, set the initial estimates of $x_i^{I}$, $x_i^{II}$ and $s$.

Subprogram 4: given $T$ and $z_i$ (or $x_i^{I}$, $x_i^{II}$), calculate activity coefficient.

Subprogram 3: given the initial estimates of $x_i^{I}$, $x_i^{II}$ and $s$, use the Newton-Raphson method for solving Eqs. (42) – (44).

Plot the binodal curve in triangle phase diagram by using the recent software such as Origin series.

Fig. 20. Flowsheet of plotting the binodal curve at liquid-liquid two-phase equilibrium.

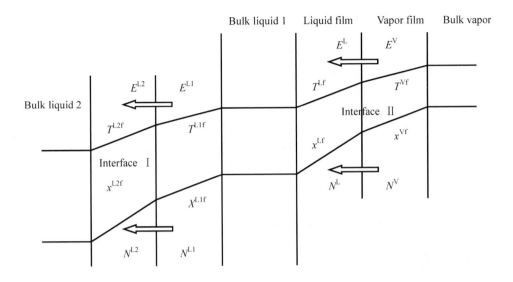

Fig. 21. Two-film model of vapor-liquid 1-liquid 2 three-phase mass and heat transfers on the tray (it is supposed that the direction of mass and heat transfers is from the vapor phase to the liquid phase and then from liquid 1 to liquid 2).

The molar transfer rates $N_i^{Vf}$, $N_i^{lf}$, $N_i^{l1f}$ and $N_i^{l2f}$ are related to the chemical potential gradients and generally solved by the famous Maxwell-Stefan equation. How to solve the Maxwell-Stefan equation is clarified in detail in chapter 1. The molar heat transfer rates, $E^{Vf}$, $E^{lf}$, $E^{l1f}$ and $E^{l2f}$, are often obtained by virtue of Reynolds analogy, Chilton-Colburn analogy or Prandtl analogy to derive the heat transfer coefficients.

### 3.2.3. Case study

On this basis, an azeotropic distillation column can be simulated by the EQ and/or NEQ stage model. Herein, for the EQ stage model, the ethanol dehydration process by azeotropic distillation with pentane as entrainer was simulated. The result coming from the references [31, 32] is shown in Figs. 22 and 23, from which we can understand the typical characteristics of azeotropic distillation column, that is, there is a steep change in temperature and composition along the column, often near the bottom or the top. As shown in Figs. 22 and 23, this range is near the bottom (between the fourth and eighth trays). The corresponding fourth and eighth trays (or a section of packing) are generally called sensitive trays. The sensitive range is marked in dashed lines. At the same time, the general assumption on constant molar flowrate adopted in the short-cut method should be cautious because flowrate change for vapor and/or liquid phases sometimes is very high, up to 25% along the column.

For the NEQ stage model, the ethanol dehydration process by azeotropic distillation with benzene as entrainer was simulated by Mortaheb and Kosuge [1]. The effect of operating conditions on the separation performance of azeotropic distillation column was studied. But it was also found that there is a steep change in temperature and composition along the column, often near the bottom or the top.

### 3.3. Multiple steady-state analysis

The term, multiple steady-state (MSS), is referred to as output multiplicities, i.e. columns with the same inputs (the same feed, distillate, bottoms, reflux and boilup molar flows, the same feed composition, number of stages and feed location) but different outputs (product compositions) and thus different composition profiles.

MSS in conventional distillation have been known from the simulation and theoretical studies dating back to 1970s and have been a topic of considerable interest in the distillation community. However, it is only recent that experimental verification of their existence has been forthcoming.

MSS is related with nonlinear analysis along the continuation path. Although the nonlinear analysis is highly mathematical and somewhat sophisticated, it is rapidly gaining acceptance by many practitioners in this field.

In order to explore the occurrence of MSS, the bifurcation parameter should be chosen among the inputs. The common bifurcation parameters are feed location, product flowrate, etc. The mathematical methods have two types: graphical method and numerical simulation.

In the graphical method for the $\infty/\infty$ case (infinite reflux and infinite number of trays or infinite long columns), the geometrical condition for the existence of MSS is as follows [33, 50-52]:

(1) As illustrated in Fig. 8, some lines parallel to the LH edge intersects the interior boundary TX more than once, or

(2) Some lines parallel to LI intersects TY more than once, or

(3) Some lines parallel to IH intersects TZ more than once.

However, the lines (e.g. $L_2H_2$, $I_2H_1$, $L_1I_1$) respectively parallel to LH, LI and IH insect TX, TY or TZ only once. Thus, no MSS is found in Fig. 8. In Fig. 24 it is supposed that there is no decanter at the top of distillation column and LLE curve isn't utilized. Fig. 24 illustrates the phenomena of MSS. The lines (e.g. $L_1L_2$) between lines $H_1H_2$ and $I_1I_2$ insect TX twice, and thus MSS exists. For the $\infty/\infty$ case, the top composition can lie in TX boundary and the bottom composition lies in LH boundary. So it is very straightforward to obtain: $B_1 = B_2$ (flowrate of bottom product) and $D_1 = D_2$ (flowrate of top product). But in these two cases, the product compositions are different, which indicates that different outputs appear. Evidently, both bottom product and top product can be used as bifurcation parameters. Moreover, MSS often appears in the vicinity of singular (fixed) points, such as the point T.

Note that the boundaries employed may be simple distillation boundaries or distillation line boundaries, depending on the column type. Furthermore, for heterogeneous azeotropic distillation with decanter at the top, the boundaries may include one part of LLE envelope.

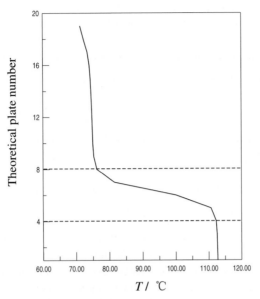

Fig. 22. Temperature profile of the ethanol / water / pentane system along the azeotropic distillation column; theoretical plate is numbered from the bottom to the top.

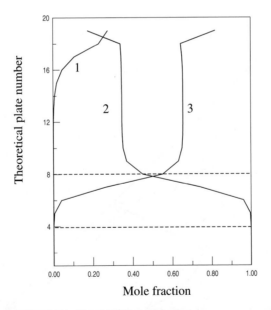

Fig. 23. Composition profile of the ethanol / water / pentane system along the azeotropic distillation column; theoretical plate is numbered from the bottom to the top; 1 - mole fraction of water in vapor phase (entrainer free), 2 - mole fraction of ethanol in liquid phase, 3 - mole fraction of pentane as entrainer in the liquid phase.

In the numerical simulation, continuation methods normally used for finding solutions to "hard" problems, where a very good initial guess is required in order for Newton's method to converge on that solution. In addition, it has proven to be a valuable tool for finding MSS in distillation columns. The underlying idea of continuation methods is that of path following, whereby a mathematical path is traced from a problem $G(x)$ for which we can obtain a solution relatively easily to the "hard" problem ($F(x)$), thereby keeping track of the changes in the variables. Such a path is implicitly defined by the homotopy equation:

$$H(x,t) = (1-t)G(x) + tF(x) \tag{52}$$

If our original problem is $n$-dimensional, this equation describes a curve in $(n + 1)$ dimensional space. Since the solution to $H(x,0) = 0$ coincides with $G(x) = 0$, we have at least one point on this curve. We now have to evaluate how the variables change along the curve, and correct each accordingly as we walk along the curve. By differentiating the homotopy equation with respect to the arc length and integrating the resulting system of differentiating equations, we can generate an estimate for a new point on the curve, which can then be corrected to give the next point on the curve. For more details about this method, the interested readers can refer to the references [53-55].

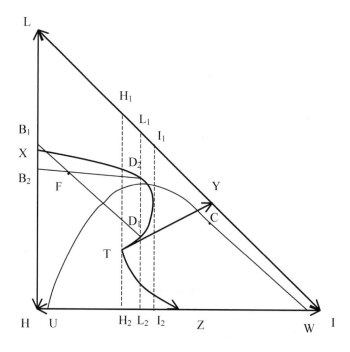

Fig. 24. MSS without decanter at the top of distillation column.

MSS are located using one of the operational specifications (e.g. reboiler duty, bottom product flowrate, feed rate, conversion) as the continuation parameter (or bifurcation parameter). This results in a system of $n$ equations with $n + 1$ variables, thereby implicitly defining a curve in a $(n + 1)$ dimensional space. If we have a converged solution for a fixed specification, we have a starting point on this curve. We can then apply exactly the same algorithm as described previously to "walk" along the curve. By doing so, we can follow the steady-state solution of the model as a function of that specification. Thus, we can obtain valuable information about steady-state column behavior and MSS regions. An added benefit of this approach is that now we are following a solution that is at any point on the curve describing a physically meaningful situation and not a strictly mathematical artifact. In the latter case, the homotopy path in itself doesn't have a physical meaning and might go through regions of negative mole fractions etc. In these regions, the solution is invalid if, for instance, in activity coefficient calculations the logarithm of the mole fractions is required.

Apart from the application of azeotropic distillation in traditional chemical engineering field, it has already been used in such front areas [12, 18, 19, 24] as preparation of nanometer-sized $ZrO_2$ / $Al_2O_3$ powders, synthesizing nanoscale powders of yttria doped ceria electrolyte, removal of byproducts in lipase-catalyzed solid-phase synthesis of sugar fatty acid esters, combining fungal dehydration and lipid extraction, and so on. It manifests that azeotropic distillation still brings out a continuous interest from the past to more recent years.

## REFERENCES

[1] H.R. Mortaheb and H. Kosuge, Chem. Eng. Process., 43 (2004) 317-326.

[2] A. Szanyi, P. Mizsey and Z. Fonyo, Chem. Eng. Process., 43 (2004) 327-338.

[3] D. Diamond, T. Hahn, H. Becker and G. Patterson, Chem. Eng. Process., 43 (2004) 483-493.

[4] V.V. Hoof, L.V.D. Abeele, A. Buekenhoudt, C. Dotremont and R. Leysen, Sep. Purif. Technol., 37 (2004) 33-49.

[5] P.A.M. Springer, S.V.D. Molen and R. Krishna, Comput. Chem. Eng., 26 (2002) 1265-1279.

[6] A.M. Eliceche, M.C. Daviou, P.M. Hoch and I.O. Uribe, Comput. Chem. Eng., 26 (2002) 563-573.

[7] P.A.M. Springer, R. Baur and R. Krishna, Sep. Purif. Technol., 29 (2002) 1-13.

[8] D.Y.C. Thong and M. Jobson, Chem. Eng. Sci., 56 (2001) 4369-4391.

[9] D.Y.C. Thong and M. Jobson, Chem. Eng. Sci., 56 (2001) 4393-4416.

[10] D.Y.C. Thong and M. Jobson, Chem. Eng. Sci., 56 (2001) 4417-4432.

[11] P.A.M. Springer and R. Krishna, Int. Commun. Heat Mass Transf., 28 (2001) 347-356.

[12] S.W. Zha, Q.X. Fu, Y. Lang, C.R. Xia and G.Y. Meng, Mater. Lett., 47 (2001) 351-355.

[13] G.Y. Feng, L.T. Fan and F. Friedler, Waste Manage., 20 (2000) 639-643.

[14] A.R. Ciric, H.S. Mumtaz, G. Corbett, M. Reagan, W.D. Seider, L.A. Fabiano, D.M. Kolesar and S. Widagdo, Comput. Chem. Eng., 24 (2000) 2435-2446.

[15] S.K. Wasylkiewicz, L.C. Kobylka and F.J.L. Castillo, Chem. Eng. J., 79 (2000) 219-227.

[16] I.L. Chien, C.J. Wang, D.S.H. Wong, C.H. Lee, S.H. Cheng, R.F. Shih, W.T. Liu and C.S.

Tsai, J. Process Control, 10 (2000) 333-340.

[17] T. Kurooka, Y. Yamashita, H. Nishitani, Y. Hashimoto, M. Yoshida and M. Numata, Comput. Chem. Eng., 24 (2000) 887-892.

[18] A.J. Tough, B.L. Isabella, J.E. Beattie and R.A. Herbert, J. Biosci. Bioeng., 90 (2000) 37-42.

[19] Y. Yan, U.T. Bornscheuer, L. Cao and R.D. Schmid, Enzyme Microb. Technol., 25 (1999) 725-728.

[20] J.M.P. King, R.B. Alcantara and Z.A. Manan, Environ. Modell. Softw., 14 (1999) 359-366.

[21] F.J.L. Castillo and G.P. Towler, Chem. Eng. Sci., 53 (1998) 963-976.

[22] M.H. Bauer and J.Stichlmair, Comput. Chem. Eng., 22 (1998) 1271-1286.

[23] C.J. Wang, D.S.H. Wong, I.L. Chien, R.F. Shih, S.J. Wang and C.S. Tsai, Comput. Chem. Eng., 21 (1997) S535-S540.

[24] H. Shan and Z.T. Zhang, J. European Ceram. Soc., 17 (1997) 713-717.

[25] M.H. Bauer and J. Stichlmair, Comput. Chem. Eng., 20 (1996) S25-S30.

[26] N. Bekiaris, G.A. Meski and M. Morari, Comput. Chem. Eng., 19 (1995) S21-S26.

[27] P. Poellmann and E. Blass, Gas Sep. Purif., 8 (1994) 194-228.

[28] H.W. Andersen, L. Laroche and M. Morari, Comput. Chem. Eng., 19 (1995) 105-122.

[29] S. Widagdo and W. D. Seider, AIChE J., 42 (1996) 96-130.

[30] N. Bekiaris, G.A. Meski, C.M. Radu and M. Morari, Ind. Eng. Chem. Res., 32 (1993) 2023-2038.

[31] R.F. Gould (ed.), Extractive and Azeotropic Distillation, American Chemical Society, Washington, 1972.

[32] E.J. Hoffman (ed.), Azeotropic and Extractive Distillation, John Wiley and Sons, New York, 1964.

[33] N. Bekiaris, G.A. Meski, C.M. Radu and M. Morari, Comput. Chem. Eng., 18 (1994) S15-S24.

[34] B. Thomas and S.K. Hans, Comput. Chem. Eng., 18 (1994) S25-S29.

[35] O.M. Wahnschaff, J.W. Kohler and A.W. Westerberg, Comput. Chem. Eng., 18 (1994) S31-S35.

[36] A. Kienle, E.D. Gilles and W. Marquardt, Comput. Chem. Eng., 18 (1994) S37-S41.

[37] E. Rev, P. Mizsey and Z. Fonyo, Comput. Chem. Eng., 18 (1994) S43-S47.

[38] J.G. Stichlmair and J.R. Herguijuela, AIChE J., 38 (1992) 1523-1535.

[39] F.A.H. Schreinemakers, Z. Phys. Chem., 39 (1901) 440.

[40] F.A.H. Schreinemakers, Z. Phys. Chem., 47 (1903) 257.

[41] R. Krishnamurthy and R. Taylor, AIChE J., 31 (1985) 456-465.

[42] R. Krishnamurthy and R. Taylor, AIChE J., 31 (1985) 1973-1985.

[43] B.P. Cairns and I.A. Furzer, Ind. Eng. Chem. Res., 29 (1990) 1383-1395.

[44] B.P. Cairns and I.A. Furzer, Ind. Eng. Chem. Res., 29 (1990) 1349-1363.

[45] B.P. Cairns and I.A. Furzer, Ind. Eng. Chem. Res., 29 (1990) 1364-1382.

[46] M.L. Michelsen, Fluid Phase Equilib., 9 (1982) 1-19.

[47] M.L. Michelsen, Fluid Phase Equilib., 9 (1982) 21-40.

[48] D.J. Swank and J.C. Mullins, Fluid Phase Equilib., 30 (1986) 101-110.

[49] J.G. Stichlmair, J.R. Fair and J.L. Bravo, Chem. Eng. Prog., 85(1) (1989) 63-69.

[50] T.E. Guttinger and M. Morari, Comput. Chem. Eng., 21 (1997) S995-S1000.

[51] T.E. and M. Morari, Ind. Eng. Chem. Res., 38 (1999) 1633-1648.

[52] T.E. and M. Morari, Ind. Eng. Chem. Res., 38 (1999) 1649-1665.

[53] T.L. Wayburn and J.D. Seader, Comput. Chem. Eng., 11 (1987) 7-25.

[54] M. Kubicek, ACM Trans. Math. Softw., 2 (1976) 98-107.

[55] A.P. Higler, R. Taylor and R. Krishna, Chem. Eng. Sci., 54 (1999) 1389-1395.

[56] M. Rovaglio, T. Faravelli, P. Gaffuri, C. Di Palo and A. Dorigo, Comput. Chem. Eng., 19 (1995) S525-S530.

[57] R. Dussel and J. Stichlmair, Comput. Chem. Eng., 19 (1995) S113-S118.

[58] M.H. Bauer and J. Stichlmair, Comput. Chem. Eng., 19 (1995) S15-S20.

[59] A.W. Westerberg, J.W. Lee and S. Hauan, Comput. Chem. Eng., 24 (2000) 2043-2054.

[60] J. Gmehling, J. Menker, J. Krafczyk and K. Fischer, Fluid Phase Equilib., 103 (1995) 51-76.

[61] J. Gmehling (ed.), Development of thermodynamic models with a view to the synthesis and design of separation processes, Springer-Verlag, Berlin, 1991.

[62] G. Schembecker and K.H. Simmrock, Comput. Chem. Eng., 19 (1995) S253-S258.

[63] G. Schembecker and K.H. Simmrock, Comput. Chem. Eng., 21 (1997) S231-S236.

[64] R.F. Martini and M.R. Wolf Maciel, Comput. Chem. Eng., 20 (1996) S219-S224.

[65] R.P. Danner, M. Hamedi and B.C. Lee, Fluid Phase Equilib., 194-197 (2002) 619-639.

[66] U. Eisele (ed.), Introduction to Polymer Physics, Springer – Verlag, New York, 1990.

[67] Z.Q. Zhu (ed.), Supercritical Fluids Technology, Chemical Industry Press, Beijing, 2001.

# Chapter 4. Catalytic distillation

Reactive distillation can be divided into homogeneous reactive distillation and heterogeneous reactive distillation (namely catalytic distillation). Since catalytic distillation is used more often than homogeneous reactive distillation, only which is emphasized in this chapter. An excellent review on reactive distillation has already been present by Taylor and Krishna [1]. This chapter may be taken on as a useful supplement to catalytic distillation. Two types of catalytic distillation, i.e. fixed-bed catalytic distillation (FCD) and suspension catalytic distillation (SCD), are discussed. In catalytic distillation, catalyst is taken on as the separating agent by reaction to promote the separation of reactants and products.

## 1. FIXED-BED CATALYTIC DISTILLATION

### 1.1. FCD advantages

*1.1.1. Introduction*

Reactive separation processes such as reactive distillation, sorption-enhanced reaction absorption, reactive extraction and reaction crystallization combine the essential tasks of reaction and separation in a single vessel. The most important example of reactive separation processes (RSP) is reactive distillation (RD). RD, the combination of chemical reaction and distillation in a single column, is one of the most important industrial applications of the multifunctional reactor concept. Since the 1980s, the research on RD is very flourishing and RD becomes a research hotspot. Up to date, there are many articles about it published every year in the international journals [1-48].

RD processes can be divided into homogeneous ones either auto-catalysed or homogeneously catalyzed, and heterogeneous processes often referred as catalytic distillation (CD) in which the reaction is catalyzed by a solid catalyst. The equipment used for homogeneous reactive distillation processes consists of bubble-cap or sieve trays with high weirs that provide the necessary liquid holdup and residence time needed for reaction. Heterogeneous reactive distillation processes use a solid catalyst, where most of the reactions take place within the catalyst particle. The equipment consists primarily of catalyst containing packing which allows for simultaneous reaction and mass transfer between vapor and liquid phases.

*1.1.2. Advantages*

RD has many advantages over sequential processes [12, 13], for instance, a fixed-bed reactor followed by a fractionating column in which the distillate or bottom of the reaction mixture is recycled to the reactor inlet. The most important advantage in use of RD for equilibrium-controlled reactions is the elimination of conversion limitations by continuous removal of products from the reaction zone. Apart from increased conversion, the following benefits can be obtained:

(1) An important benefit of RD technology is a reduction in capital investment because two process steps can be carried out in the same device. Such an integration leads to lower costs in pumps, piping and instrumentation.

(2) If RD is applied to exothermic reaction, the reaction heat can be used for vaporization of liquid. This leads to savings of energy costs by the reduction of reboiler duties.

(3) The maximum temperature in the reaction zone is limited to the boiling point of the reaction mixture so that the danger of hot spot formation on the catalyst is reduced significantly. A simple and reliable temperature control can be achieved.

(4) Product selectivities can be improved due to a fast removal of reactants or products from the reaction zone. Thus, the probability of consecutive reactions, which may occur in the sequential operation mode, is lowered.

But for catalytic distillation (CD), it has more extra advantages in comparison with homogeneous reactive distillation:

(1) An optimum configuration of the reaction and separation zones is permitted in a RD column whereas expensive recovery of liquid catalysts may be avoided.

(2) If the reaction zone in the CD column can be placed above the feed point, poisoning of the catalyst (especially ion exchange resins) by metal ions can be avoided. This leads to longer catalyst lifetime compared to conventional systems.

(3) Many catalysts used in CD column are environment-friendly such as supported molecule sieves, ion exchange resins and so on.

That is why catalytic distillation is more desirable than homogeneous reactive distillation. Table 1 lists some application of RD [13]. It can be seen that the technology of RD is applied only for etherification, esterification and alkylation (synthesis of ethylbenzene or cumene) on an industrial scale. Furthermore, the cases of catalytic distillation are larger than those of homogeneous reactive distillation. Many researchers believe that RD, especially catalytic distillation, is very promising and the potential of this technique should go far beyond today's application.

In this chapter we only pay more attention to catalytic distillation since it is used more often than homogeneous reactive distillation and similar results may hold for homogeneous reactive distillation. Catalytic distillation (CD) can be divided into two types: fixed-bed catalytic distillation (FCD) and suspension catalytic distillation (SCD). FCD is relatively conventional, but SCD is a recent development. In order to clarify the characteristics of CD, let us begin with one example of FCD.

Fig. 1 [17] shows the flow scheme of a common industrial process for the production of the fuel ethers tert-amyl-methylether (TAME) or methyl-tert-butylether (MTBE) in comparison with a possible process in which a catalytic distillation technique is employed. In the original process (Fig. 1a), in the first fixed-bed reactor (enveloped with dashed line) the main part of the overall conversion of reactive olefins is attained. This reactor is filled with an acid ion exchange resin as catalyst. For the further increase of the olefin conversion a second reactor is necessary in the common industrial process. After this, the product of TAME or MTBE is isolated via distillation from hydrocarbons and methanol. It is evident that the whole process is somewhat complicated.

180

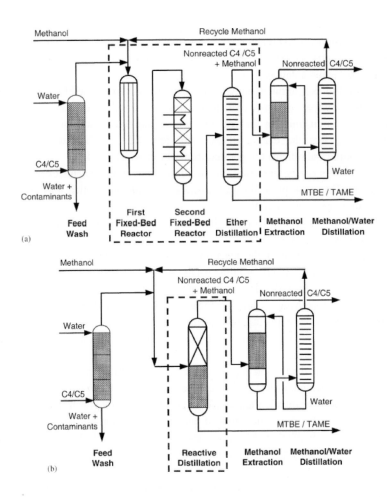

Fig. 1. The original and FCD flow scheme for the production of fuel ethers TAME or MTBE; adapted from the source [17].

It is known that the synthesis of TAME or MTBE is a chemical equilibrium-limited reaction, and thus the conversion of reactive olefins isn't complete even in case of two fixed-bed reactions are set. However, the process (in Fig. 1b) which uses a fixed-bed catalytic distillation (FCD) combines three process steps of the common process. In addition to significant saving of energy and investment costs, it is possible to achieve high purity of inert hydrocarbons and TAME or MTBE.

Note that in Fig. 1b the FCD column is composed of two zones: reaction zone in the upper part and stripping zone in the lower part, not including rectifying zone. However, another kind of FCD column is also found for producing TAME or MTBE in the references [6-10] which is composed of three zones: rectifying zone in the upper part, reaction zone in the middle part and stripping zone in the lower part (see Fig.2). The merit of this kind of FCD column is that methanol with relative high boiling point and C4/C5 with relative low boiling point is in count-current contact. The mixing condition is favorable, which possibly results in reduced catalyst amount and high conversion. But catalyst contamination from metal ions brought out by methanol can be produced, which leads to deactivation of catalyst due to ion exchange over a long time. However, it doesn't occur in Fig. 1b where the reaction zone is located above the inlet. This arrangement prolongs the catalyst lifetime. So the suitable arrangement should be selected depending on the economic consideration in industry.

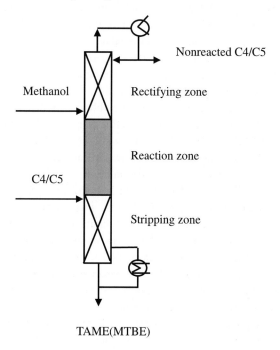

Fig. 2. The FCD column composed of rectifying zone, reaction zone and stripping zone.

## 1.2. Hardware structure

The outstanding characteristics of catalytic distillation is that solid catalyst is placed in the reaction zone of the distillation column, which is different from homogeneous reaction distillation and conventional distillation. Undoubtedly, how to design the hardware containing catalyst is crucial for catalytic distillation.

Towler and Frey [1, 74] have highlighted some of the practical issues in implementing a large-scale application of catalytic distillation. These are discussed below:

(1) Installation, containment and removal of the catalyst

It is important to allow easy installation and removal of the RD equipment and catalyst. If the catalyst undergoes deactivation, the regeneration is most conveniently done ex situ. So there must be provision for easy removal and installation of catalyst particles. Reactive distillation is often passed over as a processing option because the catalyst life would require frequent shut down. An RD device that allowed on-stream removal of catalyst would answer this concern. But for the fixed-bed catalytic distillation column, it is difficult to realize loading and downloading catalyst from time to time. From the viewpoint, the fixed-bed catalytic distillation technique is more challenging on the condition that the life of catalyst is enough long, preferably over one to three years or even longer.

(2) Efficient contacting of liquid with catalyst particles

The hardware design must ensure that the following "wish-list" is met:

(a) Good liquid distribution and avoidance of channeling. Liquid misdistribution can be expected to have a more sever effect in RD than in conventional distillation.

(b) Good radical dispersion of liquid through the catalyst bed. This is required to avoid reactor hot-spots and runaways and allow even catalyst ageing. The requirement of good radical mixing has an impact on the choice of the packing configuration and geometry.

(3) Good vapor liquid contacting in the reaction zone

If the reaction rate is fast and the reaction is equilibrium-limited, then the required size of the reactive zone is strongly influenced by the effectiveness of the vapor-liquid contacting. Vapor-liquid contacting becomes less important for slower reactions. Commonly used devices for good vapor-liquid contacting are the same as for conventional distillation and include structured packing, random packing and distillation trays.

(4) Low pressure drop through the catalytically packed reaction section

This problem arises because of the need to use small catalyst particles in the 1-3 mm range in order to avoid intra-particle diffusional limitations. The smaller are the catalyst particles, the larger pressure drop but the lower intra-particle diffusional resistance. A compromise between these two contradictory factors must be achieved. Counter-current operation in catalyst beds packed with such small-sized particles has to be specially configured in order to avoid problems of excessive pressure drop and flooding.

(5) Sufficient liquid hold-up in the reactive section

The liquid hold-up, mean residence time, and liquid residence time distribution (RTD) are all important in determining the conversion and selectivity of RD. This is in sharp contrast with conventional distillation where liquid hold-up and RTD are often irrelevant as the vapor-liquid mass transfer is usually controlled by the vapor-side resistance. For RD tray columns the preferred regime of operation would be the froth regime whereas for conventional distillation we usually adopt the spray regime.

(6) Designing for catalyst deactivation

Even though it is desirable to allow on-line catalyst removal and regeneration, such devices haven't been commercialized as yet. Catalyst deactivation is therefore accounted for in the design stage by use of excess catalyst. Besides adding excess catalyst, the reaction severity can be increased by (a) increasing reflux, leading to increased residence time and (b) increasing reaction temperature (by increase of column pressure).

Anyway, three different approaches to locate catalyst particles have ever been tried:

(1) Place beds of catalyst in a tray column, either in downcomers or above the trays

Fig. 3 shows catalyst bales above the sieve tray. Details of this catalyst bales configuration are available in [29].

Table 1
Applications of RD processes (homogeneously catalyzed (hom.); heterogeneously catalyzed (het.)); adapted from the source [13] with some replenishment

| Reaction type | Synthesis | Catalyst | References |
|---|---|---|---|
| Esterification | Methyl acetate from methanol and acetic acid | hom. | [49] |
| | Methyl acetate from methanol and acetic acid | het. | [50] |
| | Ethyl acetate from ethanol and acetic acid | het. | [51] |
| | Butyl acetate from butanol and acetic acid | hom. | [52] |
| Transesterification | Ethyl acetate from ethanol and butyl acetate | hom. | [53] |
| | Diethyl carbonate from ethanol and dimethyl carbonate | het. | [54] |
| Hydrolysis | Acetic acid and methanol from methyl acetate and water | het. | [55] |
| Etherification | MTBE from isobutene and methanol | het. | [15, 35, 56] |
| | ETBE from isobutene and ethanol | het. | [57] |
| | TAME from isoamylene and methanol | het. | [58] |
| Condensation | Diacetone alcohol from acetone | het. | [60] |
| | Bisphenol-A from phenol and acetone | - | [61] |
| Dismutation | Monosilane from trichlorsilane | het. | [62] |
| Hydration | Mono ethylene glycol from ethylene oxide and water | hom. | [63] |
| Nitration | 4-Nitrochlorobenzene from chlorobenzene and nitric acid | hom. | [64] |
| Alkylation | Ethylbenzene from ethylene and benzene | het. | [68-70, 72, 73] |
| | Cumene from propylene and benzene | het. | [59, 66, 67, 72, 73] |
| | Alkylation of benzene and 1-dodecene | het. | [23, 65, 71-73] |

There are various choices of liquid flow pass when installing them above the sieve for different requirement as shown in Fig. 4. But it is found that when operating at the high conversion, the conversion level will increase in case of decreasing the number of liquid flow passes because the liquid load per weir length increases. It is believed that the liquid hold-up is usually much higher in sieve tray columns as compared to packed columns. And this is an favorable when carrying out relatively slow, catalysed, liquid phase reactions. So in this case catalytic distillation with tray columns is advisable.

Fig. 5 shows the structure of a wire-gauze catalyst envelop above the sieve tray [30]. This structure may be as in the Katapak-s and katamax constructions of Sulzer Chemtech and Koch-Glitsch. An advantage of a catalytic sieve tray construction is that contacting on any tray is cross-current and for large diameter columns there will be a high degree of staging in the liquid phase. This is favorable from the viewpoint of selectivity and conversion.

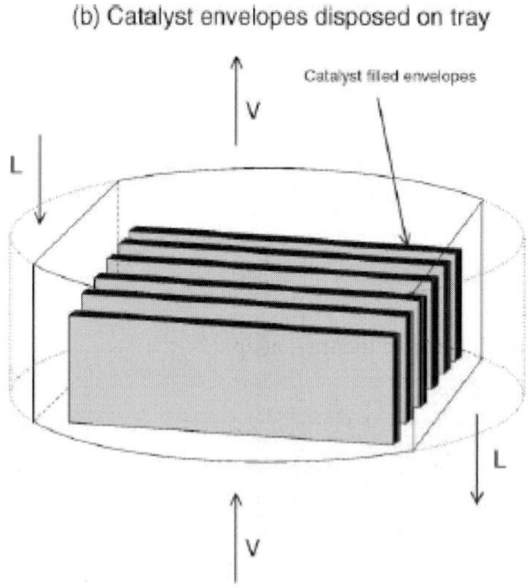

(b) Catalyst envelopes disposed on tray

Catalyst filed envelopes

Fig. 3. Catalyst bales above the sieve tray; adapted from the source [29].

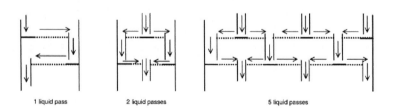

1 liquid pass    2 liquid passes    5 liquid passes

Fig. 4. Tray configuration in which catalyst bales are above the tray; adapted from the source [29].

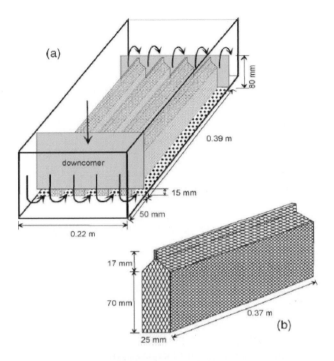

Fig. 5. A wire-gauze catalyst envelop above the tray; adapted from the source [30].

(2) Use random packing elements containing catalyst; the packings can be prepared by extrusion, impregnation of porous packings, or molding resins as random packings.

The famous active packing is the Raschig ring as shown in Fig. 6 [5, 17, 24, 29, 36]. It is made catalytically active. The catalyst rings can be prepared by block polymerization in the annular space. Their activity is quite high. However, osmotic swelling processes can cause breakage by producing large mechanical stresses inside the resin.

Another way for using random packing is just to simply blend the inactive random packing and catalytic particles which are usually in 1-3 mm in order to ensure high activity and avoid the diffusion limitations. However, the amount of catalytic particles is often small because problems of excessive pressure drop and flooding are brought out. So it is only used in laboratory, rarely in industry.

(3) Use ordered packing elements which contain catalyst particles between layers of supporting material. Elements of this type include the bale packing proposed by Smith [1, 16, 24], as well as other arrangements.

The structured active packing is very welcome and commonly used in catalytic distillation. Evidently, the excellent structured active packing should not only have high catalytic activity, but also have high mass transfer efficiency (i.e. as low HETP as possible). HETP (height equivalent to a theoretical plate) is an important index for evaluation conventional distillation packing.

(a) Catalytically active packing

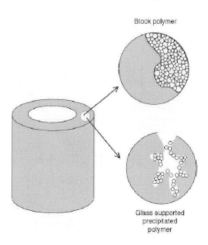

Fig. 6. The active Raschig ring proposed by Sundmacher and Hoffmann; adapted from the sources [24, 29].

The early structured active packing is proposed by Smith [1, 16, 24], in which the catalyst beads are placed in the pockets of a Fiberglas cloth belt which is rolled up using a spacer of mesh knitted steel wire ("bale-type catalytic packing"), as shown in Fig. 7.

Another kind of bale-type catalytic packing is shown in Fig. 8 [39]. It consists of metal wave mesh and catalyst particles encapsulated with Nylon cloth bags, which has been used for the hydrolysis of methyl acetate (MeOAc) in polyvinyl alcohol (PVA) production process. There are two popular kinds of catalyst particles sandwiched between corrugated sheets of wire gauze, i.e. KATAPAK-S [21] and MULTIPAK [11, 28, 41]. Both packing types combine the benefits of modern structured packings, such as low pressure drop and high throughput and offer the advantages of heterogeneous catalysis.

For KATAPAK-S type, the granular catalyst is embedded between two wire gauze sheets (see Fig. 9a).

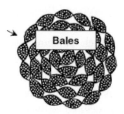

Fig. 7. A bale-type catalytic packing proposed by Smith [16]; adapted from the source [24].

The catalytic packing of MULTIPAK, as shown in Fig. 9b, has been developed in consideration of the qualities of modern structured wire gauze packings and their improvements on conventional distillation. It consists of corrugated wire gauze sheets and catalyst bags of the same material assembled in alternative sequence. This construction enables high column loads by providing open channels for the gas flow while most of the liquid is flowing downwards through the catalyst bags. Sufficient mass transfer between gas and liquid phases and radial mixing are guaranteed by the segmentation of the catalyst bags and numerous contact spots with the wire gauze sheets. The catalyst bags are generally filled with an acidic ion-exchange resin.

The important advantage of the structured catalyst sandwich structures over the catalyst bales is with respect to radial distribution of liquid. Within the catalyst sandwiches, the liquid follows a crisscross flow path to ensure sufficient mass transfer between gas and liquid phases.

Besides, one technique of structured catalyst packing is the catalytic coating of packings. However, in most cases its making is rather complex and the support of the catalyst supplier becomes necessary. This technique hasn't been put into practice in CD.

### 1.3. Mathematical models

To date there are two types of mathematical models to simulate reactive or nonreactive distillation systems, i.e. the equilibrium (EQ) stage model and the non-equilibrium (NEQ) stage model, as mentioned in chapter 2. With respect to the EQ stage model, please refer to chapter 2 (7.1 EQ stage model). In CD, like reactive extractive distillation, we need to simply add a term to take account of the effect of reaction on the mass balances. But in solving the $R$ equation (reaction rate equation), if chemical reaction is reversible, the reaction rate is assumed to be zero. That is to say, the chemical equilibrium is reached in every tray. On the contrary, if chemical reaction is irreversible, then the reaction rate is relative to the amount of catalyst used.

With respect to the NEQ stage model, please refer to chapter 2 (7.2 NEQ stage model). But the EQ and NEQ stage models in CD are a little different from those in extractive distillation in that the catalyst is involved. For example, for the material balance,

$$\frac{dM_j}{dt} = V_{j+1} + L_{j-1} + F_j - (1 + r_j^V)V_j - (1 + r_j^L)L_j + \sum_{k=1}^{r}\sum_{i=1}^{c} v_{i,k}R_{k,j}\varepsilon_j \qquad (1)$$

where $R_{k,j}$ is reaction rate, $v_{i,k}$ is the stoichiometric coefficient of component $i$ in reaction $k$, $\varepsilon$ is reaction volume or amount of catalyst, not just reaction volume as in reactive extractive distillation.

The assumptions of the EQ and NEQ stage models in CD are summarized in Table 2. Both model equations and solving procedures [88-98] are identical to reactive extractive distillation. Therefore, it is no need to describe them in detail again.

Two cases about mathematical models will be exemplified in the section of SCD, one for the EQ stage model where FCD column is involved and the other for the NEQ stage model where comparison of the EQ and NEQ stage models is made.

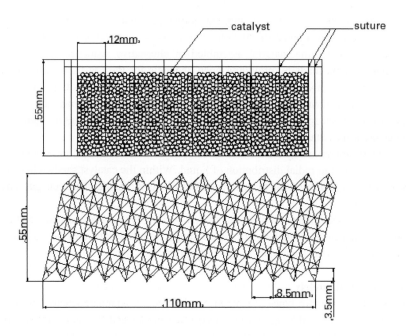

Fig. 8. Structure of a bale-type catalytic packing used for the hydrolysis of methyl acetate (MeOAc); adapted from the source [39].

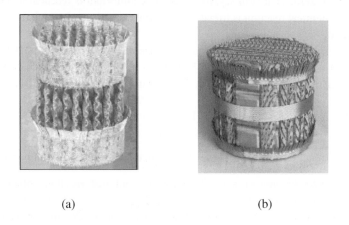

(a)                                    (b)

Fig. 9. Packing element of KATAPAK-S (left) and MULTIPAK (right); adapted from the sources [21, 28].

Table 2
Assumptions used in the EQ and NEQ stage models

| EQ stage model | NEQ stage model |
|---|---|
| 1. Operation reaches steady state. | 1. Operation reaches steady state. |
| 2. System reaches mechanical equilibrium. | 2. System reaches mechanical equilibrium. |
| 3. The vapor and liquid bulks are mixed perfectly and assumed to be at thermodynamic equilibrium. | 3. The vapor-liquid interface is uniform in each tray and assumed to be at thermodynamic equilibrium. |
| 4. Heat of mixing can be neglected. | 4. Heat of mixing can be neglected. |
| 5. Reactions take place in the liquid bulk. | 5. There is no accumulation of mass and heat at the interface |
| 6. The condenser and reboiler are considered as an equilibrium tray. | 6. The condenser and reboiler are considered as an equilibrium tray. |
| 7. If the catalyst is used, then the amount of catalyst in each tray of reaction section is equal (for fixed-bed and suspension catalytic distillations). | 7. Reactions within the interface are ignored and take place in the liquid bulk. |
| 8. If the catalyst is used, then the influence of catalyst on the mass and energy balance is neglected because the catalyst concentration suspended in the liquid phase is often low (for suspension catalytic distillation). | 8. The heat generated due to chemical reaction is neglected because heat effect is often not apparent for the reaction concerned. |
| | 9. If the catalyst is used, then the amount of catalyst in each tray of reaction section is equal (for fixed-bed and suspension catalytic distillations). |
| | 10. If the catalyst is used, then the influence of catalyst on the mass and energy balance is neglected because the catalyst concentration suspended in the liquid phase is often low (for suspension catalytic distillation). |

## 2. SUSPENSION CATALYTIC DISTILLATION

### 2.1. Tray efficiency and hydrodynamics of SCD

*2.1.1. Tray efficiency*

As we know, heterogeneous catalytic distillation is a more recent development and has attracted researchers' attention because of its inherent advantages. However, formerly catalytic distillation was only referred to FCD, until recently a new-type heterogeneous catalytic distillation process, called suspension catalysis distillation (SCD) process, has been put forward by Wen and Min [75, 76]. However, the work on the SCD process is very scarce.

In the SCD process, tiny solid particles aren't used as packing in the column but blended with liquid phase. In other words, catalyst particles are always moving in the column, not static. Comparing to conventional FCD process, it has the following unique advantages:

1. No need for structured catalytic-packing

2. No need for shutting down the unit to replace the deactivated catalyst

3. Mass and heat transfers in the fine catalyst particles and inter-phase are quicker than those in the structured catalytic-packing.

4. In a conventional FCD process, if the catalyst is quick to be deactivated, then it should be replaced by fresh catalyst from time to time. This is very tedious and difficult, especially for such reactions as synthesizing alkylbenzene with longer carbon chains. But the SCD process has no this problem and is convenient to be operated, which has been verified in our pilot plant.

It is believed that the packed column isn't suitable for the SCD process. When adding solid particles to liquid phase, the packed column is prone to be jammed. From this viewpoint, the tray columns are more suitable for the SCD process. Among the tray columns (i.e. sieve tray, valve tray and slant-hole tray), sieve tray is the best candidate because the gas-liquid-solid three phases can flow more smoothly in this tray.

As in FCD, both the EQ and NEQ stage models can be adopted to simulate the SCD process. However, developing an NEQ model for a RD process isn't as straightforward as it is for the EQ stage model in which we need to simply add a term to take account of the effect of reaction on the mass balances. As we know, NEQ model is more complicated than EQ model. In the NEQ model, the design information on the column configuration must firstly be specified so that mass transfer coefficients, interfacial areas, liquid hold-ups, etc. can be calculated. Therefore, for any new invented configuration of the column, many experiments have to be done in advance to obtain the necessary model parameters. Evidently, it is too tedious, and much time will be spent on the design of the SCD process. Fortunately, as pointed out by Lee and Dudukovic [77], a close agreement between the predictions of EQ and NEQ models can be found if tray efficiency is accurately predicted for the EQ model. On the other hand, if tray efficiency is known, it can be used for check the validity of NEQ stage model. So tray efficiency is an important variant in process simulation. An experimental study has been carried out to measure tray efficiency, as well as hydrodynamics in SCD column.

Two systems, air-water and air-water-solid particles, were adopted for measuring tray efficiency. Silica gel particles were selected as the solid phase, and it was based on the consideration that silica gel was used as the supporter of the catalyst used in the SCD process for producing cumene [75, 76]. The particle diameter distribution was obtained by screening method, and is listed in Table 3.

The experimental flowsheet is shown in Fig. 10. Air was driven by a compressor and into the measuring column from the bottom. Water (or the mixture of water and solid particles) was firstly blended with oxygen and then transported by a centrifugal pump into the top of the measuring column. So the gas from the bottom to the top and the liquid from the top to the bottom contact countercurrent in the column.

Table 3

Diameter distribution of solid particles

| Diameter / mm | Below 0.097 | 0.097-0.105 | 0.105-0.125 | 0.125-0.150 | 0.150-0.200 | 0.200-0.300 |
|---|---|---|---|---|---|---|
| wt / % | 1.99 | 2.96 | 25.77 | 20.28 | 44.57 | 4.44 |

When solid particles were added into water, more attention should be paid to preventing them from sinking down. Otherwise, the local concentration of solid particles in the liquid would be changed again and again. To avoid this, the water tank kept constantly stirred. In the pipe and column trays, solid particles were found no deposition because particles are too small and suspended by the flowing liquid.

The column having three trays, i.e. entrainment tray, measuring tray and gas distribution tray, was made up of organic glass and had 500 mm I.D., 15 mm thickness, tray distance 300 mm. In the measuring tray the big sieve hole with diameter 10 mm, as well as weir length 336 mm, downcomer area 0.0164 $m^2$ and sieve-hole area ratio 5.52%, was adopted in this investigation because it was relatively more difficult to be crowded by solid particles in this trays as compared to valve, bubble and small-hole sieve trays. The sieve holes on the tray were arranged according to the equilateral triangle manner. The section A - A of measuring tray is also shown in Fig. 10. The tray efficiency $E_{mL}$ is expressed by the following equation:

$$E_{mL} = \frac{X_i - X_o}{X_i - X_o^*} \times 100\% \tag{2}$$

where $X_i$ and $X_o$ are oxygen concentrations in the liquid phase at the inlet and the outlet, respectively, and $X_o^*$ is oxygen concentration in the water tank equilibrium with the atmosphere (or the equilibrium oxygen concentrations in the liquid phase with the air leaving the measuring tray). They were determined by oxygen gauge (type JPSJ-605). The flowrates of gas and liquid were respectively measured by orifice-plate flowmeter and rotary flowmeter, both of which had been calibrated beforehand. It took only a few of minutes to achieve the steady state for each run. In the beginning, some preliminary experiments under the same conditions were made to ensure that the data were repeatable.

The curves describing the relations of liquid flowrate with tray efficiency $E_{mL}$ at different conditions are shown in Fig. 11 through Fig. 14 in which three weight concentrations of the solid particles in the liquid, $w$ = 0%, 2% and 5%, were employed. These concentrations have covered the range of the catalyst concentration used in the SCD process, below 5 wt%. If the solid particles are in higher concentrations, such problems as sedimentation, jam and so on, may arise. It can be concluded from Figs. 11-14 that:

(1) For a given concentration of solid phase, tray efficiency decreases with increasing liquid flowrate. This is attributed to the decreasing amount of oxygen stripped from the liquid phase under a constant gas flowrate. That is to say, in Eq. (2), $X_i$ and $X_o^*$ remain constant, but $X_o$ increases, which results in the decrease of tray efficiency. But at some points, there is a slight fluctuation to this tendency. This is may be due to the unstable flow driven by centrifugal pump and thus setting a buffer tank in the outlet of centrifugal pump is advisable.

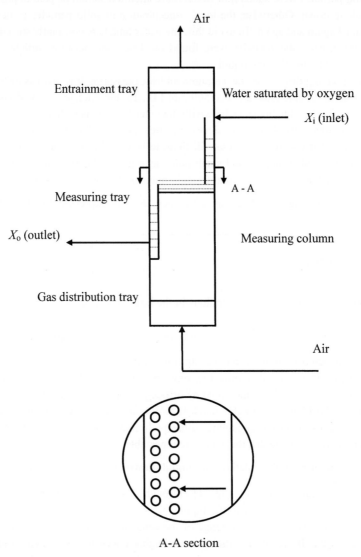

Fig. 10. Configuration of the measuring column and trays.

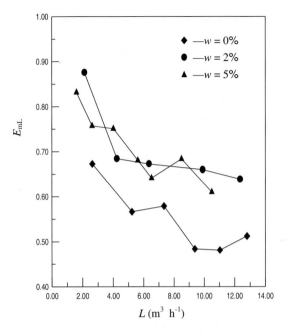

Fig. 11. Effect of liquid flowrate $L$ on tray efficiency $E_{mL}$ ($V = 359.74$ m$^3$ h$^{-1}$; weir height $h_w = 40$ mm).

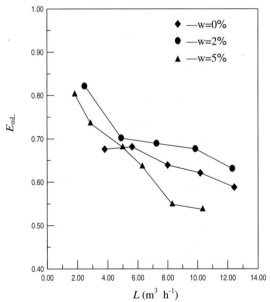

Fig. 12. Effect of liquid flowrate $L$ on tray efficiency $E_{mL}$ ($V = 440.59$ m$^3$ h$^{-1}$; weir height $h_w = 40$ mm).

194

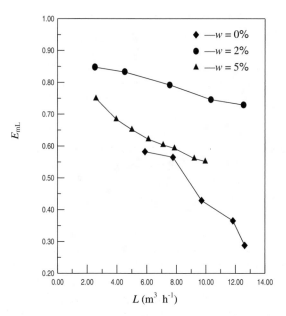

Fig. 13. Effect of liquid flowrate $L$ on tray efficiency $E_{mL}$ ($V = 359.74$ m$^3$ h$^{-1}$; weir height $h_w = 50$ mm).

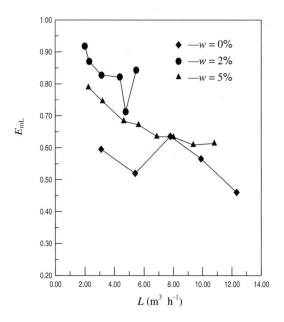

Fig. 14. Effect of liquid flowrate $L$ on tray efficiency $E_{mL}$ ($V = 440.59$ m$^3$ h$^{-1}$; weir height $h_w = 50$ mm).

(2) When raising the solid concentration, tray efficiency first increases and then decreases over a wide range of liquid flowrate. The reason may be that at low concentration the added solid particles promote the disturbance of interface between gas-liquid phases, and thus dwindle the liquid film thickness. A thinner liquid film thickness means a higher liquid-side mass transfer coefficient. Since the systems discussed in this work are just governed by liquid film, the resistance to mass transfer is mainly concentrated in the liquid film. So in terms of two-film theory tray efficiency will increase as the solid concentration rises. On the other hand, the presence of particles close to the interface may hold back the mass transfer of liquid phase and reduce the contact area between gas and liquid phases. At higher solid concentrations, this effect will be dominant. Consequently, the overall result is that tray efficiency no longer increases and even seems go down. This trend is consistent with the conclusion about the effect of solid particles on mass transfer coefficients put forwarded by Alper and Beenackers [78, 79].

It is known that the factors influencing tray efficiency are very complicated and involve physical properties of the systems, tray configuration, hydrodynamics etc. An institution of U.S.A., AIChE, has ever organized many researchers to study tray efficiency. Ultimately, a set of method recommended for estimation of tray efficiency of gas-liquid two phases flow, which is now called AIChE method [80], is put forward. Herein, this method is tested with the data obtained in this study for air-water system and then modified to predict tray efficiency of air-water-solid particles system.

Firstly, this method is used for predicting tray efficiency for the air-water system. It is found that the average relative deviation (ARD) of tray efficiency is 15.7% between this investigation and the values estimated by AIChE method for 21 points, and can satisfy the general engineering requirement. At the same time, it indicates that the obtained experimental data are reliable. The ARD is defined as

$$ARD = \sum_{i=1}^{n} \left| \frac{(E_{mL}^{\text{exp}} - E_{mL}^{\text{cal}})}{E_{mL}^{\text{exp}}} \right| / n \tag{3}$$

where $n$ is the number of data points.

Then, this method is extended to estimate the tray efficiency of the air-water-solid particles system by means of replacing the liquid physical properties of liquid with slurry density $\rho_m$ and viscosity $\mu_m$, which can be calculated from the following equations [81]:

$$\rho_m = w\rho_s + (1-w)\rho_L \tag{4}$$

$$\mu_m = \mu_L(1+4.5\phi) \tag{5}$$

where $\rho_L$ and $\mu_L$ are given by

$$\rho_L = \sum_{i=1}^{c} x_i \rho_i \tag{6}$$

$$\mu_L = \sum_{i=1}^{c} x_i \mu_i \tag{7}$$

However, in this case the ARD of tray efficiency is 13.0% between this investigation and the modified AIChE method for 50 points. It shows that the modified AIChE method can also be employed to estimate the tray efficiency of air-water-solid particles system.

Beenackers and Swaaaij [79] have studied the influence of solid particles on gas-liquid mass transfer and reported the method of calculating the volumetric liquid-side mass transfer coefficients at the gas-liquid interface $k_L a$, from which tray efficiency for gas-liquid-solid particles system can be deduced [80]. $k_L a$ is solved by

$$k_L a / (k_L a)_0 = (\mu_m / \mu_L)^{-0.42} \tag{8}$$

with index "0" indication of no solids present.

We select a three-phase system, e.g. propylene (gas) - benzene (liquid) - Silica gel (solid particles) to test the modified AIChE method. The calculated values of tray efficiency from the method of using $k_L a$ are compared with those from the modified AIChE method under the same conditions. The results are listed in Table 4 where $h_w$ = 40mm. The results of two methods are in good agreement with the difference less than 5%. It is shown that the modified AIChE method is suitable for either air-water-solid particles or organic gas-organic liquid-solid particles system.

However, it may be expected that the actual tray efficiency is greater than the calculated values coming from both the modified AIChE method and the method of using $k_L a$. It is known that $k_L a$ is the product of $k_L$ and $a$, in which $a$ (m$^2$ m$^{-3}$) is gas-liquid contact area and not changed whether or not chemical reaction takes place, but $k_L$ will increase if chemical reaction in the liquid phase is considered. The reason is that mass transfer is enhanced due to rapid chemical reaction, which can be described by enhancement factor $E_A$.

$E_A$ is defined as the ratio of $J_A$ (molar flux of component A, kmol m$^{-2}$ s$^{-1}$) with particles, to $J_A$ with the same but inert particles, in which $J_A$ is evaluated at the same overall driving force. Inert means that neither the particles nor components produced from the particles participate in the reaction as a reactant or a catalyst. Further, inert particles don't adsorb the gas phase component transported towards the bulk of the liquid phase, nor any other reactant or reaction product. Even so, it seems that the designed result is safe and conservative because we take on as the tray efficiency smaller than actual into consideration.

Table 4
Comparison of tray efficiency between the modified AIChE method and the method of using $k_L a$

| $w$ | $V / m^3 h^{-1}$ | $L / m^3 h^{-1}$ | $E_{mL}$ | |
|------|--------|--------|--------------------------|--------------------------|
| | | | The modified AIChE method | The method of using $k_L a$ |
| 0.02 | 440.59 | 2.48 | 0.8233 | 0.8278 |
| 0.02 | 440.59 | 4.90 | 0.6236 | 0.6293 |
| 0.02 | 440.59 | 7.25 | 0.5069 | 0.5123 |
| 0.02 | 440.59 | 9.85 | 0.4258 | 0.4308 |
| 0.02 | 440.59 | 12.30 | 0.3746 | 0.3792 |
| 0.05 | 440.59 | 1.82 | 0.8929 | 0.9010 |
| 0.05 | 440.59 | 2.86 | 0.7866 | 0.7984 |
| 0.05 | 440.59 | 5.00 | 0.6187 | 0.6324 |
| 0.05 | 440.59 | 6.30 | 0.5487 | 0.5623 |
| 0.05 | 440.59 | 8.30 | 0.4709 | 0.4837 |
| 0.05 | 440.59 | 10.34 | 0.4152 | 0.4271 |

One typical equation of expressing $E_A$ for rapid first-order reaction [82], constant gas phase concentration and no liquid through-flow of bulk is:

$$E_A = \frac{Ha_f}{\sinh(Ha_f)} \left\{ \cosh(Ha_f) - \frac{1}{\cosh(Ha_f) + \left[ \frac{k_r C_{si} \sinh(Ha_f)}{k_L a \bullet Ha_f} \right]} \right\} \tag{9}$$

where $C_{si}$ is the actual solids concentration in the gas-liquid film for mass transfer, described by Wimmers and Fortuin via a Freundlich isotherm still containing some empirical factors [82, 99], $k_r$ is the pseudo-homogeneous first-order reaction rate constant, and $Ha_f$ is the modified Hatta number defined by the following equation:

$$Ha_f = \frac{\sqrt{k_r C_{si} D_A}}{k_L} \tag{10}$$

So the application range of the AIChE method is extended to either for air-water-solid system or for no air-water-solid systems. By combining EQ model and tray efficiency model, much real information on the SCD column, which is necessary for the process design, is obtained. Moreover, tray efficiency can be used for checking the validity of NEQ stage model if the NEQ stage model is readily available, of course.

### 2.1.2. Hydrodynamics

It is evident that tray hydrodynamics (mainly referring to entrainment, leakage and pressure drop) plays an important role in guaranteeing normal operation of the column with gas-liquid-solid three-phase flow. Unfortunately, no research is yet available for them. The experiments of tray hydrodynamics can be done by use of the same equipment as measuring tray efficiency.

It is found that if the hydrodynamics of gas-liquid-solid three-phase is calculated with gas-liquid two-phase, the error will be great. The ARD between calculated and experimental values is 29.83% for 50 points of entrainment, and 10.25% for 50 points of pressure drop. So it is necessary to re-correlate the mathematics model of entrainment, pressure drop and leakage for gas-liquid-solid three-phase.

After correlating the experimental data, we obtain:

1. For entrainment $e_V$ on one tray,

$$e_v = \frac{0.2745}{\sigma}(\frac{w_G}{H_T - h_f})^{0.9524} \tag{11}$$

for $w = 0.02$, and

$$e_v = \frac{0.1867}{\sigma}(\frac{w_G}{H_T - h_f})^{1.2728} \tag{12}$$

for $w = 0.05$, where $w_G$ is vapor velocity through the bubble regime (m s$^{-1}$), $H_T$ is tray distance (m), and $h_f$ is bubble layer height in the tray (m).

2. For pressure drop on one tray,

$$\Delta p = \Delta p_c + \Delta p_l \tag{13}$$

$$\Delta p_c = 1000 \times \xi \frac{u_0^2}{2} \rho_V \tag{14}$$

$$\Delta p_l = \beta(h_w + h_{ow})\rho_m g \tag{15}$$

where $\Delta p_c$ is "dry" pressure drop (Pa), $\Delta p_l$ is "wet" pressure drop (Pa), $\xi = 2.1$ obtained by correlating the experiment data, $h_{ow}$ is liquid height over the weir (m), $u_0$ is vapor velocity through the hole (m s$^{-1}$), and $\beta$ is coefficient in the "wet" pressure drop.

In Eq. (15), the density of liquid-solid mixture $\rho_m$ instead of $\rho_L$ is adopted, and defined as

$$\rho_m = w\rho_s + (1-w)\rho_L. \tag{16}$$

Under this condition the ARD of pressure drop is just 5.23% for 50 points.

3. For leakage ratio $Q$ (kg kg$^{-1}$), it almost remains constant on the tray with raising the concentration of solid phase. The phenomena can be accounted for by the following equation:

$$Q = \frac{W}{L\rho_L} \tag{17}$$

$$\frac{W}{A_0} = 0.020Fr^{-1} - 0.030 \tag{18}$$

$$Fr = u_0 (\frac{\rho_G}{gh_L\rho_L})^{0.5} \tag{19}$$

where $W$ is leakage rate (kg h$^{-1}$), $A_0$ is hole area on the tray (m$^2$) and $h_L$ is liquid height on the tray (m).

Therefore, for the coupled reaction and distillation processes with gas-liquid-solid three-phase, we can safely obtain the information necessary for the equipment design on the tray column and extend to the industry, which have important practical values.

## 2.2. Alkylation of benzene and propylene

Here alkylation of benzene and propylene is studied as the example of FCD and SCD. The reaction between benzene and propylene is very popular in industry because the product, cumene, as a basic chemical material is mainly produced from this reaction. In order to accurately simulate the SCD process, it is important to know the intrinsic rate equations for this reaction system. It is known that the reactions include alkylation reaction that is irreversible and transalkylation reaction that is reversible. For transalkylation reaction, the reaction equilibrium constant can be referred in the reference [83]. But for alkylation reaction, the kinetic data for a given catalyst must be determined experimentally.

In general, the kinetic property of a chemical reaction, which is carried out in a distillation column, should be measured separately, e.g. in a fixed-bed reactor. Moreover, in the case of heterogeneously catalysed reactions, special attention should be paid to the mass and energy transport resistances inside the catalyst. Thus one has to study carefully both intrinsic and apparent kinetics of the reaction system. Herein, it is assumed that the intrinsic reaction rates are the same as global reaction rates because the employed catalyst particles in the SCD process are so tiny that mass and energy transport resistances inside and outside the catalyst particles can be neglected.

A modified $\beta$-zeolite catalyst, YSBH-01, developed from our laboratory, was used in the investigation. It has high activity and selectivity for the reaction. The kinetic performance of this catalyst was experimentally studied in a fixed-bed micro-reactor. On this basis, the equilibrium stage model (MESHR equations) is programmed. Moreover, a novel SCD process, in which alkylation and transalkylation reactions are carried out simultaneously in a single column, is suggested and the performance of SCD and FCD processes is discussed in terms of the EQ stage model.

### 2.2.1. Reaction kinetics

The involved reaction network includes alkylation and transalkylation reactions and is written as:

$$\text{B} + \text{P} \xrightarrow{\ k_1\ } \text{I}$$

$$\text{P} + \text{I} \xrightarrow{\ k_2\ } \text{D}$$

$$\text{B} + \text{D} \underset{k_4}{\overset{k_3}{\rightleftharpoons}} 2\text{I}$$

where B, P, I and D represent benzene, propylene, cumene and dialkylbenzene, respectively, and the first two are alkylation reaction, the latter transalkylation reaction.

The kinetic data of alkylation of benzene with propylene was determined in a continuous flow system (Fig. 15). The chemical materials, benzene and propylene, are all supplied by Yansan Petrochemical Corporation with over 99% purity.

The experimental system comprised three parts, a feed blending station for preparing the reaction mixture with different composition, an assembly of fixed-bed laboratory reactor-electric oven with a multi-channel temperature controller, and an off-line gas chromatogram (GC) and gas chromatogram-mass spectrometer (GC-MS).

The feed blending station consisted of two metering pumps for driving benzene and propylene, respectively, a nitrogen tube for sweeping and a mixer. The metering pumps were calibrated in advance. Feed composition was calculated based on the reading of the metering pumps and checked by gas chromatograph analysis.

The main parts of micro reactor-electric oven assembly are a fixed-bed micro-reactor of 8 mm i.d., 300 mm long and an electric oven of 800 W power. The temperature of reaction section was controlled with a temperature programmable controller and measured with a micro-thermal couple inserted in its center through a small jacket tube. The reaction section of the reactor was charged with 0.2 g catalyst of 80-100 mesh and some amount of quartz chips was loaded in both sides of the reaction section. Suitable particle size of the catalyst was determined by a preliminary experiment so as to eliminate the influence of internal diffusion. A back-pressure regulator was equipped downstream in the micro-reactor. The composition analysis system for feed and product consisted of an off-line gas chromatograph, SP 3420, equipped with a FID detector and an OV-101 capillary column (60 m long and 0.32 mm o.d.). The experiments were made under the following condition: the temperature 160 - 220 °C, the pressure 3.0 MPa, mole ratio of benzene to propylene 4 - 9, and weight hourly space velocity (WHSV) above 600 h$^{-1}$.

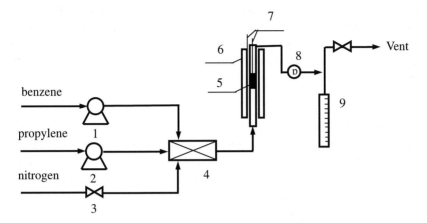

Fig. 15. The schematic diagram of fixed-bed experimental setup. 1,2 - benzene and propylene metering pump; 3 - adjusting valve; 4 - mixer; 5 - catalyst bed; 6 - temperature controller; 7 - thermocouples; 8 - pressure adjusting valve; 9 - metering vessel.

The rate equations for alkylation of benzene with propylene were obtained after correlating the experiment data by the Marquardt method and expressed as follows:

$$r_1 = 6.981 \times 10^5 \exp(-\frac{63742}{RT}) C_B^{0.96} C_P^{0.87} \tag{20}$$

$$r_2 = 4.239 \times 10^4 \exp(-\frac{79162}{RT}) C_I^{0.61} C_P^{0.92} \tag{21}$$

where $R$ is gas constant (8.314 kJ kmol$^{-1}$ K$^{-1}$), $T$ is temperature (K), and $C$ is mole concentration (kmol m$^{-3}$).

### 2.2.2. The improved SCD process

The original SCD process, in which alkylation reaction takes place in SCD column, but transalkylation reaction in another fixed-bed reactor, is shown in Fig. 16. In this work an improved SCD process, in which alkylation and transalkylation reactions take place simultaneously in a SCD column, is proposed and illustrated in Fig. 17.

In the improved SCD process, distillation and reaction always take place at the same time in the SCD column. But the upper section of this column is used for alkylation reaction and the lower section for transalkylation reaction. Transalkylation of DIPB having a high boiling point is more concentrated in the lower section than in the upper section of the column in terms of vapor-liquid equilibrium (VLE). The solid-liquid mixture leaving from the bottom is treated by a gas-liquid separator (i.e. a static sedimentary tank). The separated catalyst is recycled, and a part of them is regenerated from time to time. The separated liquid is the mixture of benzene, cumene and DIPB, which enter a benzene column (column2 ) and a cumene column (column 3 ) in series to remove unconverted benzene and side product DIPB. The ultimate product, cumene, is obtained from the top of column 3, and DIPB at the bottom of this column then flows into the lower section of column 1 for carrying out transalkylation reaction. However, in the original SCD process, only alkylation reaction is carried out in the SCD column. DIPB leaving from column 3 doesn't flow into column 1, but enters a fixed-bed reactor where only transalkylation reaction takes place. As a result, by using the improved SCD process, the equipment investment can be decreased and the process is simplified.

### 2.2.3. Simulation of the SCD column

It is evident that the SCD column is the key to the SCD process and should be paid more attention in the simulation of the SCD process. It is interesting to explore whether it is possible to integrate alkylation with transalkylation into a single SCD column by means of the simulation. In this study the equilibrium stage (EQ) model was established to simulate the SCD column. The equations that model equilibrium stages are known as the MESHR equations, into which the reaction terms including reaction rate equations and reaction heat equations are incorporated. The UNIFAC model is used for description of liquid phase nonideality, while the Virial equation of state is used for the vapor phase. The extended Antoine equation is used for calculation of the vapor pressure. The vapor enthalpy of every component is calculated by the modified Peng-Robinson (MPR) equation, and the sum of every vapor enthalpy multiplied by every mole fraction is the total vapor enthalpy. The liquid phase enthalpy is deduced from vapor phase enthalpy and evaporation heat. Thermodynamic data for this reaction system are taken from the sources [81, 84].

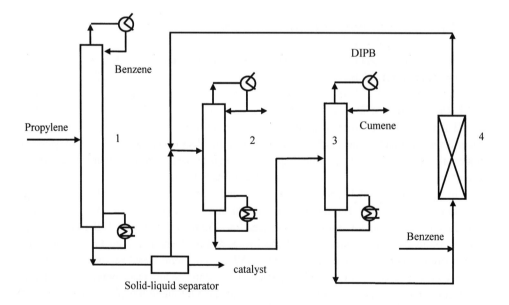

Fig. 16. Flowsheet of the original SCD process for producing cumene. 1 - SCD column; 2 - benzene column; 3 - cumene column; 4 - fixed-bed reactor.

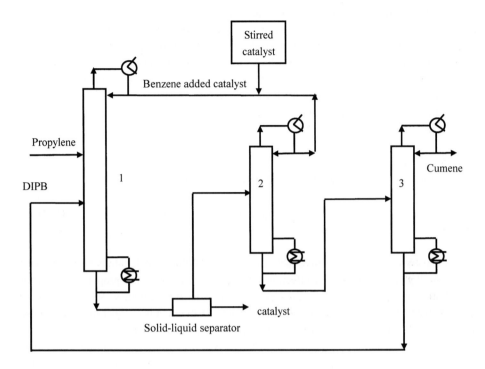

Fig. 17. Flowsheet of the improved SCD process for producing cumene. 1 - SCD column; 2 - benzene column; 3 - cumene column.

Under steady-state conditions all of the time derivatives in the MESHR equations are equal to zero. The modified relaxation method where the MESHR equations are written in unsteady-state form and are integrated numerically until the steady-state solution has been found, is used to solve the above equations.

It is known that transalkylation reaction is reversible, and in this work it is supposed to reach chemical equilibrium at each stage. The chemical equilibrium constant $K$ is written below [83]:

$$K = 6.52 \times 10^{-3} \exp(27240 / RT) \tag{22}$$

It can be seen from Figs. 16 and 17 that the distinguishable difference between a single alkylation reaction and simultaneous alkylation and transalkylation reactions in a column is that a DIPB stream is fed into the catalytic distillation column in the improved process. It is evident that the flowrate of DIPB fed into the column 1, $F1$, is an important parameter influencing the column design and operation. In the simulation of the column it is found that $F1$ can not be arbitrarily given and should be equal to the flowrate of DIPB leaving from the column, $F2$. Therefore, $F1$ should be determined in advance in a continuous production. In the simulation $F1$ is constantly adjusted until it is equal to $F2$, and thus $F1$ is determined.

It is assumed that the SCD column is divided into 41 theoretical stages, and the condenser is the stage 41 and the reboiler the stage 1. The operating pressure is 700 kPa and reflux rate 960 kmol h$^{-1}$, while the feed of benzene and propylene are 100 kmol h$^{-1}$ and 50 kmol h$^{-1}$, respectively. The relation of $F1$ and $F2$ in the SCD column is shown in Fig. 18, and the point where $F1 = F2$ can be found, which means that the process has reached steady-state, and the number of steady-states is only one. The computation has been performed on a PC (Pentium 4 1.5G). Each case needs about ten minutes for simulation.

In the case that $F1 = F2$, temperature and composition of the bottom stream in the SCD column are listed in Table 5.

Table 5
Temperature and molar composition of the bottom stream in the SCD column

| $T / K$ | Molar composition | | | |
|---|---|---|---|---|
| | benzene | propylene | cumene | DIPB |
| 514.5 | 0.0566 | 0.00 | 0.3508 | 0.5926 |

It can be seen from the above table that the SCD column with simultaneous alkylation and transalkylation reactions is effective for the reaction with final propylene conversion approximately 100%. Moreover, cumene selectivity is approximately 100% because the amount of DIPB feeding into and leaving from the SCD column is equal to each other and there is no DIPB accumulation. But it is impossible to reach this selectivity for the SCD column with only alkylation reaction because DIPB will be produced and thus both cumene and DIPB are as resultants. In order to get a higher selectivity, a fixed-bed reactor for transalkylation reaction in the original SCD process is generally set to convert the DIPB formed in alkylation reaction to cumene.

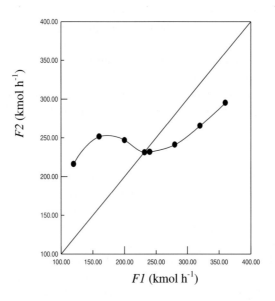

Fig. 18. The relation of $F1$ and $F2$ in the SCD column.

## 2.2.4. Comparison of the SCD and FCD columns

On the basis of the improved SCD process, it is easy to go a farther step to deduce that a fixed-bed catalytic distillation (FCD) column is also suitable for simultaneous alkylation and transalkylation reactions. It is interesting to compare these two processes. The FCD process is shown in Fig. 19, which is similar to the improved SCD process.

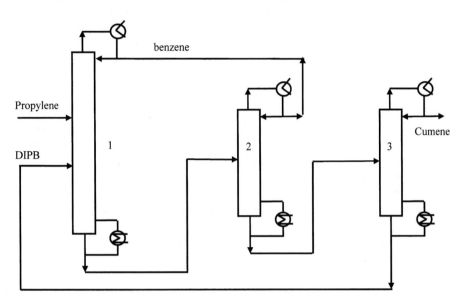

Fig. 19. Flowsheet of the FCD process for producing cumene. 1 - FCD column; 2 - benzene column; 3 - cumene column.

The FCD column is composed of three sections, named from the top to the bottom as the first reaction section, stripping section and the second reaction section. Alkylation reaction is carried out in the first reaction section, while transalkylation in the second reaction section. The mathematical model for simulating the FCD column is identical to the SCD column's model and the same intrinsic reaction rate equations are used to simulate the FCD column. Moreover, it is assumed that both the amount of employed catalyst and the column operating condition are the same in the FCD and SCD columns.

The relation of $F1$ (the flowrate of DIPB fed into the column) and $F2$ (the flowrate of DIPB leaving from the column) in the FCD column is illustrated in Fig. 20, and the point where $F1 = F2$ can be found.

In the case that $F1 = F2$, temperature and composition of the bottom stream in the FCD column are listed in Table 6.

It is found from Tables 5 and 6 that cumene concentration at the bottom is a little higher in the improved SCD process than in the FCD process. It is shown that the improved SCD, besides its unique advantages as mentioned in the introduction, is more attractive than the FCD process in view of increasing cumene yield. The reason may be that the product of cumene is a light component compared with DIPB, and is always produced along the SCD column, not just at the stripping section as in the FCD column. So it indicates that it is better to disperse the catalyst evenly along the whole column, and not to accumulate the catalyst only in some parts of the column for simultaneous alkylation and trans-alkyation reaction in a single column.

As a matter of fact, in the FCD process catalyst is generally used as structured-packing elements in the column, and thus the mass and energy transfer resistances can't be neglected. Therefore, the global reaction rates are slower than intrinsic ones. If this factor is further considered, cumene concentration at the bottom in the improved SCD process should be much higher than that in the FCD process.

Anyway, the model simulation shows that the SCD column, besides its unique advantages, is more attractive than the FCD column in view of increasing cumene yield.

It should be mentioned that although here the improved SCD process is only used for producing cumene with benzene and propylene, it isn't difficult to extend the model for production of such compounds as ethylbenzene with benzene and ethylene and so on.

Table 6
Temperature and molar composition of the bottom stream in the FCD column

| $T / K$ | Molar composition | | | |
|---------|---------|-----------|--------|--------|
| | Benzene | propylene | cumene | DIPB |
| 515.1 | 0.0498 | 0.00 | 0.3293 | 0.6209 |

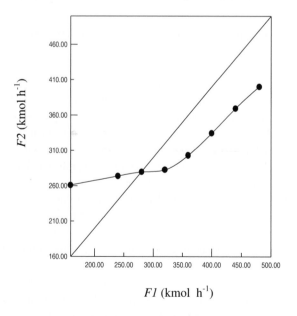

Fig. 20. The relation of $F1$ and $F2$ in the FCD column.

*2.2.5. Stead-state analysis*

Multiple steady-state (MSS) analysis in conventional distillation has been known from simulation and theoretical studies dating back to the 1970s and has been a topic of considerable interest in the FCD for synthesizing MTBE and TAME, etc. The analysis method of MSS in CD is similar as in azeotropic distillation (see chapter 3), and continuation path search under the same input conditions is commonly adopted. However, herein, we are concerned with the steady-state analysis under different input conditions. The example of alkylation of benzene with propylene is emphasized. The EQ stage model is used for the steady-state analysis of the original and improved SCD processes, respectively.

1. Steady-state analysis of the original SCD column

In the original SCD column, as shown in Fig. 21, the mixture of benzene and catalyst are evenly distributed by the stirrer and then enter the column. By varying the location to which benzene and catalyst are fed, the change of the bottom composition of benzene and cumene is investigated. The operating condition is given as follows:

(1) The SCD column has 41 theoretical stages. The total condenser is the stage 41 and the reboiler the stage 1.

(2) The feed stage of propylene is No. 20.

(3) The operating pressure is 700 kPa because under the corresponding temperature the catalyst is highly active.

(4) The reflux rate is fixed at 960 kmol h$^{-1}$ to ensure high ratio of benzene to propylene in the liquid phase and prevent polymerization of propylene.

(5) The other operating parameters are marked in Fig. 21.

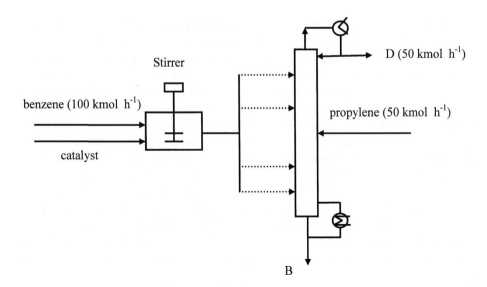

Fig. 21. The original SCD column used for the steady-state analysis.

The computation has been performed on a PC (Pentium 4 1.5G). Each case needs a few minutes for simulation. The influence of the feed location of benzene and catalyst on the bottom composition of benzene and cumene is illustrated in Fig. 22.

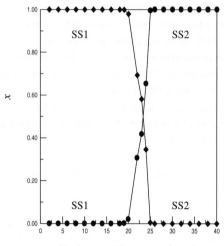

Feed stage of benzene and catalyst

Fig. 22. Two steady-states of the original SCD column, □- the bottom molar fraction of benzene; ●- the bottom molar fraction of cumene.

It can be seen from Fig. 22 that in the original SCD column there are two steady-states, that is, SS1 and SS2. When the mixture of benzene and catalyst are fed in the vicinity of the No. 22 stage, the composition change of benzene and cumene is very abrupt. But apart from this range, the composition of benzene and cumene is very stable. In SS1 the bottom molar fraction of cumene (ultimate product) is close to zero. On the contrary, in SS2 high purity of cumene can be obtained. The reason may be that in SS1, the mixture of benzene and catalyst as a heavy component is fed into the low section of the column. However, propylene as a light component is fed into the middle section of the column. It is known that the difference of normal boiling point of benzene and propylene is very high, up to 127.9 □. Therefore, benzene and propylene can't contact effectively under the distillation condition and thus alkylation reaction doesn't nearly take place, which results in the low bottom composition of cumene. In SS2, the mixture of benzene and catalyst as a heavy component is fed into the upper section of the column, and propylene as a light component is fed into the middle section of the column. Therefore, benzene and propylene can contact effectively under the distillation condition and thus alkylation reaction takes place, which results in the high bottom composition of cumene.

Since the bottom composition of cumene is higher in SS2, it is advisable to select this steady-state to operate. But it should be cautious that the feed stage of benzene and catalyst is better to be far away from No. 22 stage because it is a sensitive stage.

2. Steady-state analysis of the improved SCD column

As mentioned before, if $F1$ (the flowrate of DIPB fed into the SCD column) is equal to $F2$ (the flowrate of DIPB leaving from the SCD column) in the improved SCD column, then the process reaches steady-state.

In the simulation the operating condition is given as follows:

(1) The SCD column has 41 theoretical stages. The total condenser is the stage 41 and the reboiler the stage 1.

(2) The feed stage of propylene is No. 28.

(3) The feed stage of propylene is No. 20.

(4) The operating pressure is 700 kPa because under the corresponding temperature the catalyst is highly active.

(5) The reflux rate is fixed at 960 kmol·h$^{-1}$ to ensure high ratio of benzene to propylene in the liquid phase and prevent polymerization of propylene.

(6) The feed molar ratio of benzene to propylene is changeable.

The computation has been performed on a PC (Pentium 4 1.5G). Each case needs a few minutes for simulation. The relation of $F1$ and $F2$ at different feed molar ratio of benzene to propylene is illustrated in Fig. 23.

It can be seen from Fig. 23 that when the feed molar ratio of benzene to propylene is below or equal to 3, the curves of $F1$ and $F2$ intersect the diagonal at one point, which means that the process reaches steady-state. That is to say, in this case there is only one steady-state in the improved SCD column. However, when the ratio is higher than 3, the curves of $F1$ and $F2$ don't intersect the diagonal and there is no steady-state found in the improved SCD column.

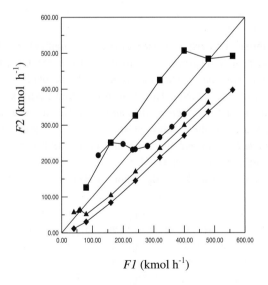

Fig. 23. The relation of $F1$ and $F2$ at different feed molar ratio of benzene to propylene, ■- feed molar ratio 1 : 1; ●- feed molar ratio 2 : 1; ▲- feed molar ratio 3 : 1; ◆- feed molar ratio 4 : 1.

In addition, it should be noted that in the design and optimization of the improved SCD column the feed molar ratio of benzene to propylene must be greater than 1 in order to prevent polymerization of propylene and inactivity of catalyst. So in the improved SCD process the feed molar ratio of benzene to propylene is better to be within the range of 1 - 3 because in this case the process can reach steady-state operation and prevent polymerization of propylene.

**2.3. Alkylation of benzene and 1-dodecene**

Linear alkylation (LAB) is an important intermediate in the detergent industry, and is usually made from alkylation of benzene and olefin with long carbon chains from C10 to C14. The catalysts for alkylation reaction are generally aluminium trichloride (AlCl3), hydrofluoric acid (HF) and solid acids. But it is known that AlCl3 and HF are to some extents hazardous to environment. Many efforts have been made to replace it with environmentally safer solid catalysts such as H-ZSM-5, H-ZSM-12, HY and lays [85, 86].

However, the common solid catalysts have the drawbacks, that is, low conversion, low selectivity and high ratio of benzene and olefin. In order to overcome those, a high-efficiency heteropoly acid catalyst which is also environment-friendly is developed [69, 70]. This catalyst is made up of 77%wt silica gel and 28%wt tungstophosphoric acid (HPW), and of high activity and selectivity for alkylation of benzene and 1-dodecene. But the kinetics equations for the reaction over any catalyst were rarely reported in open references.

Due to the unique advantages of the SCD column, it is very straightforward to combine this technology with the synthesis of C12 alkylbenzene with benzene and 1-dodecene. The kinetics equations over the heteropoly acid catalyst are firstly measured in a fixed-bed micro-reactor.

*2.3.1. Kinetic equations*

In general, the kinetic property of a chemical reaction that is carried out in a distillation column should be measured separately, e.g. in a fixed-bed reactor. Moreover, in case of heterogeneous catalytic reaction, special attention should be paid to the mass and energy transport resistance inside the catalyst. Thus both micro- and macro-kinetics of the reaction system must be studied carefully. Herein, it is assumed that intrinsic reaction rates are the same as global rates because the catalyst used in the SCD column is tiny particles and situated in strong turbulence so that the mass and energy transport resistances inside and outside the catalyst particles can be neglected.

The involved reaction network for alkylation of benzene and 1-dodecene is written as [87]:

$$B + D \xrightarrow{k_1} 2\text{-ph}$$

$$B + D \xrightarrow{k_2} 3\text{-ph}$$

$$B + D \xrightarrow{k_3} 4\text{-ph}$$

$$\text{B} + \text{D} \xrightarrow{\text{k}_4} \text{5-ph}$$

$$\text{B} + \text{D} \xrightarrow{\text{k}_5} \text{6-ph}$$

where B, D and $i$-ph ($i$ = 2, 3, 4, 5, 6) represent benzene, 1-dodecene and five different positional alkylbenzene (phenyl-dodecane isomers), respectively. It is possible that some heavier byproducts, such as multi-alkylbenzene, also form. But it is evident that those are too small to be neglected.

The kinetic equations for alkylation of benzene and 1-dodecene were determined in a fixed-bed micro-reactor as shown in Fig. 15.

The reaction was performed at temperatures between 358.15 K to 398.15 K. The series of experiments was arranged according to the principle of mathematics statistics. The pre-experiments were done to investigate the influence of external and internal diffusions. It was found that when the average diameter of catalyst particles was less than 0.21 mm (60-80 mesh) and the liquid weight hourly space velocity (WHSV) was larger than 20 $h^{-1}$, the influence of external and internal diffusions may be thought to preclude.

The kinetic equations for alkylation of benzene and 1-dodecene were obtained after correlating the kinetic data by the Marquardt method and written as follows:

$$r_{2-ph} = 402.783 \times \exp(-\frac{45734}{RT})C_B C_D \tag{23}$$

$$r_{3-ph} = 743.969 \times \exp(-\frac{46649}{RT})C_B C_D \tag{24}$$

$$r_{3-ph} = 792.109 \times \exp(-\frac{48080}{RT})C_B C_D \tag{25}$$

$$r_{4-ph} = 892.208 \times \exp(-\frac{48080}{RT})C_B C_D \tag{26}$$

$$r_{5-ph} = 217.870 \times \exp(-\frac{48225}{RT})C_B C_D \tag{27}$$

where $R$ is gas constant (8.314 J $mol^{-1}$ $K^{-1}$), $T$ is temperature (K), $C$ is mole concentration (kmol $m^{-3}$) in the liquid phase, and $r$ is the reaction rate for specified component (kmol $m^{-3}$ $kg^{-1}$ catalyst). The kinetic equations were verified by $F$-factor test, and thought to be reliable because it was consistent with the $F$-factor rules.

### 2.3.2. Comparison of experimental and simulation results

An experimental apparatus for producing C12 alkylbenzene, as shown in Fig. 24, was set up in laboratory. The SCD column comprised three parts, i.e. rectifying, reactive and stripping sections. The mixture of benzene, 1-dodecane and 1-dodecene was blend with catalyst in a mixing tank, and then into the top of the reactive section. 1-dodecane was added to dilute the concentration of 1-dodecene and prevent it from polymerization. The catalyst was evenly distributed along the reactive section. The solid-liquid mixture leaving from the bottom of the reactive section was separated by a solid-liquid separator (high-speed shear dispersing machine). The resulting catalyst was sent to be regenerated and then recycled. The resulting liquid was the mixture of benzene, 1-dodecane, 1-dodecene and $i$-ph ($i$ = 2 - 6) which entered into the stripping section.

The SCD column had 23 trays and each was regarded as one model stage for both the EQ and NEQ stage models because the accurate prediction of tray efficiency was very difficult in the case of simultaneous multi-component separation and reaction. Therefore, the column was divided into 25 model stages in which the rectifying section had 5 stages, the reactive section 10 stages and the stripping section 8 stages. The reboiler was stage 1 and the total condenser stage 25. Some structural and operating parameters are listed in Table 7. In the experiment, the column pressure was adjustable, at 101.3, 131.7 and 162.1 kPa respectively, while other operating parameters remained constant. In the simulation of the EQ and NEQ stage models, for simplifying the computer program, $i$-ph ($i$ = 2, 3, 4, 5, 6) were taken on as one component because they had similar thermodynamic and physical properties.

Figures 25, 26 and 27 show the temperature profiles along the SCD column at pressures 101.3, 131.7 and 162.1 kPa, respectively. It can be seen that the calculated values by the NEQ stage model are much closer to the experimental values than by the EQ stage model. The reason may be that in actual operation, trays rarely, if ever, operate at equilibrium. We know that for conventional distillation operation with vapor-liquid two-phase, tray efficiency can't reach 100% in most cases. However, in the reactive section of the SCD column it is vapor-liquid-solid three-phase. The addition of solid particles will further decrease tray efficiency in this case because the presence of solid particles close to the interface may hold back the mass transfer of vapor and liquid phases and reduce the contact area of two phases [77, 78].

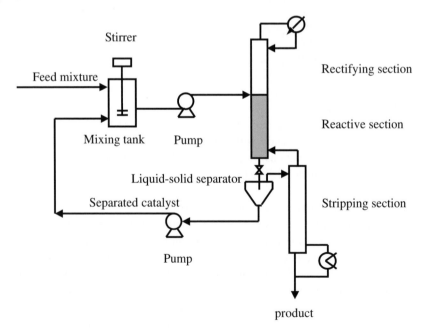

Fig. 24. Experimental flowsheet for producing C12 alkylbenzene.

Table 7
The structural and operating parameters of the SCD column

| Structural parameters | | Operating parameters | |
| --- | --- | --- | --- |
| Type of tray | Sieve | Pressure, kPa | 101.3, 131.7, 162.1 |
| Column diameter (m) | 0.05 | Reflux rate, g h$^{-1}$ | 700 |
| Total tray area (m$^2$) | 0.002 | Feed rate, g h$^{-1}$ | |
| Number of liquid flow passes | 1 | Benzene | 35 |
| Tray spacing (m) | 0.1 | 1-dodecene | 50.4 |
| Liquid flow path length (m) | 0.04 | 1-dodecane | 749.6 |
| Active area (m$^2$) | 0.002 | The concentration of catalyst in the feed mixture, g ml$^{-1}$ | 0.05 |
| Downcomer type | Round | | |
| Downcomer clearance (m) | 0.04 | | |
| Weir length (m) | 0.03 | | |
| Weir height (m) | 0.03 | | |

So the prediction of the EQ stage model isn't accurate in the case of multi-component separation and chemical reaction with vapor-liquid-solid three-phase. It is the rate of mass and heat transfer, and not the equilibrium that limits the separation.

The difference of calculated values between the EQ and NEQ stage models can also be brought out in the composition profiles along the column, which are illustrated in Figs. 28 and 29. Here the studied system is somewhat special and the boiling-point difference of the key light (benzene) and heavy (1-dodecene) components is very high, up to 133.2 ℃. This means that these two components are easy to be separated in very few equilibrium stages. So there is an abrupt change of benzene composition in the feed tray for the EQ stage model, as shown in Fig. 28. In the reactive and stripping sections of the SCD column, the benzene composition is almost zero, which leads to a higher temperature and fewer product of $i$-ph ($i$ = 2, 3, 4, 5, 6) than actual operation. However, the change of benzene composition in the feed tray in Fig. 29 for the NEQ stage model is somewhat smooth. In many places of the reactive and stripping sections, the benzene composition is away from zero in the NEQ stage model, which leads to a lower temperature and more products of $i$-ph ($i$ = 2, 3, 4, 5, 6) than in the EQ stage model. Therefore, the calculated values by the NEQ stage model are more in agreement with the experimental values.

Because the deviation between the experimental and calculated values is large for the EQ stage model, it is deduced that tray efficiency under the operation condition listed in Table 7 is low. That is to say, the assumption of EQ stage is far away from the actual. But for the NEQ stage model there is a little fluctuation of temperatures in the vicinity of the feed tray, which may be due to the great influence of the feed mixture on the mass and heat transfer rates and the limitation of accuracy of thermodynamic equations used in the NEQ stage model.

A comparison of the product composition at the bottom between the experimental and calculated values is also made in Table 8 where $x$ is the mole fraction in the liquid phase. Table 8 shows that the concentrations of both benzene and 1-dodecene at the bottom are very low. The fact of no benzene and 1-dodecene at the bottom proves that the SCD column is effective for this alkylation reaction. Furthermore, it can be seen from Table 8 that the calculated values by the NEQ stage model are more approximate to the experimental values than by the EQ stage model. So it farther proves that the NEQ stage model is reliable and can be used for the design and optimization of the SCD column.

The NEQ stage model will be preferred for the simulation of a tray reactive distillation column to the EQ stage model. However, as mentioned above, a close agreement between the predictions of EQ and NEQ stage models can be found if the tray efficiency is accurately predicted for the EQ stage model. But in the case that tray efficiency is difficult to obtain, especially for the vapor-liquid-solid three-phase system, the NEQ stage model can play a role.

In the experiment, the conversion of 1-dodecene 100% and selectivity of 1-dodecene 100% are obtained. Moreover, the weight concentration of 2-ph in the i-ph is up to 35%, which is higher than 25% reported in the fixed-bed reactor [86]. So the SCD column is positive to the selectivity of 2-ph, which is favorable in industry.

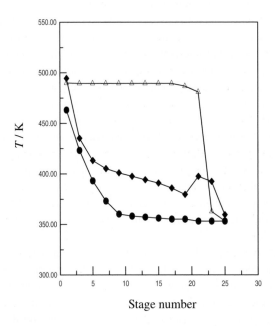

Fig. 25. Comparison of calculated and experimental values at 101.3 kPa; the tray is numbered from the bottom to the top; ● — Experimental values; ◆ — Calculated values by the NEQ stage model; △ — Calculated values by the EQ stage model.

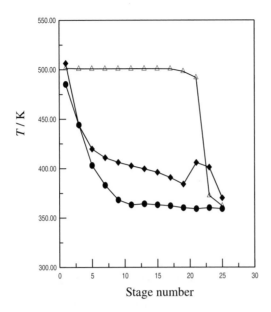

Fig. 26. Comparison of calculated and experimental values at 131.7 kPa; the tray is numbered from the bottom to the top; ● — Experimental values; ◆ — Calculated values by the NEQ stage model; △ — Calculated values by the EQ stage model.

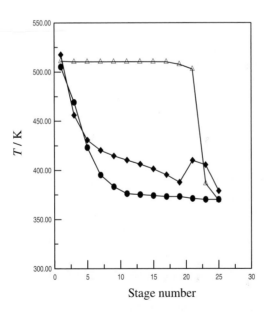

Fig. 27. Comparison of calculated and experimental values at 162.1 kPa; the tray is numbered from the bottom to the top; ● — Experimental values; ◆ — Calculated values by the NEQ stage model; △ — Calculated values by the EQ stage model.

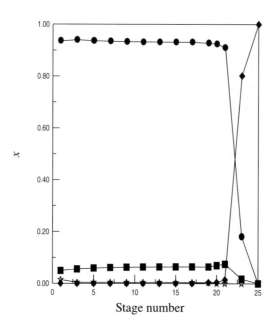

Fig. 28. Composition profile of the SCD column for the EQ stage model; the tray is numbered from the bottom to the top; ◆ —benzene; ■ — 1-dodecene; ●— 1-dodecane; ☆ — *i*-ph.

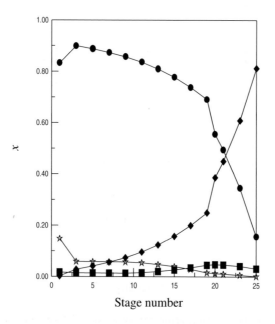

Fig. 29. Composition profile of the SCD column for the NEQ stage model; the tray is numbered from the bottom to the top; ◆ —benzene; ■ — 1-dodecene; ●— 1-dodecane; ☆ — *i*-ph.

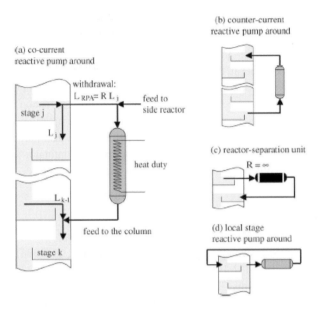

Fig. 30. Four configurations of the combination of side-reactor or external reactor and distillation column; adapted from the source [19].

Table 8
The product composition at the bottom of the SCD column at pressure 101.3 kPa

|  | $x^{exp}$ | $x^{cal}$ | |
|---|---|---|---|
|  |  | EQ stage model | NEQ stage model |
| benzene | 0.0 | 0.0 | 0.0036 |
| 1-dodecene | 0.0 | 0.0497 | 0.0113 |
| 1-dodecane | 0.8926 | 0.9361 | 0.8850 |
| C12 alkylbenzene | 0.1074 | 0.0142 | 0.1001 |

It is concluded that the NEQ stage model is more accurate than the EQ stage model. The reason may be that in actual operation, trays rarely, if ever, operate at equilibrium, especially when the solid particles are added in the SCD column. So it is advisable to select the NEQ stage model, or the combination of the EQ stage model and tray efficiency, for the design and optimization of the SCD column, which will help the future scale-up of the SCD techniques in industry.

In FCD and SCD, the requirements of high liquid or catalyst holdup aren't consistent with the requirements of good in situ separation. Apart from this, a new-type side-reactor or external reactor concept is proposed [19]. It may belong to the scope of catalytic distillation. In the side-reactor concept the reactor feed is withdrawn from the distillation column and the reactor effluent is returned back to the same column. The side reactor could be a conventional catalytic-packed bed reactor operating in liquid phase and therefore, there are no hardware design problems or conflicting requirements with regard to in situ separation. Furthermore, the reaction conditions within the side reactor (e.g. temperature) can be adjusted independently of those prevailing in the distillation column by appropriate heat exchange.

Four configurations of linking the side reactors to the distillation column can be designed, as shown in Fig. 30. In order to meet the process requirements of conversion, more than one side reactor may be necessary. Apparently, the determination of the optimum number of side reactors, along with the liquid draw-off and feed-back points to the distillation column, need careful attention and consideration. An algorithm to determine an optimum configuration of the side-reactor concept in order to maximize conversion has been developed.

## REFERENCES

[1] R. Taylor and R. Krishna, Chem. Eng. Sci., 55 (2000) 5183-5229.

[2] E. Perez-cisneros, M. Schenk and R. Gani, Comput. Chem. Eng., 20 (1996) S267-S272.

[3] K. Sundmacher, G. Uhde and U. Hoffmann, Chem. Eng. Sci., 54 (1999) 2839-2847.

[4] A.V. Solokhin and S.A. Blagov, Chem. Eng. Sci., 51(11) (1996) 2559-2564.

[5] B. Saha, S.P. Chopade and S.M. Mahajani, Catal. Today, 60 (2000) 147-157.

[6] S. Hauan, T. Hertzberg and K.M. Lien, Comput. Chem. Eng., 19 (1995) S327-S332.

[7] S. Hauan, T. Hertzberg and K.M. Lien, Comput. Chem. Eng., 21(10) (1997) 1117-1124.

[8] J. Bao, B. Gao, X. Wu, M. Yoshimoto and K. Nakao, Chem. Eng. J., 90 (2002) 253-266.

[9] S. Schrans, S. de Wolf and R. Baur, Comput. Chem. Eng., 20 (1996) S1619-S1624.

[10] S.J. Wang, D.S.H. Wong and E.K.Lee, J. Process Control, 13 (2003) 503-515.

[11] A. Hoffmann, C. Noeres and A. Gorak, Chem. Eng. Process., 43 (2004) 383-395.

[12] A. Tuchlenski, A. Beckmann, D. Reusch, R. Dussel, U. Weidlich and R. Janowsky, Chem. Eng. Sci., 56 (2001) 387-394.

[13] C. Noeres, E.Y. Kenig and A. Gorak, Chem. Eng. Process., 42 (2003) 157-178.

[14] H. Subawalla and J.R. Fair, Ind. Eng. Chem. Res., 38 (1999) 3696-3709.

[15] J.L. Degarmo, V.N. Parulekar and V. Pinjala, Chem. Eng. Prog., 88 (1992) 43-50.

[16] J.C. Gonzalez, H. Subawalla and J.R. Fair, Ind. Eng. Chem. Res., 36 (1997) 3845-3853.

[17] K. Sundmacher, G. Uhde and U. Hoffmann, Chem. Eng. Sci., 54 (1999) 2839-2847.

[18] A.C. Dimian, F. Omota and A. Bliek, Chem. Eng. Process., 43 (2004) 411-420.

[19] R. Baur and R. Krishna, Chem. Eng. Process., 43 (2004) 435-445.

[20] Y. Avramenko, L. Nystrom and A. Kraslawski, Comput. Chem. Eng., 28 (2004) 37-44.

[21] M. Kloker, E.Y. Kenig, A. Gorak, A.P. Markusse, G. Kwant and P. Moritz, Chem. Eng. Process., 43 (2004) 791-801.

[22] J. Peng, T.F. Edgar and R.B. Eldridge, Chem. Eng. Sci., 58 (2003) 2671-2680.

[23] J.F. Knifton, P.R. Anantaneni, P.E. Dai and M.E. Stockton, Catal. Today, 79-80 (2003) 77-82.

[24] R. Baur and R. Krishna, Chem. Eng. Process., 41 (2002) 445-462.

[25] A. Beckmann, F. Nierlich, T. Popken, D. Reusch, C.V. Scala and A. Tuchlenski, Chem. Eng. Sci., 57 (2002) 1525-1530.

[26] Z. Qi, A. Kolah and K. Sundmacher, Chem. Eng. Sci., 57 (2002) 163-178.

[27] E.Y. Kenig, H. Bader, A. Gorak, B. Bebling, T. Adrian and H. Schoenmakers, Chem. Eng. Sci., 56 (2001) 6185-6193.

[28] A. Kolodziej, M. Jaroszynski, A. Hoffmann and A. Gorak, Catal. Today, 69 (2001) 75-85.

[29] R. Baur, R. Taylor and R. Krishna, Catal. Today, 66 (2001) 225-232.

[30] J.M. van Baten, J. Ellenberger and R. Krishna, Catal. Today, 66 (2001) 233-240.

[31] S. Melles, J. Grievink and S.M. Schrans, Chem. Eng. Sci., 55 (2000) 2089-2097.

[32] S.P. Chopade, React. Funct. Polym., 42 (1999) 201-212.

[33] F. Bezzo, A. Bertucco, A. Forlin and M. Barolo, Sep. Purif. Technol., 16 (1999) 251-260.

[34] E. Kenig, K. Jakobsson, P. Banik, J. Aittamaa, A. Gorak, M. Koskinen and P. Wettmann, Chem. Eng. Sci., 54 (1999) 1347-1352.

[35] L.A. Smith and M.N. Huddleston, Hydro. Proc., No. 3 (1982) 121-123.

[36] K.D. Mohl, A. Kienle, E.D. Gilles, P. Rapmund, K. Sundmacher and U. Hoffmann, Chem. Eng. Sci., 54 (1999) 1029-1043.

[37] H.G. Schoenmakers and B. Bessling, Chem. Eng. Process., 42 (2003) 145-155.

[38] Z. Qi, K. Sundmacher, E. Stein and A. Kienle, Sep. Purif. Technol., 26 (2002) 147-163.

[39] J. Xiao, J. Liu, X. Jiang and Z. Zhang, Chem. Eng. Sci., 56 (2001) 6553-6562.

[40] K. Sundmacher and Z. Qi, Chem. Eng. Process., 42 (2003) 191-200.

[41] C. Noeres, A. Hoffmann and A. Gorak, Chem. Eng. Sci., 57 (2002) 1545-1549.

[42] S. Giessler, R.Y. Danilov, R.Y. Pisarenko, L.A. Serafimov, S. Hasebe and I. Hashimoto, Comput. Chem. Eng., 25 (2001) 49-60.

[43] R. Baur, R. Taylor and R. Krishna, Chem. Eng. Sci., 55 (2000) 6139-6154.

[44] Y.S. Nie, J.R. Zhang and Y.C. Du, Petrochemical Technology (China), 29 (2000) 105-110.

[45] A.V. Solokhin and S.A. Blagov, Chem. Eng. Sci., 51 (1996) 2559-2564.

[46] K. Sundmacher and U. Hoffmann, U., Chem. Eng. Sci., 51 (1996) 2359-2368.

[47] J. Tanskanen and V.J. Pohjola, Comput. Chem. Eng., 24 (2000) 81-88.

[48] Y.X. Zhang and X. Xu, Trans IchemE, 70 (1992) 465-470.

[49] V.H. Agreda, L.R.Partin and W.H. Heise, Chem. Eng. Prog., 86 (1990) 40-46.

[50] A. Gorak and A. Hoffmann, Am. Inst. Chem. Eng. J., 47 (2001) 1067-1076.

[51] H. Komatsu and C.D. Holland, J. Chem. Eng. Japan, 4 (1977) 292-297.

[52] H. Hartig and H. Regner, Chem. Eng. Tech., 43 (1971) 1001-1007.

[53] B. Davies and G.V. Jeffreys, Trans. Inst. Chem. Eng., 51 (1973) 275-280.

[54] H.P. Luo and W.D. Xiao, Chem. Eng. Sci., 56 (2001) 403-410.

[55] Y. Fuchigami, J. Chem. Eng. Japan, 23 (1990) 354-358.

[56] K. Sundmacher and U. Hoffmann, Chem. Eng. Technol., 16 (1993) 279-289.

[57] C. Thiel, Ph.D. thesis, Technical University of Clausthal, Germany, 1997.

[58] C. Oost, Ph.D. thesis, Technical University of Clausthal, Germany, 1995.

[59] J.D. Shoemaker and E.M. Jones, Hydro. Proc., 66 (1987) 57-58.

[60] G.G. Podrebarac, F.T.T. Ng and G.L. Rempel, Chem. Eng. Sci., 53 (1998) 1067-1075.

[61] S. Ung and M.F. Doherty, Chem. Eng. Sci., 50 (1995) 23-48.

[62] D. Muller, J.P. Schafer and H.J. Leimkuhler, Proceedings of the GVC, DECHEMA and

EFCE Meeting on "Distillation, Absorption and Extraction", Bamberg, Germany, 2001.

[63] A.R. Ciric and P. Miao, Ind. Eng. Chem. Res., 33 (1994) 2738-2748.

[64] D.J. Belson, Ind. Eng. Chem. Res., 29 (1990) 1562-1565.

[65] B.B. Han, Y. Wang, S.J. Han and Y. Jin, Chemical Reaction Engineering and Technology (China), 13 (1997) 190-193.

[66] C.Q. He, Z.M.Liu and Y.Z. Zhang, Journal of fuel Chemistry and Technology (China), 27 (1999) 203-208.

[67] J.Q. Fu, M.J. Zhao and J.R. Zhang, Petrochemical Engineering (China), 27 (1998) 309-313.

[68] Z.Z. Zhang, K. Zhang, J.Y. Mao, S.C. Xie and M.L. Li, Petrochemical Engineering (China), 28 (1999) 283–285.

[69] L.Y. Yang, Z.L. Wang, J.Q. Fu, J.R. Zhang and S. Chen, Petrochemical Engineering (China), 28 (1999) 347–351.

[70] L.Y. Yang, Z.L. Wang, J.R. Zhang and S. Chen, Petroleum Processing and Petrochemicals (China), 29 (1998) 54-57.

[71] Y. Xu, P.C. Wu, Z.B. Wang, Y.C. Yan and W.X. Shen, Journal of Inorganic Chemistry (China), 14 (1998) 298-302.

[72] C. Perego and P. Ingallina, Catal. Today, 73 (2002) 3-22.

[73] Z.Y. Qin, Y.X. Jiang and Y. Ning, Guangxi Chemical Technology (China), 26 (1997) 24-27.

[74] S. Kulprathipanja (ed.), Reactive Distillation, Philadelphia, 2000.

[75] L.Y. Wen and E. Min, Petrochemical Technology (China), 29 (2000) 49-55.

[76] L.Y. Wen, E. Min and G.C. Pang, J. Chem. Ind. Eng. (China), 51 (2000) 115-119.

[77] J.H. Lee and M.P. Dudukovic, Comput. Chem. Eng., 23 (1998) 159-172.

[78] E. Alper, B. Wichtendahl and W D Deckwer, Chem. Eng. Sci. 35( 1980) 217-222.

[79] A.A.C.M. Beenackers and W. P. M. V. Swaaaij, Chem. Eng. Sci., 48 (1993) 3109-3139.

[80] Committee of chemical engineering handbook (eds.), Chemical Engineering Handbook 2nd ed., Chemical Industry Press, Beijing, 1996.

[81] J.S. Tong (ed.), Fluid Thermodynamics Properties, Petroleum Technology Press, Beijing, 1996.

[82] O.J. Wimmers and J.M.H. Fortuin, Chem. Eng. Sci., 43 (1988) 313-319.

[83] Z. Gao, L.Y. He and Y.Y. Dai (eds.), Zeolite Catalysis and Separation Technology, Petroleum Technology Press, Beijing, 1999.

[84] R.C. Reid, J.M Prausnitz and T.K. Sherwood (eds.), The Properties of Gases and Liquids, McGraw-Hill, New York, 1987.

[85] J. L. G. Almeida, M. Dufaux and Y. B Taarit, JAOCS 71 (1994) 675-694.

[86] B. Vora, P. Pujado and I. Imai, Chem. Eng., 19 (1990) 187-191.

[87] Z.G. Lei, C.Y. Li, B.H. Chen, E.Q. Wang and J.C. Zhang, Chem. Eng. J., 93 (2003) 191-200.

[88] R. Baur, A. P. Higler, R. Taylor and R. Krishna, Chem. Eng. J., 76 (2000) 33-47.

[89] R. Baur, R. Taylor and R. Krishna, Chem. Eng. Sci., 56 (2001) 2085-2102.

[90] A.P. Higler, R. Taylor and R. Krishna, Chem. Eng. Sci., 54 (1999) 1389 –1395.

[91] A. P. Higler, R. Taylor and R. Krishna, Chem. Eng. Sci., 54 (1999) 2873-2881.

[92] H. Komatsu, J. Chem. Eng. Japan, 10 (1977) 200-205.

[93] H. Komatsu and C. D. Holland, J. Chem. Eng. Japan 10 (1977) 292-297.

[94] J. Jelinek and V. Hlavacek, Chem. Eng. Comm., 2 (1976) 79-85.

[95] R. Taylor, H. A. Kooijman and J. S. Hung, Comput. Chem. Eng., 18 (1994) 205-217.

[96] R. Krishna and J. A. Wesselingh, Chem. Eng. Sci., 52 (1997) 861-911.

[97] H. A. Kooijman and R. Taylor, Ind. Eng. Chem. Res., 30 (1991) 1217-1222.

[98] J. D. Seader and E. J. Henley (eds.), Separation Process Principles, Wiley, New York, 1998.

[99] O.J. Wimmers and J.M.H. Fortuin, Chem. Eng. Sci., 43 (1988) 303-312.

# Chapter 5. Adsorption distillation

Two types of adsorption distillation, i.e. fixed-bed adsorption distillation (FAD) and suspension adsorption distillation (SAD), are discussed in this chapter. In FAD the separating agent is stagnantly supported in the column as the catalyst in FCD, while in SAD the separating agent is constantly moving in the column as the catalyst in SCD. So in SAD the separating agent has the similar behavior in tray efficiency and hydrodynamics as the catalyst of catalytic distillation.

## 1. FIXED-BED ADSORPTION DISTILLATION

### 1.1. Introduction

Fixed-bed adsorption distillation (FAD) has presently been proposed by Abu Al-Rub [1-3]. It involves replacement of the inert packing material in packed-bed distillation column by an active packing material. The active packing materials used by Abu Al-Rub for separating ethanol and water are 3 Å or 4 Å molecular sieves, which are thought to be able to alter the VLE of ethanol and water considerably.

On the other hand, we think that some ion-exchange resins also can be assumed to be composed of one anion which generally has large molecular weight and one cation which is generally a single metal ion. Since salt effect may plays a role in increasing the relative volatility of aqueous solution, ion-exchange resins can also be selected as the packing material in the adsorption distillation column.

Evidently, in FAD molecular sieves or ion-exchange resins as the separating agents can be used for the separation of the mixture with close boiling point or azeotropic point. Furthermore, FAD is an extremely environment-friendly process because no extra organic solution is introduced except the components to be separated, and thus there doesn't exist solvent loss, unlike azeotropic distillation and extractive distillation.

Only was the separation of the ethanol / water system by FAD investigated because ethanol is a basic chemical material and solvent used in the production of many chemicals and intermediates [1, 2].

The experimental results obtained from the VLE of the ethanol / water system at 70 ℃ in the presence of 6.5 g of 3 and 4 Å molecular sieves showed considerable change from those without the molecular sieves. The azeotropic point for this system was eliminated. Moreover, no further improvement in this separation was achieved by increasing the weight of the molecular sieves beyond the optimum value. The alteration of the VLE is a result of the external force field exerted by the molecular sieves on the mixture's components. These results prove the feasibility of using active packing materials to alter the VLE of binary mixtures.

## 1.2. Thermodynamic interpretation

To interpret the results obtained thermodynamically, we will use the excess Gibbs energy to express the nonideality of a system. Suppose that the system is composed of component 1 (the light component), component 2 (the heavy component) and the solid (separating agent), and it is related to the interactions of the system components:

$$\Delta G^E = \Delta G^E_{12} + \Delta G^E_s \tag{1}$$

All the solid interactions are included in $\Delta G^E_s$. Thus,

$$\ln \gamma_i = \ln \gamma_{il} + \ln \gamma_{is} \tag{2}$$

or

$$\gamma_i = \gamma_{il} \gamma_{is} \tag{3}$$

where $\gamma_{il}$ is the activity coefficient in the absence of molecular sieves, and $\gamma_{is}$ is the activity coefficient due to the presence of external force fields created by the solid on the liquid. Therefore, from an experimental measurement of the activity coefficients in the absence and in the presence of external force fields, $\gamma_{is}$ can be obtained.

For the ethanol (1) / water (2) system, it shows a positive deviation from ideality in the absence or presence of molecular sieves. However, due to the effect of external force fields, the deviation from ideality is reduced, i.e. $\gamma_i < \gamma_{il}$ and $\gamma_{is} < 1$ under the same temperature and concentration.

Another factor that may affect the VLE of a binary mixture in porous media is the curvature of the gas-liquid interface inside the pore which may cause a reduction in the vapor pressure of the components to be separated [4]. But when the 3 and 4 Å molecular sieves are employed in FAD, ethanol maybe can't be adsorbed into the pore because of its relative large molecular diameter (>4 Å). Therefore, it can be assumed that only water (diameter 2.75 Å) is competitively adsorbed.

The vapor pressure of water (2) in the microporous molecular sieves is expressed by the well-known Kelvin equation [5-8]:

$$\ln\left(\frac{p_2^r}{p_2^0}\right) = -\frac{2\sigma_2 V_2}{rRT} \tag{4}$$

where $\sigma_2$ is surface tension (N m$^{-1}$), $V_2$ is molar volume of water (m$^3$ mol$^{-1}$), $r$ is a radius of curvature over a concave vapor-liquid interface, $R$ is gas constant (8.314 J mol$^{-1}$ K$^{-1}$) and $T$ is temperature (K); or

$$p_2^r = p_2^0 \exp(-\frac{2\sigma_2 V_2}{rRT}) \tag{5}$$

$$\alpha_{12}^r = \frac{y_1/x_1}{y_2/x_2} = \frac{p_1^0 \gamma_1}{p_2^r \gamma_2} = \frac{p_1^0 \gamma_1}{p_2^0 \gamma_{2l} \gamma_{2s}} \exp(\frac{2\sigma_2 V_2}{rRT}) = \frac{\alpha_{12}}{\gamma_{2s}} \exp(\frac{2\sigma_2 V_2}{rRT}) \tag{6}$$

which indicates that $\alpha_{12}^r > \alpha_{12}$ because $\gamma_{is} < 1$ and $\exp(\frac{2\sigma_2 V_2}{rRT}) > 1$. That is why the

relative volatility of ethanol to water is improved in the presence of molecular sieves.

However, there is another explanation for the reason about improvement of relative volatility [3]. It is thought that the idea of an external force field emanating in a liquid from 2 mm diameter molecular sieve beads is far removed from the VLE inside capillaries, which implies that the theory of external force field isn't valid in the liquid bulk. Rather, the addition of molecular sieves to the water / ethanol system will result in the preferential uptake of water by the sieves, thereby reducing the liquid phase concentration of water. The quantified vapor phase would then be in equilibrium with a liquid phase of lower water content, irrespectively of how capillary surfaces may alter VLE. So that is why the relative volatility of ethanol to water is improved in the presence of molecular sieves.

We think that, whatever the theory, it is the truth that molecular sieves can play a role in increasing the relative volatility and thus making it possible to separate the components with close boiling point or azeotropic point. In other words, it is anticipated that FAD, as a promising special distillation process, may replace azeotropic distillation and extractive distillation in some cases in the near future.

### 1.3. Comparison of FAD and extractive distillation

Comparison of adsorption distillation and extractive distillation for separating ethanol and water is very attractive because adsorption distillation is an attractive "new" technique and extractive distillation is an "old" technique. The most commonly used separating agents in the extractive distillation are ethylene glycol, the mixture of ethylene glycol and one salt, CaCl$_2$ and N, N-dimethylformamide (DMF) [9, 10].

Comparison of azeotropic distillation and extractive distillation has been made in chapter 2. Today, azeortopic distillation is almost replaced by extractive distillation for separating ethanol and water. So it is unadvisable to use azeortopic distillation for separating ethanol and water.

Comparison of adsorption distillation and extractive distillation is made on the basis of process experiment and VLE experiment.

#### 1.3.1. Process experiment

The experimental flow sheet of extractive distillation process with two columns (extractive distillation column and solvent recovery column) has been established in the laboratory and is shown in Fig. 1. The extractive distillation column was composed of three sections, ( I ) rectifying section of 30 mm (diameter) × 800 mm (height), ( II ) stripping section of 30 mm (diameter) × 400 mm (height) and (III) scrubbing section of 30 mm (diameter) × 150 mm (height). The solvent recovery column was composed of two sections, ( I ) rectifying section of 30 mm (diameter) × 650 mm (height) and ( II ) stripping section of

30 mm (diameter) × 300 mm (height). The two columns were packed by a type of ring-shape packing with the size of 4 mm (width) × 4 mm (height). The theoretical plates are determined by use of the system of n-heptane and methylcyclohexane at infinite reflux, having 18 theoretical plates (including reboiler and condenser) for extractive distillation column and 14 theoretical plates (including reboiler and condenser) for solvent recovery column.

In the extractive distillation process experiment, the mixture of ethylene glycol and $CaCl_2$ was used as the separating agent. The reagents, ethylene glycol and $CaCl_2$, were of analytical purity and purchased from the Beijing Chemical Reagents Shop, Beijing, PRC. Composition analyses of the samples withdrawn from the top of extractive distillation column were done by a gas chromatography (type Shimadzu GC-14B) equipped with a thermal conductivity detector. Porapark Q was used as the fixed agent of the column packing, and hydrogen as the carrier gas. The column packing was 160 ℃ and the detector 160 ℃. The data were dealt with by a SC 1100 workstation. In terms of peak area of the components, the sample compositions could be deduced.

The experimental results of extractive distillation column are given in Table 1, where the feed concentration is ethanol of 0.3576 mole fraction and water of 0.6424 mole fraction. It is shown that high-purity of ethanol (above 99.0%wt), up to 0.9835 mole fraction (almost 99.5%wt), was obtained under low solvent/feed volume ratio 2.0 and low reflux ratio 0.5, which means that the mixture of ethylene glycol and $CaCl_2$ as the separating agent is very effective for the separation of ethanol and water by extractive distillation and has been governed in industry for a long time.

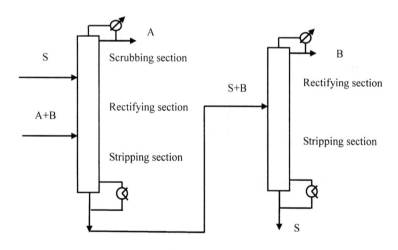

Extractive distillation column          Solvent recovery column

Fig.1. The double-column process for extractive distillation.

Table 1
Experimental results of extractive distillation column

| No. | Solvent/feed volume ratio | Reflux ratio | $T_{top}$ / K | The top composition | |
|---|---|---|---|---|---|
| | | | | $x_1$ (ethanol) | $x_2$ (water) |
| 1 | 0.5 | 0.5 | 351.1 | 0.8506 | 0.1494 |
| 2 | 1.0 | 0.5 | 351.1 | 0.9674 | 0.0326 |
| 3 | 2.0 | 0.5 | 351.1 | 0.9835 | 0.0165 |

In order to test the effect of molecular sieves and ion-exchange resins on the separation of ethanol and water in the adsorption distillation column, a set of experimental apparatus was established and shown in Fig. 2. The column was composed of two sections, ( I ) 30 mm (diameter) × 600 mm (height) and ( II ) 30 mm (diameter) × 600 mm (height). The first section was firstly filled with stainless steel ring-shape packing with the size of 3 mm width and 3 mm height, then filled with 3 Å sphere-shape molecular sieves with the diameter of 4-6 mm, and finally filled with ion-exchange resins (the type is D072 Styrene-DVB composed of the anion $RSO_3^-$ and the cation $Na^+$.) with the diameter of 1.20 - 1.30 mm. The second section was always filled with stainless steel ring-shape packing with the size of 3 mm width and 3 mm height in both experiments. In the beginning the feed mixture entered into the bottom tin with 2500 ml. Then the batch distillation column started to run. In the experiment it was assumed that the bottom and top concentration was keep constant because the operation time wasn't too long and concentration change in the bottom tin was negligible.

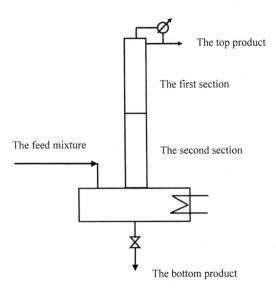

Fig. 2. The experimental apparatus used for separating ethanol and water by adsorption distillation.

The organic solvents, ethanol, was of analytical purity. Ethanol and 3 Å molecular sieves were purchased from the Beijing Chemical Reagents Shop, Beijing, PRC. Ion-exchange resins were purchased from the Chemical Plant of Nankai University. Water was purified with ion-exchange resins. Composition analyses of the samples withdrawn from the top and bottom were done by a gas chromatography (type GC4000A) equipped with a thermal conductivity detector. Porapark Q was used as the fixed agent of the column packing, and hydrogen as the carrier gas. The column packing was 160 ℃ and the detector 160 ℃. The data were dealt with by a BF9202 workstation. In terms of peak area of the components, the sample compositions could be deduced.

Under the infinite reflux condition, the experimental results are given in Table 2 where $x_1$ and $x_2$ are molar fraction of ethanol and water in the liquid phase, respectively, and $T$ is temperature (K). The fresh molecular sieves and ion-exchange resins are used and not replaced during the operation.

Table 2
The experimental results of adsorption distillation

| The top of the column | | | The bottom of the column | | |
|---|---|---|---|---|---|
| $T / K$ | $x_1$ | $x_2$ | $T / K$ | $x_1$ | $x_2$ |
| No separating agent | | | | | |
| 354.15 | 0.8057 | 0.1943 | 355.55 | 0.5930 | 0.4070 |
| 354.15 | 0.8063 | 0.1937 | 355.55 | 0.5940 | 0.4060 |
| Molecular sieves | | | | | |
| 354.55 | 0.8463 | 0.1537 | 356.05 | 0.5093 | 0.4907 |
| 354.55 | 0.8443 | 0.1557 | 356.35 | 0.5163 | 0.4837 |
| Ion-exchange resins | | | | | |
| 354.25 | 0.8205 | 0.1795 | 356.05 | 0.5093 | 0.4907 |
| 354.25 | 0.8183 | 0.1817 | 356.35 | 0.5163 | 0.4837 |

It can be seen from the above table that the enhancement of molecular sieves and ion-exchange resins on the separation of ethanol and water is very limited, and high-purity of ethanol (above 99.0%wt) can't be obtained, at most up to 0.8463 mole fraction (94.8%wt) and 0.8205 mole fraction (93.8%wt) respectively in the case study. Therefore, although adsorption distillation, in which the solid separating agent is used and the operation process is simple, may be attractive, it seems to be unattractive in the actual distillation process.

### 1.3.2. VLE experiment

In order to further compare the separation effect of different separating agents used in adsorption distillation and extractive distillation, a typical recycling vapor-liquid equilibrium cell having the volume 60 ml was utilized to measure the VLE of aqueous ethanol system. Salt and the solvent ethylene glycol were blended with the solvent/feed volume ratio 1:1 and the concentration of salt was 0.1 g (salt) /ml (solvent). In the cell the mixture with about 40 ml was heated to boil at normal pressure. One hour later, the desired phase equilibrium was achieved. At that time small samples in the vapor and liquid phases were removed by transfer pipet with 1 ml in order to make the equilibrium not to be destroyed.

Firstly we measured the equilibrium data of the ethanol (1) / water (2) system which corresponded well with the reference data [11]. It is verified that the experimental apparatus was reliable. Then the measurements were respectively made for the system ethanol (1) / water (2) / ethylene glycol (solvent/feed volume ratio is 1:1), and the system ethanol (1) / water (2) / ethylene glycol / CaCl$_2$ (solvent/feed volume ratio is 1:1 and the concentration of salt is 0.1g ml$^{-1}$ solvent) at 1 atm. The experimental VLE data are plotted in Fig. 3 where the mole fractions are on solvent free basis, along with the results from the references with 4 Å molecular sieves as the separating agent.

It can be seen from Fig. 3 that the separation effect of molecular sieves on ethanol and water is much less than that of ethylene glycol or the mixture of ethylene glycol and CaCl$_2$. Furthermore, the separation ability of the mixture of ethylene glycol and CaCl$_2$ is higher than that of single ethylene glycol, which shows that it is an effective way to improve ethylene glycol by adding a little of salt.

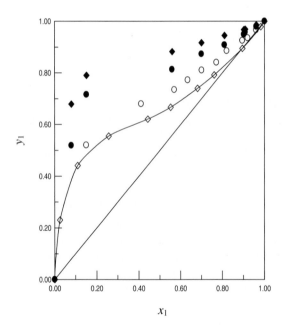

Fig. 3. VLE of ethanol and water at 1 atm for different separating agent. ◆ – ethanol (1) / water (2) / ethylene glycol / CaCl$_2$; ● – ethanol (1) / water (2) / ethylene glycol; ○ – ethanol (1) / water (2) / 300 g of 4 Å molecular sieves; ◇ – ethanol (1) / water (2).

On the other hand, it is known that the mixture of ethanol and water is very difficult to be separated when the mole fraction of ethanol in the liquid phase is over 0.895. However, the VLE change in the presence of molecular sieves isn't apparent especially when the mole fraction of ethanol in the liquid phase is in the range of 0.895 – 1.0. An important experimental operation condition should also be mentioned during measuring VLE of the ethanol-water-molecular sieves system: "fresh" molecular sieves were used for each run. It indicates that only dehydrated molecular sieves can raise the relative volatility of ethanol to water and the used "old" molecular sieves have little effect on this separation if not dehydrated in time. However, it is a pity that the actual operation condition in the adsorption distillation column is quite different from that in the VLE experiment. When molecular sieves are selected as active packing materials for separating ethanol and water in the column, high-purity of ethanol isn't anticipated because molecular sieves may be quick to be saturated by water. That is one reason why we obtained the poor result in the process experiment of adsorption distillation. To obtain high-purity of ethanol as in the extractive distillation process, besides large amount, the molecular sieves should be constantly regenerated. It is very tedious to often unload and load the molecular sieves from the column. Moreover, much energy and equipment cost will be consumed in the process of regenerating molecular sieves. Thus, we think that molecular sieves aren't good separating agent for adsorption distillation. This conclusion also holds for ion-exchange resins.

N, N- dimethylformamide (DMF) is also reported as the separating agent for separating ethanol and water [9]. Herein, DMF is improved by adding a little of salt to make into a mixture. If some factors such as solubility, price, erosion, source and so on are considered, the salt NaSCN is the prospective additive. The separation ability of DMF, the mixture of DMF and NaSCN, ethylene glycol and ion-exchange resins are compared when the composition of feed mixture in the VLE cell is ethanol (0.7861 mole fraction) and water (0.2139 mole fraction). The result is given in Table 3, where the solvent/feed volume ratio is 1.0.

Table 3
Relative volatility of different separating agent for separating ethanol and water

| The separating agent | Relative volatility $\alpha_{12}$ |
|---|---|
| No separating agent | 1.05 |
| DMF | 1.31 |
| The mixture of DMF and NaSCN (0.10 g NaSCD / ml DMF) | 1.63 |
| The mixture of DMF and NaSCN (0.20 g NaSCD / ml DMF) | 2.08 |
| Ethylene glycol | 2.23 |
| Ion-exchange resins | 1.16 |

We know that relative volatility provides a useful index for selection of suitable separating agent. According to the definition of relative volatility, obviously $\alpha_{12}$ larger than unity is desired. Thus, the following conclusions can be drawn from Table 3:

(1) Adding a little of salt to DMF can improve the relative volatility of ethanol to water.

(2) The separation ability of ethylene glycol is greater than that of single DMF, and approximate to the mixture of DMF and NaSCN (0.20 g NaSCD / ml DMF).

(3) The separation ability of ion-exchange resins is the least among all the separating agents. It isn't good that ion-exchange resins are used for separating agent for the separation of ethanol and water. The reason may be that the molecular weight of ion-exchange resins is so high that the amount of function group $Na^+$ is very little and can't effectively influence the relative volatility of ethanol to water.

Therefore, by combining the results of Fig. 3 and Table 3, it is evident that the best separating agent is the mixture of ethylene glycol and $CaCl_2$, and the corresponding separation process is just extractive distillation.

So the separation technique using active packing materials for separating ethanol and water may seem very attractive, but can't replace common extractive distillation process by now because molecular sieves and ion-exchange resins as solid separating agent are much more difficult to be regenerated than liquid separating agent, and the separation effect is also less than that of the separating agents currently used in the extractive distillation process.

## 2. SUSPENSION ADSORPTION DISTILLATION

### 2.1. Introduction

Another kind of adsorption distillation, i.e. suspension adsorption distillation (SAD), has been put forward by Cheng et al [12, 13]. It doesn't involve replacement of internals of distillation. Although SAD looks like SCD, in SAD the tiny solid particles are used as the adsorption agent and blended with liquid phase in the column. In SCD the tiny solid particles are used as catalyst.

Sometimes adsorption distillation is coupled with extractive distillation, and is applied in a same column. This kind of adsorption distillation also can be called adsorption extractive distillation. The only difference from extractive distillation is that the separating agent is the mixture of particle and liquid solvent. The whole process is unchanged, just the two-column extractive distillation process. Consequently, the merits of both extractive distillation and adsorption distillation are completely exhibited.

The most significance of employing adsorption distillation is that the gas-liquid mass transport is enhanced by adding the tiny solid particles, which correspondingly improves the tray efficiency for tray columns. However, when adding solid particles to liquid phase, the packed column is prone to be jammed. From this viewpoint, the tray column is more suitable for SAD.

However, SAD has its own drawback. For instance, it is difficult to run the distillation column again while the solid particles are dispositted after a long time. This should be noted in the operation.

## 2.2. Thermodynamic interpretation

The schematic diagram describing mass transfer of gas-liquid-solid (non-porous) three-phase is shown in Fig. 4 [14]. It is assumed that component A in the gas bulk is absorbed into the external surface of non-porous particles. So the mass transfer flux $J_A a$ (kmol m$^3$ s$^{-1}$), when adding non-porous solid particles, is:

$$J_A a = \frac{C_{AG} - C_{AiS}/m}{\dfrac{1}{k_G a} + \dfrac{1}{mk_L aE_A} + \dfrac{1}{mk_s a_p}} \tag{7}$$

and when not adding solids,

$$(J_A a)_0 = \frac{C_{AG} - C_{AL}/m}{\dfrac{1}{(k_G a)_0} + \dfrac{1}{(mk_L a)_0}} \tag{8}$$

Herein, index 0 denotes no solids present, $E_A$ is enhancement factor, $m$ is gas solubility ($m = C_L/C_G$ at equilibrium), $a$ is the specific gas-liquid contact area (m$^2$ m$^{-3}$ liquid phase), $a_p$ is the external interface area of solids per unit volume (m$^2$ m$^{-3}$ liquid phase), $C_{AG}$, $C_{AL}$ and $C_{AiS}$ (kmol m$^{-3}$) are the mole concentrations in the gas bulk, liquid bulk and near the solid surface, respectively.

$R_1$, $R_1$ and $R_3$ are defined as the resistances that control the overall mass transfer rate, i.e.,

$R_1 = 1/k_G a$     the resistance to mass transfer in the gas phase

$R_2 = 1/mk_L a$   the resistance to mass transfer in the liquid phase at the gas-liquid interface

$R_3 = 1/mk_s a_p$  the resistance to mass transfer in the liquid phase at the liquid-solid interface

The schematic diagram describing mass transfer of gas-liquid-solid (porous) three phases is shown in Fig. 5 [14]. Although inside the porous particle mass transfer and adsorption are in parallel, the overall process still can be represented by resistances in series.

A linear adsorption isotherm for adsorption of the dissolved gas A on the particles will be assumed and the concentration of adsorbed component can be expressed as [15]

$$C_{AS} = \frac{n_A}{C_S K_{ad}} \tag{9}$$

232

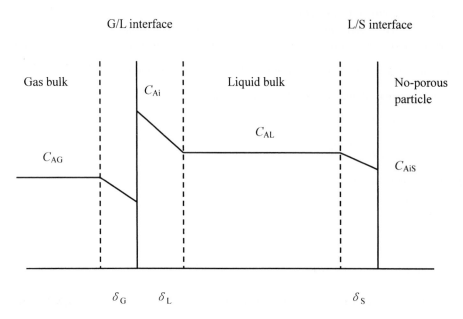

Fig. 4. The schematic diagram describing mass transfer of gas-liquid-solid (non-porous) three-phase; $\delta_G$, $\delta_L$ and $\delta_S$ are the film thickness of gas side, liquid side and solid side, respectively.

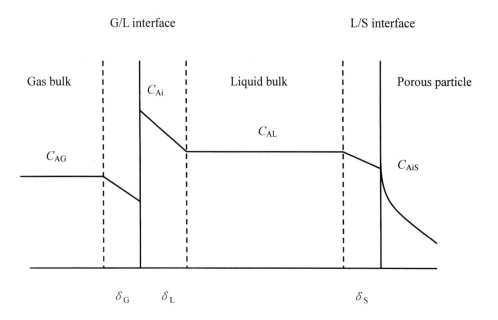

Fig. 5. The schematic diagram describing mass transfer of gas-liquid-solid (porous) three-phase; $\delta_G$, $\delta_L$ and $\delta_S$ are the film thickness of gas side, liquid side and solid side, respectively.

where $n_A$ (mol m$^{-3}$) is the number of moles A adsorbed on the particle phase, $C_{AS}$ (mol m$^{-3}$) is the concentration of A in liquid in equilibrium with actual concentration at the solid surface, $C_S$ (kg m$^{-3}$) is the concentration of solids in liquid and $K_{ad}$ (m$^3$ kg$^{-1}$ solids) is adsorption equilibrium constant. As a further simplification, the mass transfer and the adsorption on the particles will be considered to be processes in series. Thus,

$$J_A a_p = k_S a_p (C_{AL} - C_{AiS}) = k_{ad} a_p (C_{AiS} - C_{AS}) = k_p a_p (C_{AL} - C_{AS}) \tag{10}$$

with

$$\frac{1}{k_p} = \frac{1}{k_S} + \frac{1}{k_{ad}} \tag{11}$$

and

$$a_p = \frac{6C_s}{\rho_p d_p} \tag{12}$$

where $C_{AiS}$ (mol $\cdot$ m$^{-3}$ liquid) is the actual concentration of A in the liquid phase near the solid interface, $C_{AS}$ (mol m$^{-3}$ liquid) is the concentration of A in the liquid phase in equilibrium with actual concentration of A at the solid surface. The actual concentration of A at the solid surface isn't equal to $C_{AiS}$ (near the solid surface) because it is relative to per unit particle volume and A is adsorbed at the solid surface, while $C_{AiS}$ relative to per unit liquid volume and A is near the solid surface. $C_s$ (kg solids m$^{-3}$ liquid) is the concentration of solids in the liquid phase; $d_p$ (m) and $\rho_p$ (kg m$^{-3}$) are particle diameter and superficial density, respectively; $k_S$ (m s$^{-1}$), $k_{ad}$ (m s$^{-1}$) and $k_p$ (m s$^{-1}$) are liquid-to-solid mass transfer coefficient, adsorption rate constant and overall particle adsorption rate constant.

At the steady-state, the mass transfer flux $J_A a$ (kmol m$^3$ s$^{-1}$) for porous solid particles is:

$$J_A a = \frac{C_{AG} - C_{AS}/m}{\dfrac{1}{k_G a} + \dfrac{1}{m k_L a E_A} + \dfrac{1}{m k_p a_p}} \tag{13}$$

which is similar to Eq. (7) in form, only substituting $C_{AS}$ and $k_p$ for $C_{AiS}$ and $k_S$, respectively.

However, there is no information available on the influence of solid particles on $k_G a$. One might argue that such information isn't important, because for sparingly soluble gases $k_G a$ is usually not rate controlling, while for very soluble gases there is no need to add solid particles so as to promote the mass transfer due to the large numerical value of $k_G a$.

From these resistances in series, irrespective of $R_1$, the influence of solid particles is brought out mainly on $R_2$. $R_2$ might be influenced by the presence of the particles in several ways:

(1) The specific (gas-liquid) interfacial area, $a$, can be changed by the presence of the particles.

(2) $k_L$ itself can be influenced by the presence of the particles even if these are inert.

(3) In principle, in the case of dissolving particles, even the solubility of the gas phase component may change due to the presence of the dissolved particle materials and/or their products, but we neglect the effects of solubility.

When the solid particles are added into the liquid phase, in general, the mass transfer flux $J_A a$ for either non-porous or porous particles will be enhanced. Herein, enhancement factor $E_A$ is defined as [16, 17]:

$$E_A = \frac{J_A(\text{with particles adsorbing the transferred component physically})}{J_A(\text{with the same but inert particles})} \tag{14}$$

where $J_A$ is evaluated at the same overall driving force. Inert means that the particles don't participate in adsorbing the transferred component physically. But from the viewpoint of considering the difference between having particles and no particles, enhancement factor $E_A$ can also be defined as

$$E_A = \frac{J_A(\text{with particles adsorbing the transferred component physically})}{J_A(\text{with no particles and only gas - liquid two phases})} \tag{15}$$

However, this definition is rarely used due to its deficiency in distinguishing different physical contributions to enhancement factor. On the contrary, Eq. (14) is desirable. The advantage of the concept of the enhancement factor as defined by Eq. (14) is the separation of the influence of hydrodynamic effect on gas-liquid mass transfer (incorporated in $k_L$) and of the effects induced by the presence of a solid surface (incorporated in $E_A$).

There are two opposite factors affecting the mass transfer coefficient in the liquid side when adding solid particles. The presence of particles close to the interface may hold back the mass transfer of liquid phase and reduce the contact area between gas and liquid phases, which leads to decreasing the mass transfer coefficient in the liquid side. On the other hand, the added solid particles can promote the disturbance of interface between gas-liquid phases especially due to particle adsorption, and thus dwindle the liquid film thickness. A thinner liquid film thickness means a higher liquid-side mass transfer coefficient.

Either with dynamic absorption techniques or with chemical techniques, direct information on the product $k_L a$ can often be obtained in a relatively easy way. As a result, much more information is available on the product $k_L a$ than on $k_L$ and $a$ separately.

With the exception of alumina in ligroin, all results could be correlated according to [18]

$$\frac{k_L a}{(k_L a)_0} = \left(\frac{\mu_{eff}}{\mu_0}\right)^{-0.42} \tag{16}$$

with index "0" indication of no solids present. The correlation covers a superficial gas velocity up to 8 cm s$^{-1}$ and an effective viscosity range of 0.54 - 100 mPa s with a mean error of 7.7%.

Both for water and ionic aqueous solutions applying kieselguhr (7 μm), alumina (8 μm) and activated carbon (7 μm), Schumpe et al. [19] found that, provided $\mu_{eff} > 2\mu_L$,

$$\frac{k_L a}{(k_L a)_0} = \left(\frac{\mu_{eff}}{\mu_L}\right)^{-0.39} \tag{17}$$

for $u_G < 8$ cm s$^{-1}$ and $\mu_{eff}$ between 1 and 100 mPa s. It is remarkable that the relation was found to be valid for both electrolyte and non-electrolyte solutions.

Nguyen-Tien and Nigam [20, 21] observed, for three-phase fluidized beds and slurry bubble columns,

$$\frac{k_L a}{(k_L a)_0} = 1 - \frac{\varepsilon_S}{0.58} \tag{18}$$

where $\varepsilon_S$ (m$^3$ solids m$^{-3}$ liquid) is solids hold-up.

A similar relation to Eq. (18) was observed [22], although with different constants,

$$\frac{k_L a}{(k_L a)_0} = 1 - \varepsilon_s \tag{19}$$

Other relations can be found in many references. But most of them show that $k_L a$ is declined when adding inert solids. The reason may be that the presence of particles close to the interface may hold back the mass transfer of liquid phase and reduce the contact area between gas and liquid phases. The negative influence dominates on the mass transfer coefficient in the liquid side.

In order to deduce enhancement factor $E_A$, we consider the absorption of a gas phase component in a slurry with a relatively simple homogeneous model based on the Higbie penetration theory, neglecting any gas phase resistance [15].

The differential equations for the penetration of gas A into a typical surface packet are: for dissolved A:

$$\frac{\partial c_A}{\partial t} = D_A \frac{\partial^2 c_A}{\partial x^2} - k_p a_p (c_A - c_{AS})$$ (20)

(accumulation term = diffusion terms + convection term)
for adsorbed A:

$$\frac{\partial n_A}{\partial t} = k_p a_p (c_A - c_{AS})$$ (21)

(accumulation term = convection term)

where $D_A$ (m$^2$ s$^{-1}$) is the liquid diffusion coefficient which could be affected by the presence of the particles. The boundary conditions are given here for the case of initial zero loading of the slurry packet (to start with a higher initial concentration, corresponding to a higher slurry bulk loading, is of course possible):

$t = 0$; $x \geq 0$ and $t \neq 0$; $x = \infty$: $c_A = c_{AS} = n_A = 0$

$x = 0$; $t \geq 0$: $c_A = mc_{AG}$.

Then, the enhancement factor $E_A$ can be calculated from

$$E_A = \frac{\dfrac{1}{t_c} \int_0^\infty - D_A \dfrac{\partial c_A}{\partial x}\big|_{x=0}\, dt}{\left[ 2\sqrt{\dfrac{D_A}{\pi t_c}} \left( c_{A,x=0} - c_{A,x=\infty} \right) \right]}$$ (22)

Here, $t_c$ is the Higbie contact time of the slurry element at the gas-liquid surface. This system has been solved numerically by Holstvoogd et al. [15] and we consider a few interesting asymptotic solutions.

(1) Instantaneous equilibrium between particles and their surrounding liquid [23]. In this case,

$$E_A = \sqrt{1 + C_s K_{ad}}$$ (23)

(2) If the L-S mass transfer adsorption rate parameter $k_p$ is finite and the particle adsorption capacity is so large that the degree of loading remains much smaller than one under all circumstances, then the rate parameters will completely determine the rate of adsorption.

Then, the enhancement is described by the Hatta number, *Ha*, only:

$$Ha_h = \frac{\sqrt{k_p a_p D_A}}{k_L} \tag{24}$$

$$E_A = \frac{Ha_h}{\tanh(Ha_h)} \tag{25}$$

Generally, both the adsorption rate parameter and the particle capacity are important for the enhancement.

A simple heterogeneous model based on film theory is called the enhanced gas absorption model (EGAM) [24]. This model results in an analytical expression for the enhancement factor $E_A$:

$$E_A = 1 + \alpha \left[ \frac{4D_A}{d_p k_L} \left( \frac{1 - \exp(-t_p / t_s)}{t_p / t_s} \right) - 1 \right] \tag{26}$$

where $t_p / t_s$ is the ratio of the residence time of the particle at the surface and the particle saturation time $t_s$.

In contrast to homogeneous models, this heterogeneous model explains why the enhancement factor reaches a maximum as a function of $\varepsilon_s$ even if no particle saturation occurs [25]. For complete surface coverage of the gas-liquid interface we have $\alpha = 1$ and

$$E_{A,\max} = \frac{4\delta_L}{d_p} = \frac{4D_A}{k_L d_p} \tag{27}$$

For $k_L$ in aqueous systems typically $10^{-4}$ m s$^{-1}$, $D_A$ typically $10^{-9}$ m$^2$ s$^{-1}$ and $d_p = 5$ $\mu$m (which is half the film thickness), we have $E_{A,\max} = 8$ which is an important reference value for industrial applications.

Other equations of enhancement factor and mass transfer flux of gas-liquid mass transport by the tiny solid particles was proposed by Cheng et al. [12, 13] by using penetration theory and segmental linearization of adsorption isotherm curves.

The enhancement factor $E_A$ and the mass transfer flux $J_A$ under the effect of fine adsorbent particles are:

$$E_A = \frac{J_A (1 - \varepsilon_s)^2}{(k_L^0 a)(C_A^* - C_b)(1 + 0.5\varepsilon_s)^{1/2}} \tag{28}$$

$$J_A = \frac{(F_3)^{1/2}(k_L a)(C_A^* - C_{max})}{erf[z_1(F_3/D_A)^{1/2}]} \tag{29}$$

where $C_A^*$, $C_{max}$ and $C_b$ (mol m$^{-3}$) are the liquid concentration of the gas-liquid interface, the maximum and the bulk, respectively; $z_1$ and $F_3$ are adjustable parameters. In calculating diffusion coefficient $D_A$ (m$^2$ s$^{-1}$), the following apparent effective slurry viscosity $\mu_{eff}$ (Pa $\cdot$ s$^{-1}$) is adopted:

$$\mu_{eff} = \frac{\mu_L(1+0.5\varepsilon_s)}{(1-\varepsilon_s)^4} \tag{30}$$

where $\mu_L$ (Pa s$^{-1}$) is the liquid viscosity.

In Eqs. (7) and (11), relations for the mass transfer coefficient at the liquid-solid interface ($k_S$) are usually presented [26-28] in the form of

$$N_{Sh} = 2 + CN_{Re}^{n1} N_{Sc}^{n2} \tag{31}$$

with the Sherwood and Schmidt numbers defined as

$$N_{Sh} = \frac{k_S d_p}{D} \tag{32}$$

$$N_{Sc} = \mu_L/(\rho_L D) \tag{33}$$

In early attempts the Reynolds number $N_{Re}$ was often based on the hindered particle settling velocity, i.e. the actual terminal particle slip velocity in the slurry, $v_p$:

$$N_{Re} = \frac{\rho_L v_p d_p}{\mu_L} \tag{34}$$

In the end, it should be noted that the concept of adsorption distillation in this chapter means that the fields of adsorption and distillation exist simultaneously and is coupled in a same column. But while we search the website: http://isi5.isiknowledge.com/portal.cgi/wos with the title keyword "adsorption distillation", it is interesting to find that a novel adsorption-distillation hybrid scheme is proposed for propane/propylene separation [29, 30]. But this adsorption-distillation hybrid scheme is completely different from our concept of adsorption distillation (FAD and SAD) in that the separation process is carried out first by single distillation up to a propylene concentration of approximately 80% and then continuing the separation of propane from propylene by single adsorption, or, inversely, first by adsorption and then by distillation. The processes of adsorption and distillation are isolated as a unit operation. However, the suggested scheme has potential for saving up to approximately 50% energy and approximately 15-30% in capital costs as compared with current distillation technology due to eliminating energy intensive and expensive olefin-alkane distillation.

The pure component data of propane and propylene were obtained on silica gel, molecular sieve 13X and activated carbon. Although activated carbon has a greater capacity for both propane and propylene than either of the two adsorbents, it was only slightly selective for propylene. Silica gel has the greatest selectivity for propylene, which ranges from 2 to 4. The nonideality of the mixture can be attributed primarily to surface effects rather than to interactions between adsorbate molecules. This analysis is reasonable because adsorbate molecules (propylene and/or propane) themselves with similar molecular structure may form ideal mixture.

## REFERENCES

[1] A.B. Fawzi, A.A. Fahmi and S. Jana, Sep. Purif. Technol., 18 (2000) 111-118.

[2] A.A. Fahmi, A.B. Fawzi and J. Rami, Sep. Sci. Technol., 34 (1999) 2355-2368.

[3] L.M. Vane, Sep. Purif. Technol., 27 (2002) 83-84.

[4] A.A. Fahmi and R. Datta, Fluid Phase Equilib., 147 (1998) 65-83.

[5] R. Defay, I. Prigogine, A. Bellemans and D.H. Everett (eds.), Surface Tension and Adsorption, Longmans Green, London, 1966.

[6] L.M. Skinner and J.R. Sambles, Aerosol Sci., 3 (1972) 199.

[7] R.J. Hunter (ed.), Foundations of Colloid Science Vol. 1, Clarendon Press, Oxford, 1987.

[8] A.W. Adamson (ed.), Physical Chemistry of Surfaces 5th ed., Wiley, New York, 1984.

[9] G.S. Shealy, D. Hagewiesche and S.I. Sandler, J. Chem. Eng. Data, 32 (1987) 366-369.

[10] Z.G. Lei, H.Y. Wang, R.Q. Zhou and Z.T. Duan, Chem. Eng. J., 87 (2002) 149-156.

[11] J. Gmehling and U. Onken (eds.), Vapor-Liquid Equilibrium Data Collection, Dechema, Frankfurt, 1977.

[12] H. Cheng, M. Zhou, C.J. Xu and G.C. Yu, J. Chem. Ind. Eng. (China), 50 (1999) 766-771.

[13] Cheng, H., M. Zhou, Y. Zhang, C.J. Xu. J. Chem. Ind. Eng. (China), 50 (1999) 772-777.

[14] A.A.C.M. Beenackers and W.P.M. van Swaaij, Chem. Eng. Sci., 48 (1993) 3109-3139.

[15] R.D. Holstvoogd, W.P.M. van Swaaij and L.L. van Dierendonck, Chem. Eng. Sci., 43 (1988) 2181-2187.

[16] B.D. Kulkarni, R.A. Mashelkar and M.M. Sharma (eds.), Recent Trends in Chemical Reaction Engineering, Wiley-Eastern, New Delhi, 1987.

[17] H. de Lasa (ed.), Chemical Reactor Design and Technology, Martinus Nijhoff, The Hague, 1986.

[18] S.S. Ozturk and A. Schumpe, Chem. Eng. Sci., 42 (1987) 1781-1785.

[19] A. Schumpe, A.K. Saxena and L.K. Fang, Chem. Eng. Sci., 42 (1987) 1787-1796.

[20] K. Nguyen-Tien, A.N. Patwari, A. Schumpe and W.D. Deckwer, AIChE J., 31 (1985) 194-201.

[21] K.D.P. Nigam and A. Schumpe, AIChE J., 33 (1987) 328-330.

[22] H. Kojima, Y. Uchida, T. Ohsawa and A. Iguchi, J. Chem. Eng. Japan, 20 (1987) 104-106.

[23] R.I. Kars, R.J. Best and A.A.H. Drinkenburg, Chem. Eng. J., 17 (1979) 201-210.

[24] H. Vinke, G. Bierman, P.J. Hamersma and J.M.H. Fortuin, Chem. Eng. Sci., 46 (1991) 2497-2506.

[25] J.T. Tinge, K. Mencke and A.A.H. Drinkenburg, Chem. Eng. Sci., 42 (1987) 1899-1907.

[26] P.A. Ramachandran and R.V. Chaudhari (eds.), Three Phase Catalytic Reactors, Gordon and Breach, New York, 1983.

[27] W.D. Deckwer (ed.), Bubble Column Reactors, Wiley, Chichester, 1992.

[28] R.V. Gholap, D.S. Kohle, R.V. Chaudhari, G. Emig and H. Hofmann, Chem. Eng. Sci., 42 (1987) 1689-1693.

[29] R. Kumar, T.C. Golden, T.R. White and A. Rokicki, Sep. Sci. Technol., 27 (1992) 2157-2170.

[30] T.K. Ghosh, H.D. Lin and A.L. Hines, Ind. Eng. Chem. Res., 32 (1993) 2390-2399.

# Chapter 6. Membrane distillation

In membrane distillation (MD), the membrane as separating agent can be used for separating the mixture with close boiling point or forming azeotrope, but this separating agent needn't to be recycled. The separation principle of MD is based on the selectivity of membrane material to the components with different structures and properties. For example, for a hydrophobic membrane volatile organic compounds are more preferential than water to pass through the membrane. As a special distillation process, an outstanding advantage of MD is energy saving, but more work still need be done before it is extensively applied in industry.

## 1. INTRODUCTION

Separation of the mixture associated with membrane is known as membrane separation where the membrane acts as a selector that permits some components in the mixture to pass through, while other components are retained. The membrane in most cases is a thin, porous or nonporous polymeric film, or may be ceramic or mental materials, or even a liquid or gas. The selectivity of the membrane mainly depends on its structure and properties of the membrane material and the components in the mixture. Unlike conventional filtration process applied only to solid-liquid mixture, membrane separation is capable of the separation of homogeneous mixtures that are traditionally treated by distillation, absorption or extraction operations. The replacement of traditional separation processes with membrane separation has the potential to save large amounts of energy, since membrane process is mostly driven by pressure gradient or concentration gradient through the membrane. Although this replacement requires the production of high mass-transfer flux, defect-free, long-life membranes on a large scale and the fabrication of the membrane into compact, economical modules of high surface area per unit volume [1], the effectiveness and practicability of membrane separation technique have been proved by laboratory investigation and industrial production.

The membrane separation involves the process in which some components penetrate through the membrane and thus mass transfer occurs. Based on the difference in driving forces of mass transfer and effective range of separation scale (from 0.1 nanometer to 10 microns, 5 orders span), the membrane 'family' includes more than 10 members, and most of them, such as reverse osmosis (RO), gas permeation (GP), microfiltration (MF), pervaporation (PV), have been accepted as the alternatives to some conventional separation techniques in industry. The related fields of membrane separation varies from the desalination of sea water or bitter water, concentration of solutions, waste water treatment to the recovery of valuable substance from solutions, the separation of gas mixture, etc.

Membrane distillation (MD) is a new comer of the membrane family. Although the discovery of MD phenomenon can be traced back to the 1960s, it hasn't received more attention until 1980s when membrane fabrication technique gained remarkable development.

Today, MD is considered as a potential alternative to some traditional separation techniques, and is believed to be effective in the fields of desalination, concentration of aqueous solution, etc. That difference between MD and other membrane separation techniques is the driving force of mass transfer through the membrane. Unlike other members, MD is a thermally driven process. That's why it is denominated as a distillation process.

## 2. SEPARATION PRINCIPLE

### 2.1. MD phenomenon

As shown in Fig. 1, a liquid solution with high temperature is brought into contact with one side of a porous, hydrophobic membrane. The membrane which is a porous thin flexible sheet or tube acts as a barrier to separate the warm solution (called the feed side), and the permeate in either a liquid or a gas phase enters into a cooling chamber (called the permeate side). The hydrophobic nature of the microporous membrane prevents liquids / solutions from entering its pores due to the surface tension forces. As a result, a fixed interface is formed at the pores entrances. Fig. 2 [2] gives a cross sectional view of supposed straight cylindrical pores in contact with an aqueous solution to show how the vapor-liquid interfaces are supported at the pore openings. If the solution contains at least one volatile component, temperature difference at two ends of the pores produces a vapor pressure gradient within the pores. By the driving force the vapor molecules of volatile component (produced by evaporation from the feed solution at the vapor-liquid interface) migrate from the feed side to the permeate side of the membrane. At the permeate side the immigrated molecules (depending on the membrane configuration used) are either condensed or removed in vapor form from the membrane module. In this way the solution from the feed side is concentrated. In summary, membrane distillation is, by its nature, a combination of membrane separation and evaporation / condensation process, and a microporous hydrophobic membrane is employed to act as the supporter of vapor liquid interfaces.

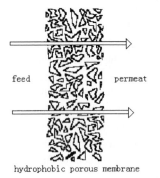

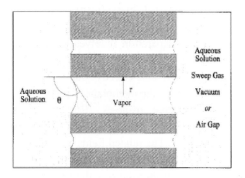

Fig. 1. MD phenomenon         Fig. 2. Vapor liquid interface in MD; adapted from the source [2].

## 2.2. Definition of MD process

Compared with such separation processes as osmotic distillation and pervaporation, the features of MD process [3] are:

(1) The membrane should be porous;

(2) At least one side of the membrane should keep in contact with the liquid to be dealt with;

(3) The driving force for each component to pass through the membrane pores is its partial pressure gradient in the vapor phase inside the pores;

(4) The membrane shouldn't be wetted by the liquid to be dealt with;

(5) No capillary condensation of vapor takes place inside the membrane pores;

(6) The membrane doesn't alter the vapor-liquid equilibrium of different components in the feed.

It is generally thought that the term "membrane distillation" arises from the similarity of this process to the conventional ordinary distillation process. Both processes depend on vapor-liquid equilibrium as the basis for separation and latent heat should be supplied to produce vapor phase. However, the remarkable difference between these two processes is that in conventional distillation the feed solution must be heated up to the temperature of its boiling point, whereas it isn't necessary for MD. It will be mentioned afterwards that MD can be performed well with the feed temperature lower than its boiling point. Moreover, the components to be separated can have close boiling point or form azeotrope. So MD belongs to a special distillation process.

## 2.3. Membrane characteristics

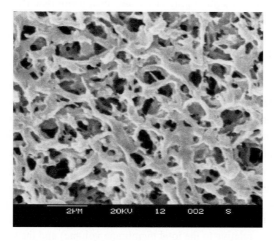

Fig. 3. Micrograph of flat sheet membrane (PVDF) by scanning electronic microcopy.

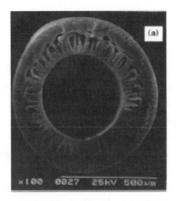

Fig. 4. Micrograph of hollow fiber membrane (PVDF) by scanning electronic microcopy; adapted from the source [56]

Although MD is still emphasized as a low cost, energy saving and potential alternative to traditional separation processes, special membranes only for this purpose haven't been provided yet. At present the membranes employed in MD are those made for microfiltration purposes, because most of the required specifications by MD processes are available from those membranes. Hydrophobic microporous membranes made of polypropylene (PP), polyethylene (PE), polytetrafluoroethylene (PTFE) and polyvinylidene fluoride (PVDF) are the commonly used membranes in MD. These membranes are fabricated in the form of either flat sheets or hollow fibers. A micrograph of a flat sheet membrane made of PVDF is shown in Fig. 3, as well as a hollow fiber in Fig. 4. By far, there are many kinds of commercially available microporous membranes maybe feasible for MD. However, as pointed out by many researchers [4], the requirement for MD membranes is: a higher permeability, lower membrane thickness, higher liquid entry pressure and low heat conductivity.

The membrane used in MD can exert its influence on transmembrane vapor flux of volatile component in three aspects [5]. Firstly, the vapor molecules move only through the pores of the membrane so that the effective area for mass transfer through the membrane is less than the total membrane area. Secondly, for most of the practical membranes, the membrane pores don't go straight through the membrane, and thus the path for vapor transport is greater than the thickness of the membrane. Thirdly, the inside wall of pores also increases the diffusion resistance by depleting the momentum of the vapor molecules.

A microporous membrane is generally characterized by four parameters, i.e. the thickness, $\delta$ (m), the mean pore size, diameter $d$ or radius $r$ (m), the porosity $\varepsilon$ (defined as the porous volume fraction relative to the total membrane volume) and the tortuosity $\tau$ (defined as the ratio of pore length to membrane thickness). Each of the parameters influences the permeability of the membrane. The relationship with the permeability can be summarized as

$$permeability \propto \frac{r^a \varepsilon}{\tau \delta}$$

where *a* may be equal to 1 or 2, depending on the predominant mass transfer mechanism within the membrane pores (about the detail see 3. transport process). It is evident that in order to obtain a high permeability or MD productivity, the membrane acting as a supporter of the vapor-liquid interface should be as thin as possible and its surface porosity and pore size should be as large as possible. Table 1 shows the usual characteristic values of hydrophobic membranes used in MD. The porosity of a membrane mainly depends on the technology of membrane preparation. Now the membrane with the porosity up to 90% is commercially available [6]. As for the thickness, there is a trade-off between high productivity and strength of the membrane. But another trade-off also exists between high productivity and high selectivity. The larger the pore size is, the more easily the pores are wetted by the feed solution.

Table 1
The usual values of characteristics of membrane used in MD

| Mean pore diameter | Porosity | Thickness | Tortuosity |
|---|---|---|---|
| 0.1 - 1.0 μm | 40 - 90% | 20 – 200 μm | 1.5 - 2.5 |

### 2.4. Membrane wetting

One requirement for MD process is that the membrane must be unwetted and only vapor is present within its pores. To avoid membrane wetting, the surface tension of the liquid in direct contact with membrane should be large and the surface energy of the membrane material should be low. A variable, called liquid-solid contact angle ($\theta$), is often used for describing liquid-solid interaction. Its definition is schematically represented in Fig. 2. The hydrophobicity means that the liquid-solid contact angle formed is greater than $90^0$ when it is in direct contact with water, which is necessary in MD. However, whether the membrane is wetted is not only associated with liquid-solid contact angle, but also associated with such factors as the liquid entry pressure (LEP). Below LEP the hydrophobicity of the membrane can't sustain water from entering its pores. The relationship of LEP with pore size, liquid-solid contact angle and liquid surface tension $\sigma$ (N m$^{-1}$) can be expressed by Cantor equation [1]:

$$LEP = -\frac{2X\sigma\cos\theta}{r} \tag{1}$$

where $X$ is a geometric factor depending on pore structure, for instance, $X = 1$ for cylindrical pore. Therefore, the membrane with larger pore size has relatively lower LEP, which indicates that it is more easily wetted during operation.

Note that the solution composition may affect the liquid-solid contact angle and liquid surface tension. Surface tension supports a pressure drop across the vapor-liquid interface up to the LEP. However, the existence of strong surfactants at the feed may reduce $\sigma|\cos\theta|$ apparently, which leads to the decrease of LEP and thus permits the feed to penetrate through the membrane pores. Therefore, it should be cautious to prevent equipment and solution from being contaminated by detergents and other surfactants.

## 2.5. The advantages of MD

In the 1960s, MD was developed to improve desalination efficiency and claimed to be a viable alternative to reveres osmosis (RO). At that time, MD was received attention mainly due to its capability of being operated with a minimum external energy requirement and a minimum expenditure of capital and land in the plant [7]. However, in comparison with RO thriving then, MD shortly lost its brightness because of its low productivity [8]. Later in the early 1980s, a renewal interest in MD was restored because development in membrane preparation technology became remarkable and membranes having porosities of as high as 80% and thickness of as low as 50 μm were available. This makes MD a more promising separation technique. Nowadays the importance of MD as a special distillation process is emphasized in the increasing number of published references. The advantages of MD over other conventional separation processes, i.e. ordinary distillation, RO, Ultra-filtration (UF), etc. are summarized as follows [2]:

(1) In principle, all non-volatile components at the aqueous feed are able to be rejected, i.e. ions, macromolecules, colloids, bacteria, etc. This is due to the hydrophobicity of the membrane. That is to say, only the vapor of volatile components are transported to the permeate side. Pressure-driven membrane separation processes, such as RO, MF and UF, can't obtain such a high rejection level.

(2) The process can be carried out under mild operation conditions. Although the feed needs to be heated to establish a temperature gradient between the two sides of the membrane, it isn't necessary to increase the feed temperature to its boiling point. In some cases MD can be performed well at the feed temperature ranging from 50–80°C with satisfactory mass flux. Therefore, low-grade waste heat energy sources, such as the cooling water from engines and the condensed water of low-pressure vapor, are good energy resources for MD. But the most attraction is the alternative energy resources such as solar and geothermal energy. MD can be coupled with these low-grade heat energy resources to constitute a high-efficiency liquid separation system. Moreover, the low operation temperature also makes MD a potential technology for the concentration of heat-sensitive substances in the field of food and pharmacy industry.

The mildness of operation condition in MD can also be brought out in the operation pressure. Unlike other pressure-driven membrane separation processes (i.e. RO and UF), MD is a thermally driven process, and pressure difference between the two sides of the membrane isn't indispensable to MD. So the pressure difference in MD may vary from zero to only a few hundred kPa. The low operation pressure results in small equipment and operation cost, low requirement for membrane mechanics and increased process safety. Besides, one important benefit from low operation pressure is the reduction of membrane fouling. Membrane fouling is a serious problem in all pressure-driven membrane separation processes and will affect mass transfer and life span. In general, membrane fouling is caused by the deposition and accumulation of undesirable materials (i.e. organic compound, inorganic compound, or a combination of both) on membrane surfaces. The lower the pressure is, the less the membrane is affected by fouling.

(3) Compared with ordinary distillation or evaporation, a large vapor-liquid interface area per unit volume is available in MD. Both MD and ordinary distillation (or evaporation) need a

large vapor-liquid interface area. In distillation column or evaporator this relies on the vapor space within the equipments. But in MD the vapor-liquid interface area is provided by the membrane which can be densely packed in the membrane module. The vapor space in the module is the pore volume of microporous membrane. As a result, the size of MD equipment is small.

## 2.6. MD configurations

There are four configurations developed to perform MD process, i.e. direct contact membrane distillation (DCMD), air gap membrane distillation (AGMD), sweeping gas membrane distillation (SGMD) and vacuum membrane distillation (VMD). The difference among these four configurations is the way in which the vapor (migrating from the vapor-liquid interface of the membrane surface at the feed side to that at the permeate side) is condensed and /or removed out of the module.

### 2.6.1. Direct contact membrane distillation (DCMD)

The DCMD configuration is illustrated in Fig. 5, where two chambers are separated by a flat sheet microporous membrane. The feed stream (with high temperature) and the permeate stream (with low temperature) flow across the two chambers respectively. Within the liquid-vapor interface at the feed side, the more volatile component vaporizes and passes through the membrane onto the liquid-vapor interface at the permeate side where it condensates. Here, direct contact means that both the feed and the permeate liquid are in direct contact with the membrane in the chambers. Among the four MD configurations, DCMD is the most extensively studied because of the convenience to setup in laboratory and the sufficiently high flux rate in comparison with other MD configurations. However, for the industrial consideration, some disadvantages are related with DCMD application. Firstly, in spite of the poor conductivity of polymeric material, the temperature difference between the feed and permeate sides not only offers a driving force for mass transfer, but also introduces a heat conduction through the membrane. Therefore, only part of the heat energy supplied to the feed is used for evaporation. In DCMD, because the permeate flow in the cooling chamber is directly contacted with the membrane, the heat loss by heat conduction is much higher than the other three configurations. So in DCMD the thermal efficiency (defined as the fraction of heat energy which is only used for evaporation) is relatively small. Secondly, in the run the operator must prepare enough permeate fluid in advance for its direct contact with membrane while flowing across the cooling chamber. Thirdly, as the condensate is mixed with the fluid in the cooling chamber, detecting the leakage or wetting of the membrane isn't easy. Even so, due to its unique advantages, DCMD is suitable for desalination or concentration of aqueous solutions such as the production of orange juice in which water is the major permeate component [8].

### 2.6.2. Air gap membrane distillation (AGMD)

The air gap membrane distillation configuration is shown in Fig. 6, where an air gap is introduced by using a plate to compart the cooling chamber from the membrane.

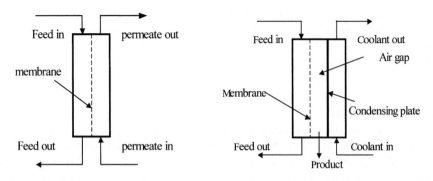

Fig. 5. DCMD configuration.                    Fig. 6. AGMD configuration.

In this configuration the migrated vapor molecules out of the membrane pores have to pass through the air gap and then condense on the plate, and the condensate formed is drained out of the air gap by gravity. In the cooling chamber the coolant is used for removing the latent heat released during the condensation of the vapor to liquid. The existence of stagnant air in the air gap presents a new resistance to mass transfer which in turn results in a low mass flux across the membrane. Because in most cases the thickness of air gap is much larger than that of the membrane, transport process through air gap is usually the controlling step in AGMD [9, 10]. However, due to the low heat conductivity of air, air gap reduces the heat loss by conduction considerably. So in AGMD the thermal efficiency is higher than in DCMD. Since the product (i.e. the condensate) can be obtained directly, it's easy to decide whether there is membrane leakage or wetting occurring by analyzing the product composition. On the other hand, compared with other MD configurations, the product of AGMD can be weighted accurately. These advantages make AGMD unique in experimental studies. Besides the application range of DCMD, AGMD can also be extended to the removal of trace volatile components from aqueous solutions.

*2.6.3. Sweeping gas membrane distillation (SGMD)*

As mentioned above, one advantage of AGMD over DCMD is the high thermal efficiency by the presence of air gap at the permeate side. However, this advantage is compromised by low mass transfer coefficient because the stagnant air gap presents a new resistance to mass transfer and transport process through air gap is the controlling step. One approach to overcome this problem is to promote the mass transfer in this region. In SGMD, this is done by blowing air into the cooling chamber and making the air flow tangentially over the surface of the membrane, instead of using a stagnant air layer comparting the membrane and the condensing surface. SGMD configuration is illustrated in Fig. 7. Compared with that in static air layer, the mass transfer in the air stream would be promoted greatly. The vapor of volatile component is taken out of the chamber by the air stream, and then condenses in an external condenser. Therefore, SGMD can be looked on as the combination of the low heat loss of AGMD and the high mass transfer coefficient of DCMD. However, by now very little work has been done in the field of SGMD. This is probably due to the fact that the permeate must be collected in an external condenser, the large sweeping gas flows is required to achieve significant permeate yield, and extra cost associated with transporting gas will be spent [11]. SGMD is suitable for removing the dissolved gas or volatile organic components from aqueous solutions.

### 2.6.4. Vacuum membrane distillation (VMD)

In VMD, the feed solution in direct contact with the membrane surface is kept at pressures lower than the minimum entry pressure (LEP); at the other side of the membrane, the permeate pressure is often maintained below the equilibrium vapor pressure by a vacuum pump. In this configuration, similar to SGMD, the vapor permeated also doesn't condense in the cooling chamber, but is taken out by vacuum and condense in an external condenser. VMD configuration is illustrated in Fig. 8. The total pressure difference between the two sides of the membrane causes a convective mass flow through the pores that contributes to the total mass transfer of VMD. On the contrary, in DCMD and AGMD there is only the diffusive flux of volatile component within the membrane pores. Therefore, mass flux of VMD is generally larger than that of other MD configurations. Another advantage of VMD comes from the negligible heat conduction through the membrane, due to the very low pressure at the permeate side of the membrane. This advantage makes VMD high thermal efficiency. Moreover, the mathematical model describing VMD becomes simple. Application of VMD is similar to that of SGMD, but more successful examples are still to be developed.

Fig. 9 shows the percentage of different MD configurations in MD references. It can be seen that DCMD has gained more attention by MD researchers, about 63% among the published references. This may be due to the simplicity in its setup and acceptable mass flux. However, it is found that VMD is the least studied one, about 8% among the published references. This may be due to the inconvenience to set up vacuum equipment and the difficulty in measuring mass flux through the membrane.

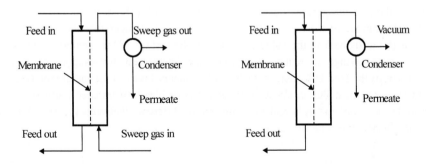

Fig. 7. SGMD Configuration.                    Fig. 8. VMD Configuration.

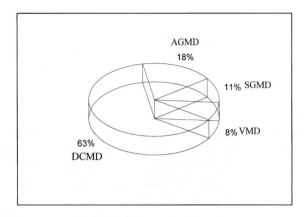

Fig. 9. Percentage of different MD configurations in MD references.

## 3. TRANSPORT PROCESS

### 3.1. Heat transfer

MD is a complicated physical process in which both heat and mass transfers are involved. According to the classical theory of heat transfer, a thermal boundary layer will be formed when a fluid is in direct contact with a solid surface, as long as the temperatures of these two objects are different. The thermal boundary layer is adjacent to the solid surface, and it is assumed that only in this region does the fluid exhibit its temperature profile. This viewpoint is adopted to depict the MD process. Within the MD module, two fluids with different temperatures are separated by a microporous membrane (with the thickness of $\delta$), so two thermal boundary layers appear at the feed side (with the thickness of $\delta_f$) and the permeate side (with the thickness of $\delta_p$) of the membrane respectively, as shown in Fig. 10. Within the boundaries, the feed temperature decreases from the value of $t_f$ (K) at its bulk to the value of $t_{fm}$ (K) at the surface of the membrane, while the permeate temperature increases from the value of $t_p$ (K) at its bulk to the value of $t_{pm}$ (K) at the surface of the membrane. Since MD relies on phase change to realize separation, the latent heat for evaporation must be transferred from the feed bulk, across its thermal boundary layer, to the membrane surface at the feed side. The heat flux $q_f$ (W m$^{-2}$) depends on the film heat transfer coefficient in the boundary layer $h_f$ (W m$^{-2}$ K$^{-1}$) and the temperature difference between the feed bulk and membrane surface. It can be written as

$$q_f = h_f \left( t_f - t_{fm} \right) \tag{2}$$

At the membrane surface of the feed side, the volatile component vaporizes. The produced vapor passes through the membrane, and then condenses on the vapor-liquid interface at the permeate side. By this means the latent heat is transferred from the feed side to the permeate side with the flux:

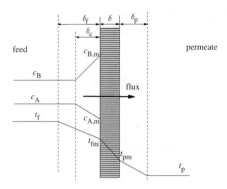

Fig. 10. Schematic representation of temperature and concentration profiles in the boundary layer adjacent to the membrane.

$$q_v = N \cdot \Delta H \tag{3}$$

where $N$ (kmol m$^{-2}$ K$^{-1}$) is the mass transfer flux through the membrane, and $\Delta H$ (kJ kmol$^{-1}$) is the latent heat of the volatile component. Here $q_v$ (W m$^{-2}$) is only a fraction of the heat energy transferred to the membrane surface and through the membrane in the form of latent heat. Besides the heat flux represented by $q_v$, due to the temperature difference between the two surfaces of the membrane, heat is also conducted through the membrane material and the gas that fills the pores with the flux:

$$q_m = h_m\left(t_{fm} - t_{pm}\right) \tag{4}$$

where $h_m$ (W m$^{-2}$ K$^{-1}$) is the heat transfer coefficient of the membrane, and can be determined by

$$h_m = \frac{\lambda_m}{\delta_m} \tag{5}$$

$\lambda_m$ (W m$^{-1}$ K$^{-1}$) is the average heat conductivity of membrane material and the gas that fills the pores:

$$\lambda_m = \varepsilon \cdot \lambda_s + (1 - \varepsilon)\lambda_g \tag{6}$$

where $\varepsilon$ is the porosity of the membrane, and $\lambda_s$ and $\lambda_g$ (W m$^{-2}$ K$^{-1}$) are the heat conductivity of membrane material and the gas that fills the pores, respectively.

The heat transfer coefficient of the membrane depends on its characteristics (i.e. the thickness and porosity) and the materials of which the membrane is fabricated. These factors are implied in Eqs. (5) and (6). $h_m$ is related with the heat conductivity of the membrane material and the air or vapor trapped within the membrane pores. The heat conductivity of some materials or gas involved in MD is listed in Table 2 [12]. It is shown that temperature has only a minor effect on the heat conductivity. The heat conductivity of crystalline polymers is almost not affected by temperature, as the polymer chains in crystalline structure hardly move or vibrate when energy is absorbed. And it is believed that for air and water vapor the change of thermal conductivities to temperature isn't sensitive.

Table 2
Heat conductivity of some materials or gas involved in MD

| $T / K$ | PVDF / $\text{W m}^{-1}\cdot\text{K}^{-1}$ | PTFE / $\text{W m}^{-1}\cdot\text{K}^{-1}$ | PP / $\text{W m}^{-1}\cdot\text{K}^{-1}$ | Air / $\text{W m}^{-1}\cdot\text{K}^{-1}$ | Water vapor / $\text{W m}^{-1}\cdot\text{K}^{-1}$ |
|---|---|---|---|---|---|
| 296 | 0.17 - 0.19 | 0.25 - 0.27 | 0.11 - 0.16 | 0.026 | 0.020 |
| 348 | 0.21 | 0.29 | 0.20 | 0.030 | 0.022 |

Fortunately, there is just a small difference between the thermal conductivities of water vapor and air. Thus, it is reasonable to take on the gas mixture in the pores as one component.

Owing to the contribution of both evaporation and conduction, the total heat flux transferred through the membrane can be written as

$$q_m = q_v + q_{m1} = N \cdot \Delta H + h_m \left( t_{fm} - t_{pm} \right) \tag{7}$$

Within the thermal boundary layer at the permeate side, heat flux is produced in the similar manner as at the feed side, i.e.

$$q_p = h_p \left( t_{pm} - t_p \right) \tag{8}$$

where $h_p$ ($\text{W m}^{-2}\ \text{K}^{-1}$) is the film heat transfer coefficient in the boundary layer.

In summary, the heat transfer process in MD includes three steps, and the corresponding regions are the thermal boundary layer of the feed side, the membrane itself and the thermal boundary layer of the permeate side. The heat fluxes for these regions are schematically represented in Fig. 11 in an electrical analog.

### 3.2. Mass transfer

In general, the mass transfer in MD consists of two steps: one is across the boundary layer at the feed side; the other is across the membrane. The latter is somewhat complicated and includes several basic mechanisms. Fig. 12 illustrates the relationship of all the possible basic mass transfer mechanisms in an electrical analogy. Note that the surface diffusion mechanism, which occurs in the process of gas mass transfer through porous medium, isn't included because it has little influence on the whole process due to the weak molecule-membrane interaction. Herein, it should be mentioned that the occurrence and the weight of a mechanism in mass transfer process rest on many factors, such as the composition of the feed, the fluid dynamics, the operation temperature and pressure, the membrane characteristics, the structure of the membrane module, etc.

If pure water (or other pure volatile solvent) is used as the feed in MD, the concentration of volatile component at the membrane surface is equal to that of the bulk fluid, i.e., no concentration profile exists at the feed side. In this case no mass transfer resistance is associated with the feed, and all mass transfer resistance is concentrated on the membrane itself.

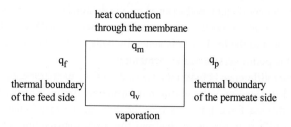

Fig. 11. Schematic representation of heat transfer in MD.

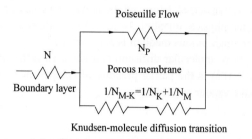

Fig. 12. Schematic representation of mass transfer in MD.

For another case, if a solution, which contains one non-volatile component and at least one volatile component, is used as the feed in MD, then within the mixture the mass transfer resistance to volatile component exists. Due to this resistance, the evaporation of volatile component at the membrane surface will result in its depletion and/or the buildup of the non-volatile component near the membrane surface. The region near the membrane surface, where the concentration profile of volatile and non-volatile components is established, is called concentration boundary layer. As shown in Fig. 10, the thickness of concentration boundary layer is $\delta_c$ (m). In this layer the concentration of non-volatile component increases from $c_B$ to $c_{B,m}$ (kmol m$^{-3}$), while that of volatile component decreases from $c_A$ to $c_{A,m}$ (kmol m$^{-3}$). The depletion or buildup of components in the concentration boundary layer due to the mass transfer resistance is referred to as concentration polarization. For a given bulk concentration, the presence of concentration boundary layer reduces the driving force for the volatile component to pass through the membrane, and thus decreases the transmembrane mass flux.

According to the theory of mass transfer in boundary layer, a mass balance in the feed side boundary layer yields to the relationship of the concentration of non-volatile component at the membrane surface and at feed bulk:

$$N_A = k_f c \ln \frac{c_B}{c_{B,m}} \tag{9}$$

where $N_A$ (kmol m$^{-2}$ s$^{-1}$) is the volatile component mass flux across the boundary layer, $k_f$ (m s$^{-1}$) is the mass transfer coefficient, and $c$ (kmol m$^{-3}$) is the total concentration at the feed.

As long as no wetting occurs, the membrane used in MD can 100% rejects the non-volatile component at the feed, so it is generally believed that no concentration boundary layer is formed at the permeate side of the membrane.

The temperature difference between the two sides of the membrane, $t_{fm}$ - $t_{pm}$, gives rise to a vapor pressure difference of volatile component across the membrane, $P$ ($t_{fm}$) - $P$ ($t_{pm}$), which acts as the driving force for mass transfer through the membrane. The hydrophobic membrane used in MD is a porous media, and the mass transfer through such medium is regulated by three kinds of basic mechanisms: Knudsen diffusion, molecular diffusion and Poiseuille flow. But the actual process is sometimes regulated by the combination of them known as transition mechanism. Two important factors affecting mass transfer are the mean free path of the gas molecule transferred $\lambda$ (m) and the mean pore diameter of the membrane $d$ (m). In accordance with the physical quantity, $\lambda / d$, defined as Knudsen number ($Kn$), the mass transfer mechanism through porous medium is summarized in Table 3, where P, M and K represent the Poiseuille flow, molecular diffusion and Knudsen diffusion, respectively, $P$ (Pa) is the total pressure, $P_i$ (Pa) is the partial pressure of component $i$, and $y_i$ is the mole fraction of component $i$ in the vapour or gas phase.

Table 3
The influence of Knudsen number on mass transfer through porous medium

| Driving force | $Kn < 0.01$ | $0.01 < Kn < 1$ | $Kn > 1$ |
|---|---|---|---|
| Single gas $\nabla P \neq 0$ | P | P – K transition | K |
| Gas mixture $\nabla P = 0$, $\nabla P_A \neq 0$ or $\nabla y_A \neq 0$ | M | M - K transition | K |
| Gas mixture $\nabla P \neq 0$ | P - M | P – M - K transition | K |

### 3.3. Mechanism of gas transport in porous medium
(1) Mass transfer of single gas

For mass transfer of single gas, the resistance caused by molecular - molecular collision within the pores is neglected. The gas would permeate through the membrane if only a total pressure gradient is imposed on the membrane. The mass transfer resistance may come from the collision of the gas molecules with the inside wall of the pores and the viscosity of the gas. The weight of these two resistances depends on Knudsen number. In the case of $Kn > 1$ (or $\lambda > d$), the molecules - wall collision predominates in the mass transfer process, and the resistance coming from the viscosity of the gas can be neglected. Since the mass transfer within membrane pores is regulated by Knudsen diffusion [13], it can be written as

$$N_K = -\frac{4}{3} d \frac{\varepsilon}{\tau} \sqrt{\frac{1}{2\pi RMT}} \nabla P \tag{10}$$

where $\nabla P$ (Pa) is the pressure gradient within the membrane pores, $R$ (J mol$^{-1}$ K$^{-1}$) is universal gas constant, $M$ (kg kmol$^{-1}$) is molecule weight of the gas, and $T$ (K) is the gas temperature.

In the case of $Kn < 0.01$ (or $\lambda < 0.01d$), the pore size is much larger than the free path of the molecules. In this case the molecules-wall collision can be neglected, and the resistance coming from the gas viscosity predominates in the mass transfer process. The mass flux describing the gas passing through the pores by Poiseuille flow mechanism [13] is:

$$N_P = -\frac{\varepsilon}{\tau}\frac{d^2}{32\,\eta}c\nabla P \tag{11}$$

where $\eta$ (Pa s) is the gas viscosity, and $c$ (kmol m$^{-3}$) is the total concentration of the gas.

In the case of $0.01 < Kn < 1$, neither of the resistances can predominate exclusively in the mass transfer process, and in this case both Knudsen diffusion and Poiseuille flow have noticeable contribution to mass transfer. Therefore, it is advisable to use the combination of these two mechanisms, called Poiseuille flow – Knudsen diffusion transition, to describe mass transfer. The whole mass flux can be written:

$$N_{P-K} = N_P + N_K \tag{12}$$

where $N_P$ and $N_K$ (kmol m$^{-2}$ s$^{-1}$) represents the contribution of Poiseuille flow and Knudsen diffusion to mass transfer, respectively.

(2) Mass transfer of gas mixture with no total pressure gradient

For simplicity, it is assumed that the gas mixture within membrane pores is binary, consisting of components A and B. When total pressure gradient is zero, there is no resistance caused by gas viscosity in the pores. Only if there exists concentration gradient of A within the pores, mass transfer will occur. The actual mechanism is also associated with the value of $Kn$. In the case of $Kn > 1$, Knudsen diffusion predominates in the mass transfer process and the corresponding mass flux can be written as

$$N_K = -\frac{4}{3}d\frac{\varepsilon}{\tau}\sqrt{\frac{1}{2\pi RMT}}\nabla P_A \tag{13}$$

where $\nabla P_A$ represents the partial pressure gradient of volatile component A. However, in the case of $Kn < 1$, the possibility of molecule-molecule collision of components A and B within the pores increases. In other words, molecule diffusion plays a role. In the case of $Kn < 0.01$, molecule-molecule collision overwhelms molecule-pore wall collision so that the mass transfer of component A should be governed by molecule diffusion [14]. The corresponding mass flux can be written as

$$N_A = -\frac{\varepsilon D_{AB}}{\tau RT}\nabla p_A + y_A N_t \tag{14}$$

where $D_{AB}$ (m$^2$ s$^{-1}$) is the diffusivity of component A relative to B (B as the inert component trapped within the membrane pores), $y_A$ is the mole fraction of component A and $y_A N_t$ (kmol m$^{-2}$ s$^{-1}$) represents the contribution of bulk flow to the mass transfer within the pores.

When $0.01 < Kn < 1$, the corresponding mass transfer mechanism is Kndusen-molecule diffusion transition, in which both of the resistance of molecule-pore wall collision and molecule-molecule collision have noticeable influence on mass transfer, but neither of them can predominate exclusively. The total mass flux is related with $N_M$ and $N_K$:

$$\frac{1}{N_{M-K}} = \frac{1}{N_M} + \frac{1}{N_K} \tag{15}$$

which is the result of the competition between these two kinds of collisions.

(3) Gas mixture with total pressure gradient

When $Kn > 1$, mass transfer is controlled by the resistance of molecule-pore wall collision, and mass flux can be expressed by Eq. (13). When $Kn < 0.01$, however, with total pressure gradient, Poiseuille flow has noticeable contribution to mass transfer. Moreover, the contribution from molecule diffusion can't be neglected if only concentration gradient of component A exists within the pores. So in this case the mass transfer mechanism is Poiseuille flow-molecule diffusion transition, and mass flux is the sum of contributions of Poiseuille flow and molecule diffusion.

$$N_{P-M} = N_P + N_M \tag{16}$$

In the case of $0.01 < Kn < 1$, in principle, all the three basic mechanisms have their influence on mass transfer, and the total mass flux is:

$$N_{P-M-K} = N_P + N_{M-K} \tag{17}$$

However, if the pressure within the pores is small enough (e.g. in VMD process), the free path of the gas molecule is larger than the pore size. In this case the contribution of molecule diffusion $N_M$ can be neglected in transition mechanism.

### 3.4. Determining the characteristics of porous membrane

Knowledge of membrane characteristics, such as its porosity, thickness, and mean pore size etc., are important in the studies of membrane fabrication and membrane separation process. One of the most popular methods to determine the characteristics of porous membrane is by gas permeation (GP) experiment. The experiment materials involve the porous membrane to be investigated and a kind of single gas such as nitrogen. The single gas is driven through the membrane by a total pressure difference exerted between the two sides of the membrane. As mentioned above, the permeation of single gas driven by a total pressure across a porous membrane will be regulated by Knudsen diffusion – Poiseuille flow mechanism. The total mass flux can be obtained from Eqs. (10) – (12):

$$N_{K-P} = \left( -\frac{8}{3} \frac{r\varepsilon}{\tau} \sqrt{\frac{1}{2\pi RMT}} - \frac{\varepsilon}{\tau} \frac{r^2}{8\eta} c \right) \nabla p \tag{18}$$

Integrate this differential equation over the membrane thickness, and the steady-state gas permeation flux is obtained:

$$N_{K-P} = \left( \frac{8}{3} \frac{\varepsilon r}{\tau\delta} \sqrt{\frac{1}{2\pi RMT}} + \frac{\varepsilon r^2}{\tau\delta} \frac{1}{8\eta} \frac{p_m}{RT} \right) \Delta p \tag{19}$$

In another form,

$$J_{K-P} = A_0 + B_0 p_m \tag{20}$$

where $J_{\text{K-P}} = \dfrac{N_{\text{K-P}}}{\Delta p}$, $A_0 = \dfrac{8}{3}\dfrac{\varepsilon r}{\tau\delta}\sqrt{\dfrac{1}{2\pi RMT}}$ and $B_0 = \dfrac{\varepsilon r^2}{\tau\delta}\dfrac{1}{8\eta RT}$.

$P_{\text{m}}$ (Pa) is the average pressure within the membrane pores. In order to get $A_0$ and $B_0$, gas permeation experiment is carried out at various $P_{\text{m}}$, while keeping the pressure difference across the membrane constant. Under this condition, the gas permeation fluxes $N_{\text{K-P}}$ (kmol m$^{-2}$ s$^{-1}$) through the membrane are measured. In terms of Eq. (20), by plotting the curve of $J_{\text{K-P}}$ (kmol m$^{-2}$ s$^{-1}$ Pa$^{-1}$) with $P_{\text{m}}$, the intercept with $y$-axis is $A_0$, and the slope is $B_0$. On this basis, the membrane characteristics can be obtained from $A_0$ and $B_0$ by the following equations:

$$\frac{\varepsilon r}{\tau\delta} = \frac{3}{8}A_0\sqrt{2\pi RMT}, \quad \frac{\varepsilon r^2}{\tau\delta} = 8B_0\eta RT \tag{21}$$

$$r = \frac{16}{3}\frac{B_0}{A_0}\sqrt{\frac{8RT}{\pi M}}\eta, \quad \frac{\varepsilon}{\tau\delta} = \frac{8\eta RTB_0}{r^2} \tag{22}$$

The flowsheet of a GP experiment is diagrammed in Fig. 13. Nitrogen is introduced from the cylinder pump into the membrane module, in which the flat sheet membrane is put on a porous sintered body made of stainless steel or ceramic as its supporter. The pressures at two sides of the membrane are regulated by valves, and measured with pressure sensors. A soap flowmeter is used to measure the flowrate of the gas that permeates through the membrane. The tank locates between the membrane module and the soap flowmeter, and acts as a buffer to get a stable gas stream.

Fig. 14 shows the experimental result about the relationship of $J_{\text{K-P}}$ with $P_{\text{m}}$ for a flat sheet PTFE membrane under a constant pressure difference of 100368 Pa. The data are correlated well with a straight line. In terms of Eqs. (21) and (22), the membrane characteristics can be obtained from the slop and the intercept of this line. For the membrane investigated $r$ is 0.172 μm, and $\varepsilon/\tau\delta$ is 2333.8 m$^{-1}$. Similarly, under other pressure differences, the respective values of $r$ and $\varepsilon/\tau\delta$ can also be obtained, and the results are listed in Table 4, from which it can be seen that membrane characteristics change slightly with pressure difference. The reason may be due to the compaction of membrane during experiment. The increase of pressure difference makes the membrane with constant polymer volume become thinner, which leads to the decrease of the porosity and pore size.

## 4. MATHEMATICAL MODEL

Mathematical model of MD is necessary for understanding the process and acting as an infrastructure for industrial design. It focuses on the description of the heat and mass transfers. In what follows, how to deduce the model equations and analyze the MD performance according to the solution of these equations are clarified. Moreover, for simplification, the feed liquid is assumed to be pure water if without specially mentioned. A number of successful attempts to model the MD process on the ground of confirmed experimental results or theoretical analysis can be found in literatures. The primary purpose of these models was to predict the mass flux through the membrane and its dependence on design and process variables. In the modeling of MD process, the effect of both the temperature and concentration polarizations should be accounted for.

Table 4
Membrane characteristic under various pressure difference

| $\Delta P$ (Pa) | $A_0$ | $B_0$ | $r$ ($\mu$m) | $\varepsilon / \tau\delta$ (m) |
|---|---|---|---|---|
| 31280 | 4.98E-5 | 2.41E-10 | 0.211 | 1829.9 |
| 48824 | 5.07E-5 | 2.16E-10 | 0.186 | 2108.1 |
| 64736 | 5.12E-5 | 2.07E-10 | 0.177 | 2247.6 |
| 82552 | 5.17E-5 | 2.06E-10 | 0.175 | 2295.5 |
| 100368 | 5.19E-5 | 2.05E-10 | 0.172 | 2333.8 |

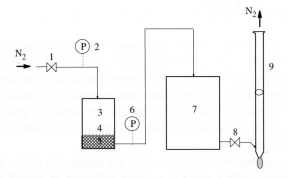

Fig. 13. Flowsheet of gas permeation experiment: 1, 8 - regulating valve; 2, 6 - pressure gauge; 3 - membrane module; 4 - membrane; 5 - membrane supporter; 7 - tank as buffer; 9 - soap flowmeter.

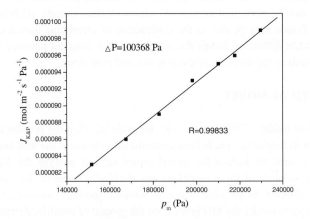

Fig. 14. The relationship of $J_{K-P}$ with $P_m$ obtained from gas permeation experiment.

## 4.1. Mathematical model of DCMD

Among the four MD configurations, DCMD is the one whose mathematical model is mature and the easiest to understand. Therefore, firstly we will show the derivation of model equations for the heat and mass transfers in DCMD, and then take insight into DCMD by analyzing the solutions of these equations.

In DCMD the pressure difference at two sides of the membrane will be zero when both the feed and permeate flows are under atmospheric pressure. In this case the contribution of Poiseuille flow to mass transfer can be neglected. At the typical membrane temperature of 60 ℃, the mean free path of water vapor is 0.11 $\mu$m and the mean pore diameter of membranes is 0.1 - 0.5 $\mu$m. Therefore, in the system where water (component A) is used as volatile component $Kn$ may vary from 0.2 to 1.0. As air (component B) is trapped in the pores, the permeation of water vapor through the membrane is regulated by the Knudsen-molecule diffusion transition mechanism. For molecule diffusion, the contribution of bulk flow to the permeation of component A through the membrane should be considered, regarded as the term $y_A N_t$ in Eq. (14). The total flow is:

$$N_t = N_{tA} + N_{tB} \tag{23}$$

As the air trapped in the pores is static, no net flux for component B crosses through any plane parallel to the membrane surface. The total flux of component B is caused only by diffusion. For a binary gas system, the diffusion flux of the two components is equal. So it becomes:

$$N_{tB} = \frac{\varepsilon D_{BA}}{\tau RT} \nabla P_B = -\frac{\varepsilon D_{AB}}{\tau RT} \nabla P_A \tag{24}$$

and

$$N_{tB} = y_B N_t = (1 - y_A) N_t \tag{25}$$

In terms of Eq. (25), it can be deduced that

$$N_t = -\frac{1}{1 - y_A} \frac{\varepsilon D_{AB}}{\tau RT} \nabla P_A \tag{26}$$

Substituting this equation into Eq. (14) gives rise to

$$N_{A,M} = -\frac{1}{1 - y_A} \frac{\varepsilon D_{AB}}{\tau RT} \nabla P_A \tag{27}$$

Substituting Eq. (27) and Eq. (13) into Eq. (15) produces the following differential equation of mass flux for component A in the Knudsen - molecule diffusion transition mechanism:

$$N_{K-M} = -\frac{\varepsilon}{RT\tau} \left( \frac{P - P_A}{P D_{AB}} + \frac{3}{4d} \sqrt{\frac{2\pi M}{RT}} \right)^{-1} \nabla P_A \tag{28}$$

where $P$ (Pa) is the total pressure within the pores. Integrate Eq. (28) over the membrane thickness where the corresponding partial pressure of vapor deceases from $P_{fm}$ (Pa) to $P_{pm}$ (Pa), and the steady-state mass flux of water vapor is obtained:

$$N_{\text{K-M}} = \frac{\varepsilon}{\tau\delta}\frac{PD_{AB}}{RT_m}\ln\frac{\dfrac{P-P_{\text{pm}}}{PD_{AB}}+\dfrac{3}{4d}\sqrt{\dfrac{2\pi M_A}{RT_m}}}{\dfrac{P-P_{\text{fm}}}{PD_{AB}}+\dfrac{3}{4d}\sqrt{\dfrac{2\pi M_A}{RT_m}}} \tag{29}$$

where $T_m$ (K) is the mean temperature within the pores. $D_{AB}$ (m$^2$ s$^{-1}$) at the temperature of 273 - 373 K is estimated from the following empirical equation [15]:

$$PD_{AB} = 1.895\times10^{-5}T^{2.072} \tag{30}$$

where the unit of $PD_{AB}$ is Pa·m$^2$·s$^{-1}$.

On the other hand, the resistances in the heat transfer process of DCMD consist of three parts: the resistance of boundary layer at the feed side, of the membrane and of the boundary layer at the permeate side. These resistances are connected in series, and at steady-state the heat flux is a constant, i.e. $q_f = q_m = q_p$. Thus, from Eqs. (2), (7) and (8) we obtain:

$$t_{\text{fm}} = \frac{h_m\left(t_p + t_f\cdot h_f/h_p\right)+h_f\cdot t_f - N\cdot\Delta H}{h_m + h_f\left(1+h_m/h_p\right)} \tag{31}$$

$$t_{\text{pm}} = \frac{h_m\left(t_f + t_p\cdot h_p/h_f\right)+h_p\cdot t_p + N\cdot\Delta H}{h_m + h_p\left(1+h_m/h_f\right)} \tag{32}$$

Eqs. (29), (31) and (32) form a non-linear algebraic equation group in which $t_{\text{fm}}$, $t_{\text{pm}}$ and $N$ are variables. An explicit secant method [16] is recommended to solve it. However, when applying these equations to predict mass flux of water vapor in DCMD, it should be mentioned:

(1) Membrane characteristics (including $r$ and $\varepsilon/\tau\delta$) which can be determined by either gas permeation experiment or scanning electron microscopy and image analysis program.

(2) Membrane heat transfer coefficient $h_m$ which can be determined by Eqs .(5) and (6), or by empirical estimation. In general, the mean heat conductivity of porous membrane, $k_m$, is in the magnitude of $10^{-2}$ W m$^{-1}$·K$^{-1}$, the thickness of $10^{-4}$ m. Thus, the heat transfer coefficient is in the magnitude of $10^{2}$ W m$^{-2}$·K$^{-1}$.

(3) Heat transfer coefficient in the boundary layers, $h_f$ and $h_p$, which is affected by many factors such as the structure of the membrane module, flow velocities of the feed and permeate in the module, the physical properties of feed and permeate, the operation temperature and so on. Because of the geometrical similarity of flow channel in MD membrane module and heat exchangers, many researchers like to extend the applicable range of the correlations from heat exchangers to MD module. Table 5 lists some correlations for laminar and turbulent flow that may be useful in MD. But it is wise to use heat transfer correlations directly from the MD experiment because the boundary conditions at the membrane surface aren't exactly as same as at the heat transfer wall in heat exchangers [17].

Table 5
Correlations of heat transfer coefficient that may be useful in MD, adapted from the source [12].

| Laminar flow | Turbulent flow |
|---|---|
| $Nu = 1.86\left(\dfrac{\text{Re Pr}}{L/D}\right)^{1/3}$ | $Nu = \left(1+\dfrac{6D}{L}\right)\left[\dfrac{(f/8)\text{Re Pr}}{1.07+12.7(f/8)^{1/2}\left(\text{Pr}^{2/3}-1\right)}\right]$ |
| $Nu = 4.36+\dfrac{0.0036\,\text{Re Pr}(D/L)}{1+0.0011[\text{Re Pr}(D/L)]^{0.8}}$ | $Nu = \left(1+\dfrac{6D}{L}\right)\left[\dfrac{(f/8)(\text{Re}-1000)\text{Pr}}{1+12.7(f/8)^{1/2}\left(\text{Pr}^{2/3}-1\right)}\right]$ |
| $Nu = 0.13\,\text{Re}^{0.64}\,\text{Pr}^{0.38}$ | $Nu = 0.023\left(1+\dfrac{6D}{L}\right)\text{Re}^{0.8}\,\text{Pr}^{1/3}$ |
| $Nu = 1.95\left(\dfrac{\text{Re Pr}}{L/D}\right)^{1/3}$ | $Nu = 0.036\,\text{Re}^{0.8}\,\text{Pr}^{1/3}\left(\dfrac{D}{L}\right)^{0.055}$ |
| $Nu = 0.097\,\text{Re}^{0.73}\,\text{Pr}^{0.13}$ | $Nu = \left(1+\dfrac{6D}{L}\right)\left[\dfrac{(f/8)\text{Re Pr}}{1.2+13.2(f/8)^{1/2}\left(\text{Pr}^{2/3}-1\right)}\right]$ |
| $Nu = 3.66+\dfrac{0.104\,\text{Re Pr}(D/L)}{1+0.0106[\text{Re Pr}(D/L)]^{0.8}}$ | $Nu = 0.027\left(1+\dfrac{6D}{L}\right)\text{Re}^{0.8}\,\text{Pr}^{1/3}\left(\dfrac{\mu}{\mu_w}\right)^{0.14}$ |
| $Nu = 11.5(\text{Re Pr})^{0.23}(D/L)^{0.5}$ for cooling | |
| $Nu = 15(\text{Re Pr})^{0.23}(D/L)^{0.5}$ for heating | |

Note: $f = [0.79\ln(\text{Re})-1.64]^{-2}$ for all correlations

(4) The vapor pressures of volatile component, $P_{fm}$ (Pa) and $P_{pm}$ (Pa) which is often determined by the vapor-liquid equilibrium. The driving force of MD is vapor pressure difference across the membrane imposed by a temperature difference across the membrane, or by a vacuum or a sweeping gas. An important assumption adopted in modeling MD is that the kinetic effects at the vapor-liquid interface are negligible. In other words, the vapor and liquid are assumed to be in the equilibrium state at the temperature of the membrane surface and at the pressure within the membrane pores. According to this assumption, vapor-liquid equilibrium equations can be applied to determine the partial vapor pressures of each component at each side of the membrane. For pure solvent, the partial vapor pressure is equal to the saturation pressure. According to the famous Antoine equation,

$$p_i^0 = \exp\left(a-\frac{b}{c+T}\right) \tag{33}$$

where $p_i^0$ (Pa) is the saturation pressure, $T$ (K) is the temperature, and $a$, $b$ and $c$ are substance constants and are readily available in the references [53-55]. For water, $a = 23.1964$, $b = 3816.44$, $c = -46.13$.

### 4.2. Performance of DCMD

By means of mathematical model, we can know the performance of DCMD and find out the main factors affecting the process. Some results are obtained based on a specified microporous flat sheet membrane with the characteristics: $d = 0.5 \ \mu m$, $\varepsilon/\tau\delta = 3000 \ m^{-1}$ and $h_m = 300W \ m^{-2}\cdot K^{-1}$. The heat transfer coefficient is calculated by the following correlation obtained from the experiment:

$$Nu = 0.19 \, Re^{0.68} \, Pr^{0.33} \qquad (34)$$

#### 4.2.1. The effect of temperature

In DCMD, the driving force for mass transfer of a volatile component through the membrane is its vapor pressure difference caused by the corresponding temperature difference between the two sides of the membrane. In principle, mass flux through the membrane can be improved by either increasing the feed temperature or decreasing the permeate temperature. Fig. 15 shows the change of mass flux with feed temperature under constant permeate temperature. The influence of feed temperature is remarkable in DCMD. As shown in Fig. 15, mass flux increases with feed temperature in an exponential way, reflecting the exponential increase of vapor pressure with temperature for the volatile components.

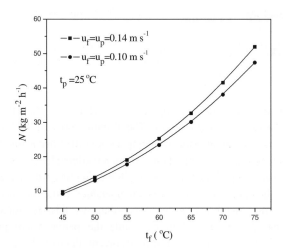

Fig. 15. Change of transmembrane flux with feed temperature in DCMD.

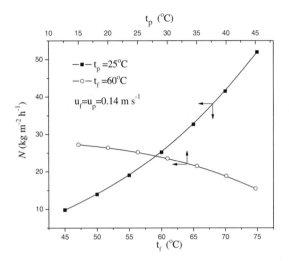

Fig. 16. Comparison of the effect of $t_f$ and $t_p$ on mass flux in DCMD.

A comparison on the effect of increasing feed temperature under constant permeate temperature and decreasing permeate temperature under constant feed temperature is made in Fig. 16. Compared with feed temperature, the permeate temperature has less effect on mass flux. This attributes to the different degree that vapor pressure changes with temperature. At the high temperature (i.e. above 45℃) the vapor pressure changes significantly with the increase of feed temperature, while relatively slowly with the increase of temperature within the range of permeate temperature.

Although increasing the feed temperature can effectively enhance mass flux in DCMD, the large temperature difference may also improve heat conduction through the membrane, which isn't good for heat transfer. From this viewpoint, there is a trade-off between mass flux and thermal efficiency.

### 4.2.2. The effect of flow velocity

There are two thermal boundary layers at two sides of the membrane respectively. Due to the heat transfer resistances within the layers, the temperature difference between the two membrane surfaces is, more or less, lower than that imposed on the bulks of the feed and permeate. This phenomenon is called as temperature polarization. The relative value of the boundary layer resistances to the total heat transfer resistance is defined as the temperature polarization coefficient (TPC) [2]:

$$\text{TPC} = \frac{t_{fm} - t_{pm}}{t_f - t_p} \tag{35}$$

TPC approaching unity means that the fluid dynamics of the system is in good condition and the process is controlled by mass transfer through the membrane; otherwise, TPC approaching zero means that the system is designed poorly and the process is controlled by heat transfer through boundary layer. Another physical meaning of TPC is that it is the fraction of driving force used for mass transfer across the membrane. For most of the systems, TPC is in the range of 0.4 to 0.8. The factors affecting heat transfer coefficient, i.e. properties of the fluids, flow velocity, the module geometrical shape, etc. also have influence on TPC.

The effect of flow velocities of the feed and permeate over the membrane on mass flux in DCMD is shown in Fig. 17. It can be seen that mass flux is improved by increasing flow velocity. The reason is that the increase of flow velocity leads to the reduction of the heat transfer resistance within the boundary layers, and thus the effect of temperature polarization is mitigated. As a result, $t_{fm}$ increases and $t_{pm}$ decreases, which causes a larger driving force for mass transfer through the membrane.

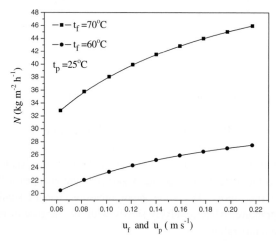

Fig. 17. The effect of flow velocity on mass flux in DCMD.

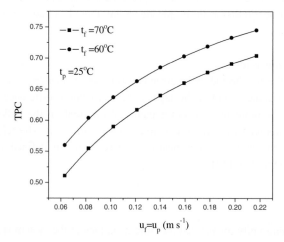

Fig. 18. Change of TPC with flow velocity.

Fig. 18 is plotted according to Eqs. (31), (32) and (35), showing the obvious change of TPC with flow velocity. It indicates that temperature polarization has a considerable effect on DCMD. In the 1960s when hydrophobic porous membrane preparation technology wasn't well developed, it is difficult to observe this phenomenon.

Increasing flow velocity isn't the only way to mitigate the effect of temperature polarization in DCMD. Turbulent promoters are another way to improve heat and mass transfers in membrane processes. Net-type spacers, as illustrated in Fig. 19, are often placed in the flow channels in such membrane processes as reverse osmosis and untrafiltration to improve mass transfer and to reduce the effect of concentration polarization and fouling. When they are applied to MD, it is found to be very effective in improving mass flux, up to approximately 50% enhancement in DCMD [12]. The spacers act as the turbulent promoter in flow channels, destabilize the flow and create eddy currents in the laminar regime. Therefore, momentum, heat, and mass transfers are enhanced. The reduced heat transfer resistance within the thermal boundary layer improves TPC so that the driving force for mass transfer across the membrane is improved.

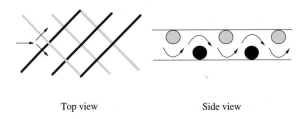

Top view                                    Side view

Fig. 19. Spacer packed in the membrane module as turbulent promoter; adapted from the source [12].

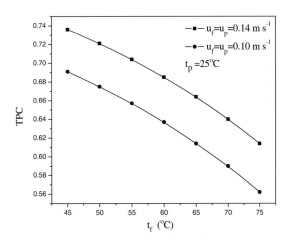

Fig. 20. The effect of temperature on TPC.

One important factor affecting TPC is the temperatures of the feed and permeate, as shown in Fig. 20. It can be seen that the feed temperature has a negative effect on TPC. When a high feed temperature is adopted in operation, mass flux increases. At the same time, more heat flux penetrates through the thermal boundary layer, and thus according to Eqs. (4) and (8), the corresponding temperature gradient is aggrandized.

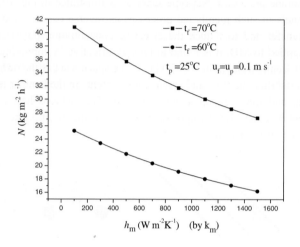

Fig. 21. The influence of $h_m$ on the transmembrane flux in DCMD.

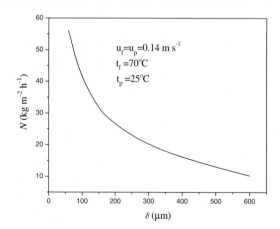

Fig. 22. Change of mass flux with membrane thickness.

*4.2.3. The effect of membrane heat transfer coefficient and membrane thickness*

Fig. 21 shows the influence of heat transfer coefficient of the membrane, $h_m$, on mass flux under a given membrane characteristics. It is known that $h_m$ has relationship with membrane material and its thermal conductivity. The higher the thermal conductivity is, the more the heat loss by conduction is. This means high thermal conductivity would result in a low mass flux. In other words, the fraction of the heat energy used for evaporation becomes less if a membrane with high thermal conductivity is employed in DCMD. So in this case thermal efficiency of the system is reduced.

However, thermal conductivity of membrane material isn't the only factor that affects $h_m$. Eq. (5) indicates a linear decrease of $h_m$ with the thickness of the membrane $\delta$. So a membrane with a large thickness should show a high thermal efficiency in DCMD.

On the other hand, Eq. (29) indicates that mass flux is inversely proportional to the membrane thickness. The total influence of membrane thickness on mass flux is shown in Fig. 22, which is plotted by calculation according to Eqs. (29), (31) and (32). Here, the sharp decline of mass flux with membrane thickness is attributed to the lengthened path within the membrane for the vapor to pass through. However, it can be seen that the change of mass flux with membrane thickness isn't always in an inversely proportional manner. When the decline of mass flux is up to a tenfold, the corresponding increase of membrane thickness is only by six folds. This further proves that thermal efficiency of the system is enhanced when a membrane of large thickness is used. In summary, there is always an optimum membrane thickness that would result in a more thermally efficient system with an acceptable mass flux.

Another aspect concerning membrane thickness is its mechanical strength. Although DCMD is usually performed under pressures close to the atmosphere pressure, the thinner the membrane is, the more likely it is to be damaged.

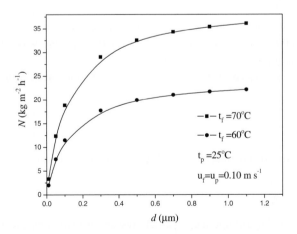

Fig. 23. The effect of pore diameter on the transmembrane flux in DCMD.

## 4.2.4. The effect of pore size

For the transport of a gas through porous medium, the pore size will influence the permeability of the medium. The effect of pore size (pore diameter) on mass flux of DCMD is theoretically predicted based on the mathematical model discussed above, and the results are shown in Fig. 23. It can be seen that the pore size has almost no influence on mass flux when the pore diameter is larger than 0.4 $\mu$m. A considerable mass flux change can be observed only when the pore diameter is in the range of being less than 0.2 $\mu$m.

As mentioned before, mass transfer through the membrane in DCMD is often regulated by Knudsen-molecular diffusion transition. These two resistances to mass transfer are combined in series. The free path of water vapor at 60 ℃ is about 0.11 $\mu$m, and when the pore diameter is less than this value, the molecular-wall collision (or Knudsen diffusion) will dominate the mass transfer process. Therefore, pore size is an important factor (as indicated in Eq. (13)). But with the increase of pore size, the molecular-wall collision (or Knudsen diffusion) becomes less important in mass transfer. The larger the pore size is, the less the opportunity for the vapor molecule to collide with the inside wall of the pores within the membrane is. In this case Knudsen diffusion mechanism will become less important. On the other hand, the reductive molecule-wall collision gives prominence to molecule-molecule collision within the pores. However, the molecular diffusion mechanism has no relation with pore size, as indicated in Eq. (14), so the two curves in Fig. 23 tend to plateau when the pore size becomes large enough.

The mass transfer process in DCMD includes two steps: the volatile component passes through the concentration boundary layer on the feed side of the membrane and through the microporous membrane itself. For mass transfer through the membrane, because of its complexity, many researchers adopted an empirical approach to describe this process, assuming that mass flux is proportional to the vapor pressure difference across the membrane:

$$N = C\left[P\left(t_{fm}\right) - P\left(t_{pm}\right)\right] \tag{36}$$

where $C$(kmol m$^{-2}$ s$^{-1}$ Pa$^{-1}$) is the membrane distillation coefficient (MDC), and can be determined from MD experiments. $P(t_{fm})$ and $P(t_{pm})$ (Pa) are the vapor pressures at the membrane surface as a function of temperature.

Schofield [18] suggested that for the Knudsen-diffusion mechanism, MDC can be calculated by

$$C_K = 1.064 \frac{r\varepsilon}{\tau\delta}\left(\frac{M}{RT_m}\right)^{0.5} \tag{37}$$

and for the molecular-diffusion mechanism, MDC can be calculated by

$$C_M = \frac{1}{P_{B,m}}\frac{PD_{AB}\varepsilon}{\tau\delta}\left(\frac{M}{RT_m}\right) \tag{38}$$

where $P_{B,m}$ (Pa) is the logarithmic mean partial pressure of inert component at two sides of the membrane.

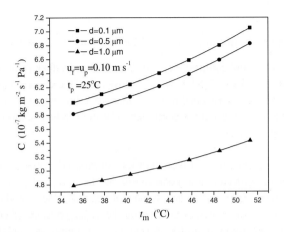

Fig. 24. Change of MDC with mean temperature within the membrane.

Eqs. (37) and (38) indicates that for Knudsen-diffusion mass transfer process, MDC will decrease proportionally with $T_m^{0.5}$, while for molecule-diffusion mass transfer process, MDC will increase with $T_m$ (by Eq. (30), $D_{AB}$ is proportional to $T_m^2$). MDC derived from Knudsen-molecule diffusion transition model is shown in Fig. 24. The fact that MDC increases with the rising of $T_m$ indicates that molecule diffusion plays a considerable role in mass transfer through the membrane. Even for the membrane with only 0.1 $\mu$m pore diameter, MDC shows no decrease with the rising of $T_m$, indicating a noteworthy mass transfer resistance caused by molecule-molecule collision. The influence of molecule-molecule collision on mass transfer is also embodied by the difference between $C_K$ and MDC. For a typical MD, $C_K$ is usually larger than $1.0 \times 10^{-6}$ kg m$^{-2}$ s$^{-1}$ Pa$^{-1}$, but MDC is lower than $8.0 \times 10^{-7}$ kg m$^{-2}$ s$^{-1}$ Pa$^{-1}$, as shown in Fig. 24.

### 4.2.5. The influence of solute at the feed

For simplicity, the above analysis is based on the assumption that the feed is pure water. In the experiment, the pure water is usually used as the feed in order to measure the fluid dynamics in the module, evaluate the membrane permeability, etc. In practice, however, the liquid feed, more or less, contains some solutes maybe volatile or non-volatile. The existence of solutes at the feed can affect the volatility of the solvent in two ways:

(a) Due to the dilution effect, the vapor pressure of the solvent in the mixture is lower than that of the pure solvent.

(b) If the mixture is non-ideal, the activity coefficient of each component in the solution should be away from 1.0.

For non-ideal binary mixture consisting of a volatile and a non-volatile components, the partial pressures of volatile component is:

$$P_i = (1-x)\gamma_i p_i^0 \tag{39}$$

where $x$ is the liquid mole fraction of non-volatile component, $\gamma_i$ is the activity coefficient of volatile component in the solution, and can be calculated either from activity coefficient models, such as NRTL, Wilson, UNIQUAC, UNIFAC, etc. (see chapter 1), or from empirical correlations. Here, the feed is assumed to be an aqueous solution of sodium chloride. For the mixture of water and NaCl, the activity coefficient of water can be estimated by [2]

$$\gamma_i = 1 - 0.5x - 10x^2 \tag{40}$$

Note that the molar fraction $x$ in Eq. (39) and (40) is relative to the membrane surface, i.e. $x_{fm}$. Due to the effect of concentration polarization at the feed side, the concentration of non-volatile component at the membrane surface at the feed side is higher than that in the bulk of the feed. This causes a negative effect on the vapor pressure of volatile component. From Eq. (9), the concentration profile within the concentration boundary layer can be described by

$$x_{fm} = x\exp\left(\frac{N}{k_f\rho}\right) \tag{41}$$

where $x$ is the mole fraction of non-volatile component in the bulk, $x_{fm}$ is at the membrane surface, $N$ (kmol m$^{-2}$ s$^{-1}$) is the mass transfer flux across the membrane, $\rho$ (kmol m$^{-3}$) is the density of the solution, and $k_f$ (m s$^{-1}$) is the mass transfer coefficient usually obtained from the analogy between heat and mass transfers. Therefore, the existence of NaCl in water makes the mathematical model a little complicated. To obtain the partial pressure of water vapor at the membrane surface $P_{fm}$, Eqs. (39) – (41) are incorporated into the mathematical model.

The difference of water vapor pressures between the bulk and the membrane surface may be attributed to three factors. Firstly, the temperature polarization (TP) causes a temperature difference between these two locations. Secondly, the vapor pressure reduction (VPR) introduced by dilution effect or the "non-ideality" of the solution causes the water vapor pressure of the solution to be lower than that of the pure water. Thirdly, the concentration polarization (CP) makes the concentration of water at the membrane surface lower than that in the bulk. The influence of these three factors can be estimated by solving model equations.

The influence of CP and VPR on mass flux in DCMD is shown in Fig. 25 and Fig. 26, where an aqueous solution of NaCl is used as the feed. It can be seen that mass flux may change remarkably for each factor. If all of these three factors are considered in the mathematical model, mass flux $N_{TVC}$ is the lowest and decreases remarkably with the increase of solute concentration. If the factors of TP and VPR are considered, mass flux $N_{TV}$ is negatively affected but with less extent than $N_{TVC}$. In contrast to these two cases, if only the factor of TP is considered, it can be observed that mass flux $N_{TP}$ would be enhanced with the increase of solute concentration! But it has been verified by experiments that this conclusion isn't correct. This contradiction manifests the importance of considering all factors affecting mass flux in mathematical model. On the other hand, the enhancement in mass flux with high feed temperature may be also affected by these factors. It can be seen from Fig. 26 that the exponential increase in mass flux with feed temperature would be negatively shifted. The higher the feed temperature is, the more the difference among $N_{\text{TVC}}$, $N_{\text{TP}}$ and $N_{\text{TV}}$ is.

The large difference between $N_{\text{TV}}$ and $N_{\text{TP}}$ indicates that VPR has negative influence on mass flux, and this influence becomes more and more remarkable with the increase of the concentration of NaCl. The influence of CP on mass flux is embodied by the difference between $N_{\text{TV}}$ and $N_{\text{TVC}}$. CP exhibits its negative contribution to mass flux not considerably when very diluent solution is used as the feed. But at the high concentration, e.g. 7% or more, the influence of CP is obvious, as shown in Fig. 25.

Fig. 27 shows the experimental results of a batch DCMD process in which an aqueous solution of NaCl is used as the feed. As a time-dependent process, the feed concentration increases with time as a result of the continuous evaporation of water from the feed side to the permeate side. As time goes on, the water vapor pressure at the feed side decreases because the salt (NaCl) concentration increases.

So the effect of CP and VPR becomes more prominent with time. The changes of the physical properties of the feed, such as viscosity, density, thermal conductivity, etc., and heat transfer coefficient with time also contribute to the decrease of mass flux. In the beginning of the batch experiment, the concentration of NaCl is 5.26% (weight fraction of NaCl) and mass flux about 39 kg m$^{-2}$ h$^{-1}$. In the end the concentration of NaCl is 20.8% (the solution is close to saturation) and mass flux 25.5 kg m$^{-2}$ h$^{-1}$. That is to say, the feed concentration increases by fourfold, but the flux decreases only 35.6%. However, for RO process, when the concentration of 10% (weight of NaCl) is used as the feed, it requires a pressure difference of above 10 MPa to perform this process, whereas the DCMD process is able to be carried out at atmospheric pressure and at a temperature below the boiling point of water (here it is 70.3 ℃). This example demonstrates the advantage of DCMD in the treatment of high concentrated feed solution without decreasing mass flux considerably.

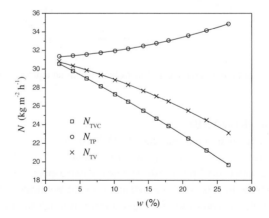

Fig. 25. The influence of CP and VPR on mass flux vs concentration in DCMD.

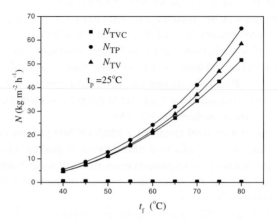

Fig. 26. The influence of CP and VPR on mass flux vs feed temperature in DCMD.

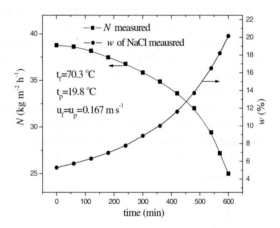

Fig. 27. Experimental results of a batch DCMD process in which an aqueous solution of NaCl is used as the feed.

## 4.3. Mathematical model of VMD

Recently, vacuum membrane distillation (VMD) has received much attention in concentrating the aqueous solution or removing trace amount of volatile organic species from water. The difference between VMD and DCMD lies in two aspects: how the driving force for mass transfer through the membrane is exerted and how the permeated vapor is condensed or removed. In DCMD, the driving force for mass transfer through the membrane is the vapor pressure difference of volatile component across the membrane, which is caused by the temperature difference between the feed and the permeate liquid. And the permeated vapor is condensed in the permeate liquid stream within the membrane module. In VMD, however, the process is driven by both the temperature difference and the total pressure difference between two sides of the membrane. The total pressure difference arises from the vacuum pulled by a vacuum pump in the cold chamber of membrane module. The permeated vapor is sucked out of the cold chamber by the vacuum pump and condensed in an external heat exchanger.

If hot water is used as the feed and flows in the direction tangent over the membrane, the temperature difference at two sides of the membrane brings about a water vapor pressure gradient within the membrane pores. This driving force produces a mass flux through the membrane, and in this case both Knudsen and molecule diffusion are involved in this process. However, since VMD generally operates at the total pressure of the order of 10 to 30 kPa which is usually below the vapor pressure of water, only trace amount of air exists in the membrane pores. Therefore, the mass transfer resistance caused by molecule-molecule collision can be neglected and the diffusion process within the pores is dominated by Knudsen diffusion. According to Eq. (13), the mass flux related to this mechanism is:

$$N_K = -\frac{8}{3}\frac{r\varepsilon}{\tau}\sqrt{\frac{M}{2\pi RT}}\nabla P_A \tag{42}$$

where $\nabla P_A$ is the water vapor pressure gradient within the membrane pores. In addition to Knudsen diffusion, the total pressure difference arising from the vacuum causes a convective mass transport through the pores, known as Poiseuille flow. Becuase almost no air exists within membrane pores, the water vapor in the pores is pure gas and thus the total pressure difference acting as the driving forece for Poiseuille flow also is $\nabla P_A$. According to Eq. (11), the mass flux related to this mechanism is:

$$N_P = -\frac{\varepsilon}{\tau}\frac{r^2}{8\eta}\frac{P_A M}{RT_m}\nabla P_A \tag{43}$$

where $\nabla P_A$ is the total pressure gradient within the membrane. In VMD these two mechanisms contribute in parallel to the mass transfer, and the total mass flux is:

$$N = N_K + N_P = \left(-\frac{8}{3}\frac{\varepsilon r}{\tau}\sqrt{\frac{M}{2\pi RT}} - \frac{\varepsilon r^2}{\tau}\frac{P_A M}{8\eta RT_m}\right)\nabla P_A \tag{44}$$

Integrating this equation over the membrane thickness gives rise to the total mass flux through the membrane:

$$N = \left(\frac{8}{3}\frac{\varepsilon r}{\tau\delta}\sqrt{\frac{M}{2\pi RT}} + \frac{\varepsilon r^2}{\tau\delta}\frac{P_A M}{8\eta RT_m}\right)\left(P_{fm} - P_{pm}\right) \tag{45}$$

where $P_{fm}$ (Pa) is the water vapor pressure at the membrane surface of the feed side, and $P_{pm}$ (Pa) is that of the permeate side. In VMD the pressure at the permeate side is so low that almost no air presents at this side and $P_{pm}$ is assumed to be equal to the vacuum pressure of the cooling chamber.

The heat energy needed for the water to vaporize into the membrane pores is provided by the heat transfer through the thermal boundary layer at the feed side. The heat flux is given by

$$q_f = h_f\left(t_f - t_{fm}\right) \tag{46}$$

Because of the very low pressure at the permeate side of the membrane, heat conduction across the membrane is negligible. That is to say, all the heat energy transferred to the membrane surface is used for the water evaporation. So Eq. (46) can be rewritten as [19]:

$$h_f \left( t_f - t_{fm} \right) = N \cdot \Delta H \tag{47}$$

where $\Delta H$ (kJ kmol$^{-1}$) is the latent heat of vaporization for water.

The model equations of VMD can be used to solve two types of problems: one is to determine the heat transfer coefficient $h_f$ of the membrane module by VMD experiments; the other is to predict mass flux of VMD by using $h_f$.

A dimensionless Nusselt number $Nu$ is commonly used to relate $h_f$ with other factors that affect heat transfer in the boundary layer, i.e.

$$\frac{h_f d_e}{\lambda} = Nu = a \, Re^b \, Pr^c \tag{48}$$

$$\log Nu / Pr^{1/3} = \log a + b \log Re \tag{49}$$

where $d_e$ (m) is the hydraulic diameter of the flow duct, $\lambda$ (W m$^{-1}$ K$^{-1}$) the thermal conductivity of the feed flow. Usually, the exponent of Prandtl number $c$ is considered to be 1/3. To determine $a$ and $b$, the VMD experiment is performed to measure mass flux under various feed flow rates (constant $t_f$ and vacuum degree). In terms of the experimental result, the temperature of the membrane surface $t_{fm}$ can be obtained from Eq. (45). Eq. (45) is an implicit non-linear equation which is suitable for solving by secant method. $t_{fm}$ and $P_{fm}$ may be related by Antoine equation. Due to the non-conductivity of the membrane in VMD, $t_m$ is assumed to be equal to $t_{fm}$ [19].

The experimental result is listed in Table 6, where $t_f$, $N$, $u_f$ (the flow velocity of the feed in membrane module, m s$^{-1}$) and vacuum degree can be directly obtained from VMD equipment, whereas $t_{fm}$, $h_f$, $Re$, $Nu$ and $Pr$ are obtained from a calculation program.

Table 6
Experimental result in VMD at the vacuum degree of 94 kPa

| No. | $t_f$ (°C) | $u_f$ (m s$^{-1}$) | $N$ (kg m$^{-2}$ h$^{-1}$) | $h_f$ (W m$^{-1}$ K$^{-1}$) | $Re$ | $Nu$ | $Pr$ |
|------|------|-------|-------|--------|---------|-------|------|
| 1 | 59 | 0.172 | 65.43 | 6068.63 | 690.24 | 18.17 | 3.04 |
| 2 | 59 | 0.210 | 67.11 | 6206.85 | 848.40 | 18.57 | 3.04 |
| 3 | 59 | 0.249 | 68.75 | 6469.46 | 1005.38 | 19.35 | 3.04 |
| 4 | 59 | 0.288 | 71.18 | 6921.04 | 1162.88 | 20.70 | 3.04 |
| 5 | 59 | 0.326 | 72.95 | 7129.44 | 1323.05 | 21.32 | 3.04 |
| 6 | 59 | 0.365 | 74.31 | 7348.07 | 1481.84 | 21.97 | 3.04 |

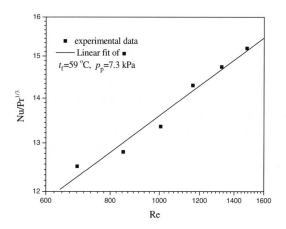

Fig. 28. $Nu / Pr^{1/3}$ vs $Re$ obtained from VMD experiments.

As indicated in Eq. (49), plotting $Nu / Pr^{1/3}$ vs $Re$ in logarithm coordinates will get $a$ as intercept and $b$ as the slop of the straight line. For instance, plotting $Nu / Pr^{1/3}$ vs $Re$ from the data listed in Table 6 in logarithm coordinates get a straight line as shown in Fig.28, with an intercept of 0.317 and a slop of 0.272. Consequently, the correlation for the heat transfer in VMD is:

$$Nu = 0.124 \, Re^{0.658} \, Pr^{0.33} \tag{50}$$

Due to the advantage of excluding molecular diffusion and heat conduction through the membrane, VMD is often used for studying heat and mass transfers, especially for deriving the heat transfer correlation. Note that the assumptions implied in Eqs. (45) and (47) will be reasonable only when the operated vacuum is high enough. From Eqs. (45) and (47), the mass flux in VMD can be calculated.

Substituting Eq. (42) into Eq. (40) gives rise to

$$\frac{h_f}{\Delta H} = \frac{8}{3} \frac{\varepsilon r}{\tau \delta} \sqrt{\frac{M}{2\pi RT}} \Delta P_A + \frac{\varepsilon r^2}{\tau \delta} \frac{p_A M}{8\eta RT_m} \Delta P \tag{51}$$

where $t_{fm}$ is implicitly expressed. The secant method can also be used for solving $t_{fm}$ from this non-linear algebraic equation. $h_f$ can be obtained from the correlation of heat transfer coefficient, e.g. Eq. (50). Once $t_{fm}$ is determined, mass flux can be calculated from Eq. (47). Therefore, for a given membrane characteristics, the performance of VMD can be simulated. Fig. 29 through Fig. 34 show the calculated results for the membrane characteristics: $r = 0.116 \, \mu m$, $\varepsilon / \tau \delta = 8621 \, m^{-1}$.

## 4.4. Performance of VMD

*4.4.1. The effect of feed temperature*

The change of mass flux with feed temperature in VMD under different conditions is shown in Fig. 29. Similar to DCMD, the increase of mass flux with feed temperature is the result of corresponding increase of vapor pressure of water. However, compared with DCMD, a higher mass flux is obtained in VMD. This increase may be attributed to the elimination of molecule-molecule resistance to mass transfer within the membrane pores by vacuum operation or the additional convective flow caused by the total pressure difference (i.e. Poiseuille flow). To find out the relative importance of these two mechanisms in VMD, the following fraction is used to identify the relative weight of Poiseuille flow and Knudsen diffusion in the mass transfer of VMD:

$$fr = \frac{\dfrac{\varepsilon r^2}{\tau \delta} \dfrac{P_A M}{8\eta R T_m} \Delta P}{\dfrac{8}{3} \dfrac{\varepsilon r}{\tau \delta} \sqrt{\dfrac{M}{2\pi R T}} \Delta P_A + \dfrac{\varepsilon r^2}{\tau \delta} \dfrac{P_A M}{8\eta R T_m} \Delta P} \tag{52}$$

which is derived from Eq. (45). The calculated values of $fr$, which is corresponding to the mass flux in Fig. 29, are shown in Fig. 30. It can be observed that although $fr$ increases with feed temperature and feed flow velocity, Poiseuille flow contributes less than 5% to the total flux of VMD under the conditions investigated. As indicated in Eq. (52), this result is due to the small pore size of the membrane. With the increase of membrane pore size, Poiseuille flow contributes more to the total mass flux. For instance, according to the results of Lawson et al. [19], for the membrane with pore diameter larger than 0.7 $\mu$m, Poiseuille flow contributes more than 10% to the total mass flux in VMD.

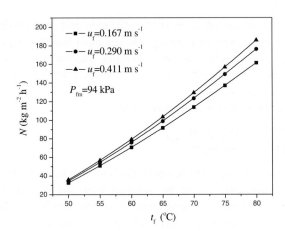

Fig. 29. Change of mass flux with feed temperature in VMD.

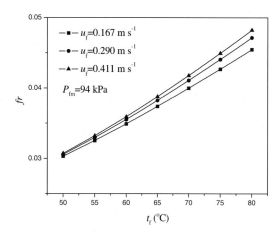

Fig. 30. The fraction of Poiseuille flow vs feed temperature.

## 4.4.2. The effect of membrane pore size

Fig. 31 shows the effect of membrane pore size on mass flux in VMD. Unlike DCMD, the pore diameter is an important factor affecting VMD. Mass flux shows a significant change within the pore diameter range of 0.1 to 0.7 $\mu$m. Note that this pore diameter range is often adopted in MD. The reason why the performance of VMD significantly depends on the pore diameter is attributed to the mechanisms that regulate the mass transfer through membrane.

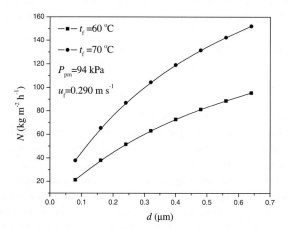

Fig. 31. The influence of membrane pore diameter on transmembrane flux in VMD.

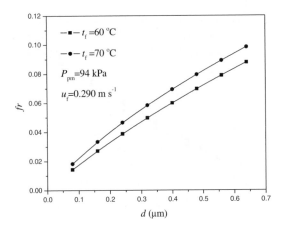

Fig. 32. Change of the fraction of Poiseuille flow with the membrane pore diameter.

In DCMD the pore size is of little effect on mass flux, reflecting that molecule-molecule collision at the large pore diameter predominates. On the contrary, in VMD, the larger the pore size is, the smaller the resistance to Knudsen diffusion and Poiseuille flow imposed by the inside wall of the membrane pores is. So the effect of pore size is different in the degrees for these two mechanisms. Eq. (45) indicates that the contributions from Knudsen diffusion and Poiseuille flow are proportional to $d$ and $d^2$, respectively. The effect of membrane pore size can also be embodied in the change of $fr$ with pore diameter, as shown in Fig. 32. Fig.32 is obtained under the same conditions as Fig. 31. It can be seen that the contribution of Poiseuille flow to the total mass flux of VMD increases with membrane pore size, and its influence cannot be neglected if a membrane with pore diameter larger than 0.5 $\mu$m is employed.

Despite the advantage of high mass flux brought out by membranes with large pore size, the possibility of risking wetting is also high in VMD because of the low LEP associated with such membranes.

### 4.4.3. The influence of vacuum (or vacuum degree)

Eq. (45) implies that increasing the vacuum can increase the driving force linearly for both Knudsen diffusion and Poiseuille flow, and thus mass flux increases linearly with vacuum. The calculated results are shown in Fig. 33. However, this tendency may be changed if VMD is operated at low vacuum. Fig. 34 shows mass flux under various vacuums obtained from the experiment. It can be seen that at the vacuum of 80 kPa a transition appears. This phenomenon can't be explained by Eq. (45) because there is an implicit relationship between the operating temperature (60 °C) and absolute pressure (about 20 kPa) relative to this vacuum. It is at this vacuum that the water of 60 °C is boiling. When the feed water boils at the entry of the membrane pores, the trapped air within the membrane pores will be displaced by the increasing amount of water vapor at the entry of the membrane pores at the feed side.

As time goes on, only a trace amount of air exists within the pores. Thus, the mass transfer resistance caused by molecule-molecule collision is eliminated. But at the temperature below boiling point, molecule-molecule collision resistance plays a role and makes mass flux low. So it should be cautious that Eq. (45) is only used for the VMD process operated at the vacuum that enables the feed boil (or the saturated pressure at the feed temperature).

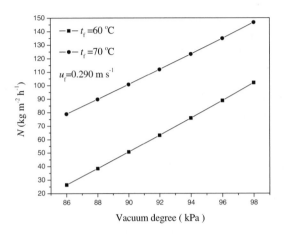

Fig. 33. The influence of vacuum on transmembrane flux in VMD.

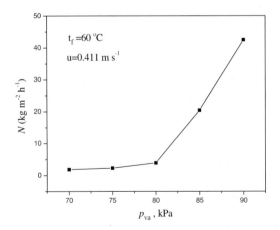

Fig. 34. The influence of vacuum on mass flux in VMD.

## 4.5. Mathematical model of AGMD

In air gap membrane distillation (AGMD), the volatile component at the feed vaporizes at the entry of membrane pores. The vapor has to pass through two static air layers: one is that of trapped air within the membrane pores; the other is introduced by using a plate to separate the cooling chamber from the membrane, i.e. air gap. The exotic air gap makes AGMD quite different from DCMD in the operation and simulation.

In AGMD the heated feed liquid brings heat flux, which has to transport through six regions connected in series, as illustrated in Fig. 35. The mathematical model of AGMD mainly focuses on determining the heat flux in each of these regions.

(1) Within the thermal boundary layer where the temperature decreases from $t_f$ to $t_{fm}$. The heat flux penetrates through this region in two ways: one is the heat conduction, which can be expressed by Fourier law: $-\lambda_f (dt/dy)$ where $dt/dy$ represents the temperature gradient within this region, $\lambda_f$ (W m$^{-1}$ K$^{-1}$) the thermal conductivity of the feed; the other is associated heat energy with mass flux within this region and in this case the heat flux is $Nc_{p,f}t$. The heat balance for an element of this region is given by the following differential equation:

$$\lambda_f \frac{d^2 t}{dy^2} - Nc_{p,f} \frac{dt}{dy} = 0 \tag{53}$$

with the boundary conditions: $y = y_0$, $t = t_f$; $y = y_1$, $t = t_{fm}$.

Integrating this equation gives the temperature profile within the thermal boundary layer:

$$t(y) = t_f + (t_{fm} - t_f) \frac{\exp(Xy/\Delta y_0) - 1}{\exp(X) - 1} \tag{54}$$

where $X = Nc_{p,f}/h_f$, $h_f = \lambda_f/\Delta y_0$, $\Delta y_0 = y_1 - y_0$.

The heat flux caused by heat conduction at the interface of feed bulk and thermal boundary layer where $y = y_0$ or $t = t_f$ can be obtained from Eq. (54):

$$q_c = -\lambda_f \left. \frac{dt}{dy} \right|_{y=y_0} = -h_f (t_{fm} - t_f) \frac{X}{\exp(X) - 1} \tag{55}$$

If mass flux is small enough, heat flux can be rewritten as [20]

$$\lim_{N \to 0} q_c = h_f (t_f - t_{fm}) - \frac{1}{2} Nc_{p,f} (t_f - t_{fm}) \tag{56}$$

Thus, the total heat flux at $y = y_0$ is:

$$q = q_c + Nc_{p,f} (t_f - t_{fm}) = h_f (t_f - t_{fm}) + Nc_{p,f} \left( \frac{t_f + t_{fm}}{2} \right) \tag{57}$$

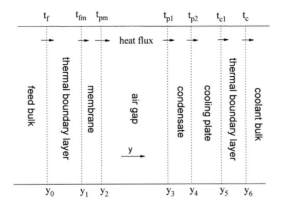

Fig. 35. Regions for heat transfer in AGMD.

(2) Within the membrane where the temperature decreases from $t_{fm}$ to $t_{pm}$, heat flux at the membrane surface of the feed side where $y = y_1$ or $t = t_{fm}$ is:

$$q = N\left[c_{p,f}t_{fm} + \Delta H(t_{fm}) + \frac{1}{2}c_{p,g}(t_{pm} - t_{fm})\right] + \frac{\lambda_m}{\delta}(t_{fm} - t_{pm}) \tag{58}$$

(3) Within the air gap where the temperature decreases from $t_{pm}$ to $t_{p1}$, heat flux at the membrane surface of the permeate side where $y = y_2$ or $t = t_{pm}$ is:

$$q = N\left[c_{p,f}t_{fm} + \Delta H(t_{pm}) + c_{p,g}\left(\frac{t_{pm} + t_{p1}}{2} - t_{fm}\right)\right] + \frac{\lambda_g}{b}(t_{pm} - t_{p1}) \tag{59}$$

where $b$ (m) is the thickness of the air gap.

(4) For the region of condensate layer where the temperature decrease from $t_{p1}$ to $t_{p2}$, heat flux at the surface of condensate where $y = y_3$ and $t = t_{p2}$ is:

$$q = h_p(t_{p1} - t_{p2}) + Nc_{p,p}\left(\frac{t_{p1} + t_{p2}}{2}\right) \tag{60}$$

(5) For the region of cooling plate where the temperature decreases from $t_{p2}$ to $t_{c1}$, heat flux across the cooling plate is:

$$q = \frac{\lambda_s}{\delta_s}(t_{p2} - t_{c1}) \tag{61}$$

where $\lambda_s$ (W m$^{-1}$ K$^{-1}$) is the heat conductivity of the cooling plate, $\delta_s$ (m) is the thickness of the plate.

(6) For the thermal boundary layer of coolant where the temperature decreases from $t_{c1}$ to $t_c$, heat flux across the surface of the plate where $y = y_5$ or $t = t_{c1}$ is:

$$q = h_c(t_{c1} - t_c) \tag{62}$$

If the heat transfer process in AGMD reaches steady-state, heat flux across each of the regions in Fig. 35 is equal. Therefore, Eqs. (57) – (62) form a linear equation group in which the five interface temperatures, $t_{fm}$, $t_{pm}$, $t_{p1}$, $t_{p2}$ and $t_{c1}$, are solved under a given operation condition. However, it should be mentioned that mass flux $N$ is included in Eqs. (57)- (60). This implies that mass transfer equation, e.g. Eq. (29), has to be incorporated to determine these temperatures, which makes the problem somewhat complicated and an iteration program is needed to deal with it.

### 4.6. Performance of AGMD

By solving the mathematical model, the temperature profile within the module for AGMD is plotted in Fig. 36, as well as the temperature profile of DCMD module under the approximate operation conditions. It can be seen that the temperature difference through the membrane in AGMD is lower than in DCMD, which means that the driving force for mass transfer is small in AGMD. Moreover, it can go a further step to deduce that mass flux in AGMD is also lower than in DCMD. But to our surprise, there is an abrupt temperature decline in the region of air gap. That is to say, the temperature difference imposed externally by heater and cooler in AGMD membrane module is mostly depleted in this region, not effectively utilized for mass transfer through the membrane.

Fig. 37 shows the comparison of mass flux between AGMD and DCMD obtained from the experiment. As mentioned before, compared with DCMD, AGMD utilizes only a small fraction of temperature difference imposed externally for mass transfer, so AGMD has lower mass flux than DCMD. It can be seen from Fig. 36 that the sensitivity of mass flux to the feed temperature is different between AGMD and DCMD because of the different slope of the two curves. Compared with DCMD, the response of mass flux to feed temperature is less rapid in AGMD than in DCMD. In AGMD, besides the gas layer within the membrane pores, the water vapor has to pass through the static air gap, which is usually tenfold or more thicker than the former. It is the static air gap that acts as the main mass transfer resistance in AGMD. So the way to increase the feed temperature to improve mass flux is limited in AGMD.

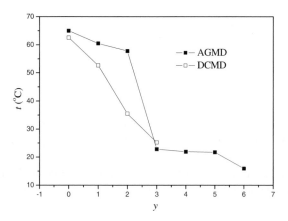

Fig. 36. The temperature profile within the modules of AGMD and DCMD.

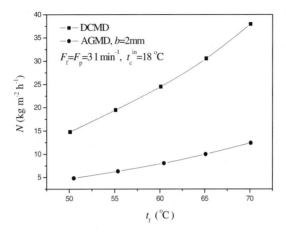

Fig. 37. Comparison of mass flux between AGMD and DCMD.

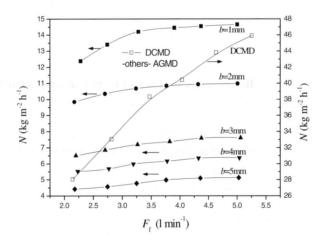

Fig. 38. The influence of feed flowrate on mass flux in AGMD and DCMD.

Fig. 38 shows the effect of feed flowrate on mass flux in AGMD and DCMD. For DCMD, the degree in which the feed flowrate affects mass flux is great, indicating an obvious temperature polarization. In AGMD, however, there is little influence caused by feed flowrate. This also holds for the effect of feed temperature, as shown in Fig. 37. This tendency becomes more and more obvious with the increase of the thickness of air gap. However, in AGMD the feed flowrate, as well as feed temperature, isn't associated with the air gap, so its improvement is almost of no use for reducing the mass transfer resistance.

In summary, the AGMD process can't be enhanced by improving operation conditions such as feed flowrate and feed temperature. But the thickness of the air gap can affect mass flux apparently. This is verified by the experimental results of both DCMD and AGMD, as shown in Fig. 39 where $b = 0$ is referred to DCMD configuration. The tendency of the left curve implies that the crucial factor affecting mass flux in MD is its configuration, similar as Figs. 37 and 38. The tendency of the right curve implies that air gap is the dominant factor affecting mass flux in AGMD.

As mentioned before, in MD only a fraction of the heat energy supplied to feed is utilized for the evaporation process, while the remaining are consumed by heat conduction through the membrane to the permeate side. The thermal efficiency of MD process is defined as the fraction of heat energy utilized for evaporation of the feed.

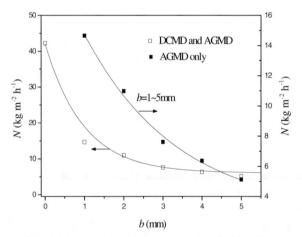

Fig. 39. The influence of the thickness of the air gap on mass flux in AGMD.

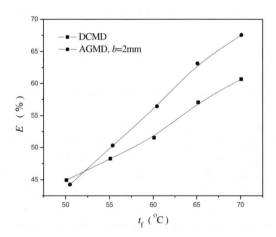

Fig. 40. Comparison of thermal efficiency between AGMD and DCMD.

The amount of the energy utilized for evaporation of the feed can be determined according to mass flux and latent heat of water. The total energy supplied is determined by the enthalpy change of the feed liquid in and out of the membrane module. Fig. 40 shows the comparison of thermal efficiency between AGMD and DCMD, which is obtained from the experiment. It is evident that AGMD is advantageous over DCMD in thermal efficiency. This should be attributed to the existence of air gap in the membrane module. It is the air gap that forms an additional heat transfer resistance by conduction from the feed to the permeate. As a result, the thermal efficiency of AGMD is improved. In addition, for both AGMD and DCMD, the thermal efficiency becomes higher with the increase of feed temperature. As we know, mass flux increases with feed temperature in an exponential way, whereas the heat loss through the membrane by conduction is linearly proportional to the temperature difference between the two sides of the membrane. Therefore, at a high feed temperature, the amount of heat energy utilized for evaporation is more than that transferred to the permeate side by conduction through the membrane.

Although AGMD is thermally efficient, the low mass flux makes it unlikely to be accepted in industry. On the other hand, while DCMD shows a relatively low thermal efficiency, this disadvantage can be overcome by setting an external heat exchanger to recover heat energy from the permeate.

## 5. MODULE PERFORMANCE

In previous sections of this chapter we focus on the performance of various MD configurations in lab-used membrane module. Because of small membrane area, there is almost no change in the fluid field when the fluid flows through the ducts of the module. Such parameters as the fluid temperature, velocity and mass flux through the membrane are maintained constant within the membrane module. Such module can be considered as a lumped system, and algebraic equations, as derived in previous sections, are suitable for describing this process. However, the membrane area of an industrially applied membrane module may be hundreds or even thousands of times larger than that of the lab-used module. Therefore, these variables may change remarkably within the scale-up membrane module, and thus the module will show some features that are different from the lab-used module. From this viewpoint, differential equations are needed to describe the MD process in the scale-up membrane modules. This section mainly concerns about the performance of these modules, especially focuses on their differences from lab-used membrane modules

### 5.1. Performance of flat sheet membrane module

As we know, both flat sheet and hollow fiber membrane can be employed for MD. Although the hollow fiber membrane module is less affected by the temperature polarization [18], the non-uniform fiber packing in the module results in serious flow maldistribution at its shell side. For this reason, a large portion of membrane area is ineffective. Even more, the flow maldistribution makes the heat recovery rate less than that of an ideal module. So it is difficult to perform the latent heat recovery in a heat exchanger. Flat sheet membrane module is generally used in experimental study of MD. The fact that there is a low temperature polarization coefficient (TPC) in a flat sheet membrane module seems to prevent it from being extensively applied. However, Lawson's studies [8] showed that TPC in the well-designed flat sheet membrane module was higher than 0.85 under various temperatures, and the water flux up to 75kg m$^{-2}$ h$^{-1}$ was obtained. In general, the pore size of flat sheet membrane is larger than that of hollow fiber membrane, but its thickness is lower than that of hollow fiber membrane. Accordingly, its mass transfer resistance is expected to be lower than that of hollow fiber membrane. Now we investigate a scale-up flat sheet membrane module that may be used in industry.

### 5.1.1. The pilot plant membrane module

The flowsheet of a DCMD pilot plant, consisting of a flat sheet MD membrane module, two heat exchangers and a heating source, is illustrated in Fig. 41.

In the MD module, the hot feed (the brackish water) and the cold permeate (the produced pure water) are pumped from the feed and permeate tanks separately, and contact in a counter-current mode. The mass and energy are transferred across the membrane from the feed to the permeate side. The two streams out of the module enter into the heat exchanger where energy is returned to the feed stream from the permeate stream. The heat recovery is embodied by the feed temperature improvement when feed streams pass through this heat exchanger. It should be noted that only when the permeate stream is hotter than the feed stream, heat recovery is possible in the heat exchanger. As the energy can't be 100% recovered in this heat exchanger, another heat exchanger, i.e. the cooler, is set to maintain the permeate temperature at a low level. In order to supply the energy sufficient to perform MD in the module, the feed is heated by a heating source.

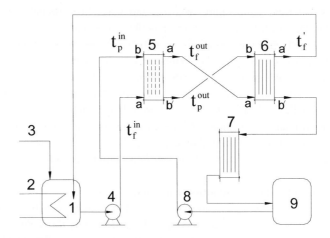

Fig. 41. Flowsheet of a membrane distillation process; 1 - feed tank; 2 - heater for the feed; 3 - water pipe; 4 - feed pump; 5 - membrane module; 6 - recovery heat exchanger; 7 - cooler for the permeate; 8 - permeate pump; 9 - permeate tank.

The pilot plant investigated is applied for desalination through flat sheet membrane module. This module is illustrated in Fig. 42. It is very similar in configuration to a plate heat exchanger. A number of flat sheet membranes are assembled in parallel, forming a number of flow channels for the feed and permeate to flow at two sides of the membranes. Each flow stream within the channels is in direct contact with one side of two membranes. Plastic spacers are packed within the flow channels to improve the flow field and reduce the effect of temperature polarization.

### 5.1.2. Mathematical model
(1) Membrane module

Mathematical model is established by investigating the feed and the permeate elements of the module. Considering a feed element, in which the $iF^{th}$ feed stream passes through the channel formed by the $im^{th}$ and $im+1^{th}$ membranes. An enthalpy conservation for this element gives a differential equation with respect to the temperatures of the $i_F^{th}$ feed stream alongside the module:

$$\frac{dt_f^{i_F}}{dA_m} = \frac{\left[N^{2i_F-1}\Delta H + h_m^{2i_F-1}\left(t_{fm}^{2i_F-1} - t_{pm}^{2i_F-1}\right)\right] + \left[N^{2i_F}\Delta H + h_m^{2i_F}\left(t_{fm}^{2i_F} - t_{pm}^{2i_F}\right)\right]}{F_f^{i_F} c_{p,f}^{i_F}} \tag{63}$$

with the boundary condition:

$$A_m = 0, \ t_f^{i_F} = t_f^{in} \tag{64}$$

where $i_F = 1,2,\cdots,n_F$, and $n_F$ is the total number of the feed streams in the module, $A_m$ (m$^2$) is the membrane area for each flow channel, $F_f^{i_F}$ (kmol h$^{-1}$) is the flow rate of the $i_F$ feed flow duct in the module, and $c_{p,f}^{i_F}$ (kJ kmol$^{-1}$ K$^{-1}$) is the heat capacity, $t_f^{in}$ (K) is the feed temperature at the module inlet.

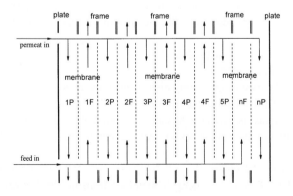

Fig. 42. Schematic representation of the flat sheet membrane module.

Similarly, another differential equation is obtained with respect to the temperatures of the $i_p{}^{th}$ permeate stream alongside the module:

$$\frac{dt_p^{i_p}}{dA_m} = \frac{\left[N^{2i_p-2}\Delta H + h_m^{2i_p-2}\left(t_{fm}^{2i_p-2} - t_{pm}^{2i_p-2}\right)\right] + \left[N^{2i_p-1}\Delta H + h_m^{2i_p-1}\left(t_{fm}^{2i_p-1} - t_{pm}^{2i_p-1}\right)\right]}{F_p^{i_p} c_{p,p}^{i_p}} \tag{65}$$

with the boundary condition:

$$A_m = A, \quad t_p^{i_p} = t_p^{in} \tag{66}$$

where $t_p^{in}$ is the permeate temperature at the module inlet, $i_p = 2, \cdots, n_p - 1$, and $n_p$ is the total number of the permeate streams in the module. Therefore, the total number of the flat sheet membranes in the module is $n_F + n_P - 1$.

The two plates at both ends of the module are used to host (as a supporting structure for this module) the whole sandwiched flat sheet membranes (with flow channels) together, and their inner surfaces facing the first and last membranes form two additional flow channels. In terms of the enthalpy conservation for the elements in the two channels, the following differential equations are obtained with respect to the temperatures of the $1^{th}$ and $n_p{}^{th}$ permeate streams alongside the module:

$$\frac{dt_p^1}{dA_m} = \frac{\left[N^1 \Delta H + h_m^1 \left(t_{fm}^1 - t_{pm}^1\right)\right]}{F_p^1 c_{p,p}^1} \tag{67}$$

$$\frac{dt_p^{n_p}}{dA_m} = \frac{\left[N^{2n_p-2}\Delta H + h_m^{2n_p-2}\left(t_{fm}^{2n_p-2} - t_{pm}^{2n_p-2}\right)\right]}{F_p^{n_p} c_{p,p}^{n_p}} \tag{68}$$

Eqs. (63) – (68) form an ordinary differential equation group with splitting boundary values, and thus a shooting method is suitable for solving them. At the same time, the values of $N$, $t_{fm}$ and $t_{pm}$ in each differential element, which are obtained from the solution of the nonlinear algebraic equation group made up of Eqs. (29), (31) and (32), are needed.

Some assumptions are adopted in the mathematical model of this flat sheet membranes module:
(1) There is no heat lost from the module;
(2) Axial diffusion is neglected;
(3) The streams in the module are evenly distributed at the cross section.
(4) The flowrates of the feed and permeate streams maintain constant within the membrane module. This assumption is based on the consideration that the difference of the feed flowrate at the inlet and outlet of the module is very small (generally no more than 5%).

(2) The heat recovery exchanger
This is used to recover the energy transferred from the feed streams to the permeate streams in the membrane module. The heat recovery ratio (HRR) is defined as the percentage of the recovered heat transferred in the module:

$$HRR = \frac{t_f' - t_f^{out}}{t_f^{in} - t_f^{out}} \tag{69}$$

where $t_f^{out}$ (K) is the feed temperature at the module outlet, $t_f'$ (K) is the feed temperature at the outlet of the heat exchanger. For counter-current flow [21], $t_f'$ can be calculated by

$$t_f' = \frac{(B_1 - B_2)\exp B_1}{B_2 \exp B_2 - B_1 \exp B_1} t_f^{out} + \frac{B_1(\exp B_2 - \exp B_1)}{B_2 \exp B_2 - B_1 \exp B_1} t_p^{out} \tag{70}$$

where $B_1 = KA/F_f n_F c_p$, $B_2 = KA/F_p n_p c_p$. $KA$ (W K$^{-1}$) is the product of total heat transfer coefficient multiplied by the heat transfer area of the heat exchanger, and usually represents the capacity of a heat exchanger. $F_f n_F$ and $F_p n_P$ (kg h$^{-1}$) are the total flow rate of the feed and permeate streams respectively. Evidently, the maximum $t_f'$ is $t_p^{out}$ (the permeate temperature at the module outlet) with the corresponding maximum HRR. But in this case the capacity of this heat exchanger is infinite.

(3) The feed tank

In the feed tank the feed water and heat energy are input. The former comes from the makeup water line, and the latter is offered by a heating source. The temperature of the feed tank can be derived from the enthalpy conservation for the tank at steady-state. Apparently, it is equal to the temperature of the feed stream at the inlet of membrane module. So we have

$$t_f^{in} = \frac{NAc_p t_0 + F_f n_F c_p t_f' + Q_a}{F_f n_F c_p} \tag{71}$$

where $Q_a$ (W) is the amount of heat energy added to the tank by the heater.

For simplifying the above equations, it is assumed that the temperature in the permeate tank is kept constant. For instance, in this case it is 25 °C. In addition, the feed stream is recycled as a loop in the process, and this loop should be teared to solve these equations. Here, we select the line between the feed tank and the membrane module to be teared. That is to say, the temperature of feed stream at the module inlet $t_f^{in}$ is taken as the iterative variable. The method of partial substitution is applied in the iteration program where a relaxation factor is introduced to expedite convergence. To avoid the possible error in calculating the physical properties of the feed stream, both the feed and the permeate are taken on as pure water.

*5.1.3. Simulation results*

It is generally believed that scale-up is very straightforward for membrane module, and simply adding membrane modules can increase the production capacity. Therefore, the simulation results of flat sheet membrane module can be extended to a large-scale plant. The following situations are considered in the simulation:

(1) The influence of the amount of energy supplied by the heater

It is known that the module flux increases exponentially with the rising of feed temperature. This is verified by the experimental, and can be explained by the exponential relationship between the water vapor pressure and its temperature. Eq. (71) indicates that feed temperature $t_f^{in}$ may be affected by the sensible heat introduced by water added to the feed tank, denoted by $NA_m c_p t_0$, and by the energy supplied by the heater, denoted by $Q_a$. However, since the latter is great, the relationship between $t_f^{in}$ and $Q_a$ is almost linear. The simulation results are shown in Fig. 43, and it seems that the mass flux of the module exponentially increases with $Q_a$.

However, Fig. 43 shows that $N$ may be linearly related with $Q_a$. The reason is due to the long-span of feed and permeate temperature within the scale-up membrane module investigated as shown in Fig. 44. The total membrane area assembled in the module is 3.2 m$^2$, 0.26 m$^2$ for each duct. Only 0.8 or 1.0 l min$^{-1}$ of the feed or permeate passes through. So the residence time of the feed and permeate in this membrane module may be much larger than that in a laboratory-scale one. As we know, the more the residence time of the feed and permeate in the module, the greater the temperature difference of the two streams. Because of the great temperature change in the module, the module performance will not be subjected to stream's temperatures at the module inlet. For a laboratory-scale membrane module, however, the two streams experience little temperature change, so the module performance is regulated by the temperature at the module inlet. This analysis can be used for determining the amount of energy consumed by MD system.

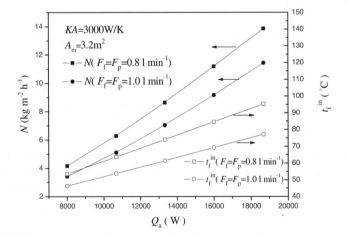

Fig. 43. The influence of $Q_a$ on the module flux and feed temperature.

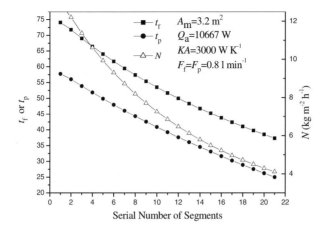

Fig. 44. Temperature and mass flux profile in the module.

(2) Performance enhancement by heat recovery

The heat recovery is carried out in a heat exchanger where the feed and permeate streams out of the membrane module contact counter-current again. In general, the capacity of a heat exchanger is represented by $KA$. For a recovery heat exchanger, the influence of $KA$ on the module flux under various feed and permeate flowrates are shown in Fig. 45, where $KA = 0$ means that no recovery heat exchanger is set in the plant. It can be seen that heat recovery has a significant influence on the module flux. The plant with heat recovery always has an obvious larger module flux than that without heat recovery under various feed and permeate flowrates. The remarkable enhancement of module flux by heat recovery is due to the corresponding increase of feed temperature $t_f^{in}$. Fig. 46 shows how the heat recovery rates

(HRR, defined by Eq. (69)) of a plant and $t_f^{in}$ are influenced by $KA$. The tendency of HRR

and $t_f^{in}$ vs $KA$ is very similar to that of module flux vs $KA$ because it is by the recovery heat

exchanger that HRR makes sense, leading to high $t_f^{in}$ and module flux.

In addition, Figs. 45 and 46 also indicate that there may exist an optimum $KA$ for given feed and permeate flowrates. Below this value, the remarkable increment of production capacity is anticipated with the rising of $KA$; above this value, however, the module flux tends to plateau. This is due to the decline of the heat transfer driving force in recovery heat exchanger with the rising of $KA$. As mentioned before, the role of heat recovery exchanger is

to raise the feed temperature from $t_f^{out}$ to $t_f'$. Nevertheless, due to the thermodynamic

restriction, the maximum $t_f'$ is $t_p^{out}$, and the corresponding maximum HRR is $HRR_{max}$. As

shown in Fig. 47, with the rising of $KA$, HRR approaches to its maximum value more and more and at the same time $t_f'$ approaches to $t_p^{out}$.

The above analysis is important for determining the size of the heat recovery exchanger. Under specified amount of heat energy supplied into the plant, the heat exchanger with optimum $KA$ should be coupled with the membrane module.

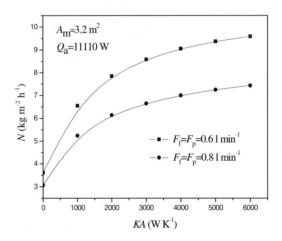

Fig. 45. Improvement of module flux by heat recovery.

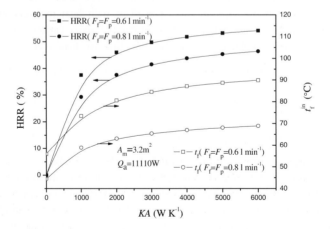

Fig. 46. Change of HRR and $t_f^{in}$ with $KA$.

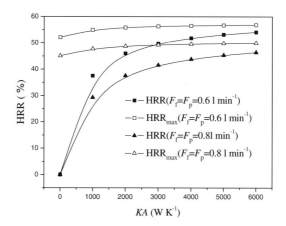

Fig. 47. The difference between HRR and HRR$_{max}$.

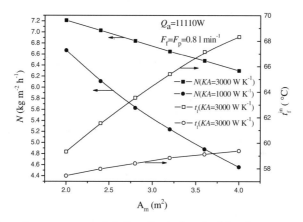

Fig. 48. Effect of membrane area on module flux (by the length of ducts).

(3) Module flux enhancement by increasing membrane area

It is very straightforward to figure out that increasing the production capacity can be achieved by increasing the membrane area. If it is done by means of using more layers of flat sheet membrane in one module or using more modules, the production capacity is increased linearly, accompanied by extra energy consumption by the pumps. Another way to increase membrane area is prolonging the ducts in the module at a constant number of membrane layers.

Fig. 48 shows the influence of membrane area on the module flux, where the membrane area increases with the rising of the length of flow ducts in the module. But the module flux is decreased with the rising of membrane area. This is due to the prolonged residence time of the feed and permeate in the ducts, which leads to a low temperature difference between the two sides of the membrane.

Fig. 49 shows the influence of membrane area on the production rate, in which the membrane area is also raised by membrane length. It can be seen that as long as a heat recovery exchanger is employed in the plant, increasing the membrane area enhances the plant's capacity. When the length of the flow channels becomes longer, the residence time of both the feed and permeate streams within the membrane module is increased, which in turn results in a lower feed outlet temperature and a higher permeate outlet temperature. This means that more energy can be recovered in the recovery heat exchanger, and thus the production rate is improved.

However, when increasing the membrane area, the heat recovery exchanger should be properly designed simultaneously; otherwise, unfavorable results may be obtained. As shown in Fig. 49, with a membrane area up to 4.0 m$^2$ coupled with a suitable heat recovery exchanger of $KA$ = 3000 W K$^{-1}$, a remarkable increase in production rate is obtained. On the other hand, using a heat recovery exchanger of $KA$ = 1000 W K$^{-1}$ will result in a slow increase in production rate. If no heat recovery exchanger is employed, i.e. $KA$ = 0, increasing the membrane area is no sense for increasing production rate. This conclusion reveals that the effectiveness of increasing production rate by increasing membrane area largely depends on the heat recovery system, and the larger $KA$, the higher production rate.

(4) The influence of flowrate on module flux

It is generally believed that the transmembrane water flux can be improved by increasing the flowrates of the feed and permeate under the condition that the feed temperature is maintained constant by a heating source. This is attributed to two aspects that have a positive effect on the driving force: firstly, increasing the flowrates will impair the temperature polarization; secondly, the temperature at the sides of the feed and permeate is almost not affected.

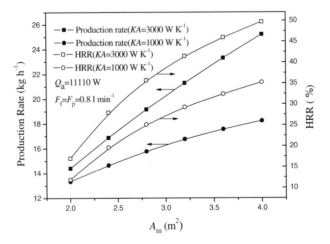

Fig. 49. Effect of membrane area on production rate.

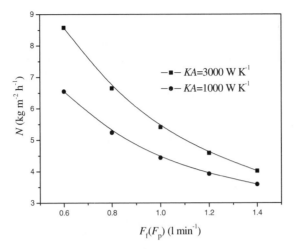

Fig. 50. The influence of the flowrates on module flux.

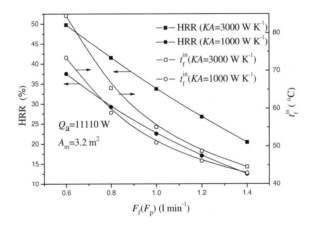

Fig. 51. The influence of flowrate on HRR and $t_f^{in}$.

Fig. 50 shows the decline of the flux of scale-up module with the increase of feed and permeate flowrates. For specific amount of heat energy added, the heat recovery exchanger can "yield" extra energy. Consequently, the feed temperature is partly determined by the heat recovery rate. The small flowrates of the feed and permeate will produce a large temperature difference (the driving force of heat transfer) in heat recovery exchanger, and thus a relatively high HRR is gotten. For this reason, the small flowrates make the MD process operate at high feed temperature. This has been proved by the change of $t_f^{in}$ and HRR with flowrates, as shown in Fig. 51. Although severe temperature polarization may take place under small flowrates, this can be offset by the positive influence from the increased feed temperature because the transmembrane water flux increases exponentially with feed temperature.

Note that the above analysis is based on the assumption that $KA$ is constant, but the flowrates may be changed. In practice, the total heat transfer coefficient $K$ will become lower when the flowrates decreases. If this factor is considered, e.g. $K \propto F^{0.8}$, the tendency of module flux vs the flowrates is very similar as that in Fig. 50 except that the slop is a little small.

The simulation is relative to the scale-up flat sheet membrane module, and the results may be different from those of small module used in laboratory. But the mathematical model is common, and the interested readers can try to calculate for different conditions.

## 5.2. Performance of hollow fiber membrane module
### 5.2.1. Introduction

The hollow fiber membrane module is a bundle of porous hollow fibers packed into a shell in configuration similar to a tube and shell heat exchanger. The fibers are often randomly packed in the shell. Let's consider a cylinder in Fig. 52. It contains many small microporous hollow filaments (the number is $n_f$). At the shell side of the cylinder is a fluid flowing past the tubes. Inside the tubes (lumen) flows a fluid that is immiscible to the shell side fluid. The two phases readily form an interface in the pores of the fiber. If chemical potentials are unequal, some components will get across the interface and thus mass transfer occurs. So this device can be used as a contactor to achieve gas/liquid and liquid/liquid mass transfer [22]. The main advantages of hollow fiber membrane module over flat sheet membrane module are:

(1) High-efficiency of heat and mass transfer.

Among the configurations such as plate-and-frame, stirred tank and hollow fiber, hollow fiber is less affected by temperature polarization in MD process [18].

(2) Large interfacial area per unit volume.

Hollow fiber module is commonly used in many fields, besides MD, such as liquid-liquid extraction, artificial kidney, desalination and wastewater treatment.

Mass transfer and fluid dynamics in hollow fiber module have been extensively studied [23]. For the convenience of calculation, some assumptions are adopted by many researches. For example, it is assumed that all the fibers in one module have the same inner diameter and that they are arranged in a nonrandom way, or they are evenly distributed at the shell side of the module. That is to say, the flow is uniformly distributed at both sides of the module.

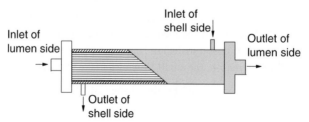

Fig. 52. Hollow fiber membrane module.

In practice, however, these modules often exhibit unexpected performance, and flow maldistribution is considered to be only one of the most possible factors. Some of the possible factors include the polydispersity of fiber inner diameter, the uniformity of fiber packing, fiber movement during operation and the effect of module inlet and outlet [24]. The first two factors are often the main reasons of flow maldistribution in the lumen and shell respectively. Flow maldistribution isn't a serious problem in the laboratory scale but becomes a critical issue in a scale-up.

Since the reasons of flow maldistribution are somewhat complicated, the ideal hollow fiber module is firstly discussed and then the more practical one.

*5.2.2. Mathematical model of ideal hollow fiber membrane module*

The configuration of a hollow fiber module is similar to a heat exchanger and the fluids flows countercurrent at two sides. The ideal hollow fiber membrane module is postulated to have the following features:

(1) There is no polydispersity of fiber inner diameter, i.e. all fibers in the module have the same inner diameter and the flow is evenly distributed into each channel at the lumen side of the module;

(2) The fibers are evenly and regularly distributed at the shell-side of the module, so the shell side flow is evenly distributed into each channel;

(3) The effect of module inlet and outlet is neglected.

In summary, an ideal hollow fiber membrane module is the one having no flow maldistribution at either the shell or lumen side along with no inlet and outlet effects.

The following assumptions are adopted in establishing the heat balance equations for an ideal hollow fiber module used in MD:

(1) There is no heat loss from the module;

(2) Axial diffusion is negligible;

(3) The streams at both sides of the membrane are in the fully developed laminar flow in hydrodynamics and thermodynamics.

As an example, the case that two streams of pure water are fed into the lumen and shell duct of a hollow fiber membrane module respectively in countercurrent flow is considered.

In the hollow fiber module, the temperature and transmembrane flux vary axially alongside the module. Considering a differential element in the module, the following differential equations can be obtained by the enthalpy conservation for the element, describing the temperature of the feed and permeate streams alongside the module. At the lumen side,

$$\frac{dt_f}{dz} = \frac{\left[ N \cdot \Delta H + \frac{\lambda_m}{\delta}\left(t_{fm} - t_{pm}\right)\right] n_f \pi d_m}{F_f c_{p,f}} \tag{72}$$

with the boundary condition:

$$z = 0, \quad t_f = t_f^{in} \tag{73}$$

and at the shell side,

$$\frac{dt_{\mathrm{p}}}{dz} = \frac{\left[N \cdot \Delta H + \frac{\lambda_{\mathrm{m}}}{\delta}\left(t_{\mathrm{fm}} - t_{\mathrm{pm}}\right)\right] n_{\mathrm{f}} \pi d_{\mathrm{m}}}{F_{\mathrm{p}} c_{\mathrm{p,p}}} \tag{74}$$

with the boundary condition:

$$z = L, \quad t_{\mathrm{p}} = t_{\mathrm{p}}^{\mathrm{in}} \tag{75}$$

where $d_{\mathrm{m}}$ (m) is the logarithmic mean diameter of the hollow fiber, $L$ (m) is the effective length of fiber, and $n_{\mathrm{f}}$ is the number of fibers in the module.

Eqs. (72) – (75) form an ordinary differential equation group with splitting boundary values, and thus a shooting method is suitable for solving these equations. The initial permeate temperature at $z = 0$ ($t_{\mathrm{p}}^{\mathrm{out}}$) is guessed and the boundary condition in Eq. (73) is used. The solution searches forward by using Eqs. (72) and (74) until $z = L$, and then the guessed value of $t_{\mathrm{p}}^{\mathrm{out}}$ is modified by using an implicit secant method. This procedure is repeated until the boundary condition in Eq. (75) is satisfied.

While solving the differential equations, the values of $N$, $t_{\mathrm{fm}}$ and $t_{\mathrm{pm}}$ in each differential element must be known beforehand. Like the flat sheet membrane module, they are obtained from the solution to the nonlinear algebraic equation group made up of Eqs. (29), (31) and (32).

### 5.2.3. Flow maldistribution

In fact, the hollow fiber membrane module isn't ideal at all and often subjects to flow maldistribution, which may bring on low-efficiency of the module. Therefore, flow maldistribution should be considered in predicting the performance of hollow fiber module used in MD.

Formerly, Noda et al [25] studied the shell-side flow maldistribution in dialysis process. They proposed a model incorporating flow maldistribution by assuming that the shell-side of a hollow fiber module can be divided into two categories: well distributed across the fiber bundle and bypass the bundle. But the later doesn't contribute to the overall mass transfer. Tompkins et al [26] studied the solvent extraction process using axial flow hollow fiber module. The mass transfer performance from the experiment were about 10% deviation from the calculation, and this was attributed to fluid maldistribution at the shell side and thus reducing the proportion of active membrane area. Lemanski et al [27] studied the effect of residence time distribution (RTD) at the shell side on the mass transfer performance. They pointed out that it was plug flow in an ideal hollow fiber module, but in a real shell-side flow the distribution of fluid across the fiber bundle tended to broaden the RTD.

Another way to describe the effect of shell-side flow is to establish empirical equations to predict the overall mass transfer coefficient [28-30]. The general equation form is: $Sh = f(\phi, \mathrm{Re}, Sc)$ where $\phi$ is the packing fraction of the hollow fiber module (the fraction of the volume occupied by fibers to the total shell volume). The introduction of $\phi$ may imply the influence of shell-side flow maldistribution.

Recently, Voronoi tessellation has been used to determine the distribution of fibers and shell-side flow in randomly packed fiber bundles assuming flow parallel to the fiber [24, 31, 32]. By calculating the theoretically exponential distribution of areas associated with each sub-division of the fiber bundles and local friction factor, the distribution of the shell-side flow and mass transfer coefficient are determined.

Flow distribution at the lumen side is also non-uniform in some cases. Using high-speed photography and dye tracer, Park et al [33] found that flow distribution depends on the inlet manifold type and size, tube length, fiber inner diameter, shell diameter, fiber packing fraction and Reynolds number. Crowder and Wickramasinghe et al. [34, 35] found that for polydisperse fiber inner diameter, the average mass transfer coefficient is reduced due to the polydispersity.

In what follows, it is assumed that the maldistribution at the lumen side is due to the polydispersity of fiber inner diameter and at the shell side due to non-uniform fiber packing.

### 5.2.4. Modeling the lumen side flow distribution

Now we investigate the effect of polydispersity of fiber inner diameter. Here the size distribution of fiber inner diameter is taken into account. But it is assumed that fibers are packed uniformly, so the flow distribution at the shell side of the module is completely even.

In general, the fiber inner diameters are non-uniform, following a Gauss distribution [23, 30, 35]. The probability density function for this random variable is:

$$g(x) = \frac{1}{\sqrt{2}\sigma} e^{-\frac{(x-d_{i,m})^2}{2\sigma^2}} \tag{76}$$

The probability for inner diameters between $x_1$ and $x_2$ is:

$$P\{x_1 < x \le x_2\} = \int_{x_1}^{x_2} \frac{1}{\sqrt{2}\sigma} e^{-\frac{(x-d_{i,m})^2}{2\sigma^2}} \, dx \tag{77}$$

where $d_m$ (m) is the average fiber inner diameter, and $\sigma$ (mm) is standard deviation. In theory, the analysis should be done for each fiber in the module, but in practice a histogram describing the fiber diameter distribution is constructed. This histogram consists of $M$ categories, each category represents a size interval of the fiber inner diameter, and the probability of a category can be calculated from Eq. (77).

Thus, a hollow fiber module is correspondingly divided into $M$ regions, and each one is regarded as an ideal hollow fiber module. These regions have the same packing fraction, but the fiber number and fiber inner diameter are different each other. Therefore, there exists a flow distribution at the lumen side of the module. Because the flowrate in a duct is proportional to the quartic of hydraulic diameter for laminar flow, the flowrate at the lumen side of $k$th region $F_{t,k}$ (kg s$^{-1}$) can be determined by:

$$F_{t,k} = \frac{n_k d_{i,k}^4}{\sum\limits_{j=1}^{M} n_j d_{i,j}^4} F_t \tag{78}$$

where $n_k$ is the number of fibers in the $k$th region, $d_{i,k}$ (m) is the fiber inner diameter of this region, and $F_t$ (kg s$^{-1}$) is the total flowrate at the lumen side of the module. With the assumption that the shell is uniformly packed, the flowrate at the shell side of $k$th region is proportional to the number of fibers it contains:

$$F_{s,k} = \frac{n_k}{\sum\limits_{j=1}^{M} n_j} F_s \tag{79}$$

where $F_s$ (kg s$^{-1}$) is the total flowrate at the shell side, and $F_{s,k}$ (kg s$^{-1}$) is the flowrate at the shell side of $k$th region.

If it is assumed that the hollow fiber bundle in each region is ideal, Eqs. (72) – (75) can be used to describe the MD process. The equations are solved over each of these regions individually by using the feed and permeate temperatures obtained from Eqs. (31) and (32), as well as membrane area. The total mass flux of the module is the sum of the production rates of all regions divided by the total membrane area:

$$N = \frac{\sum N_j A_j}{\sum A_j} \tag{80}$$

The outlet temperatures of the feed and permeate is calculated by

$$t_f^{out} = \frac{\sum t_{f,j}^{out} F_{f,j}}{F_f}, \quad t_p^{out} = \frac{\sum t_{p,j}^{out} F_{p,j}}{F_p} \tag{81}$$

Now we investigate the effect of polydispersity of fiber inner diameter.

### 5.2.5. Modeling the shell side flow distribution

In principle, the fibers in a bundle can be spaced regularly across the shell of a module, and arranged in either square or triangular array, as in tube and shell heat exchanger. In the most industrial modules, however, the fibers are often packed randomly in the shell. The duct sizes and shapes in the shell aren't same, and the module shows a certain extent variation of the local packing fraction, as illustrated in Fig. 53. Since the hydraulic resistance of each duct for fluid to pass through is different, the randomness of arrangement leads to flow maldistribution at the shell side. Now we investigate how to determine the flow maldistribution in the shell and its influence on MD. Firstly, all the fibers at a randomly packed module are divided into a number of groups (or regions) of different packing fraction. And it is assumed that all fibers in a group are evenly distributed. Then the performance for these regions is simulated independently by solving Eqs. (72) – (75), and the results obtained are incorporated into Eqs. (80) – (81) to evaluate the overall performance of a module.

(1) The distribution of local packing fraction

In order to characterize the local difference of fiber packing in the hollow fiber module, subdivision of the module into areas associated with each fiber is necessary. Voronoi tessellation is a useful mathematical tool to describe the subdivision of space between randomly packed points, and has been proved to be in a good agreement with the ball packing experiments. This tool has been applied by many researches [24, 31-32] to divide hollow fiber bundle into a number of polygonal cells, as illustrated in Fig. 54. In their work, fiber cross-section was described using probability function to determine cell size distribution. The probability density function $f$ obtained by Chan [36] was an exponential one, and applied to approach the probability that there is no other fiber within a polygonal area surrounded by $s$ nearest fibers.

$$f(\psi_i) = \frac{s^s}{\overline{\psi}^s} \frac{\psi_i^{s-1}}{(s-1)!} e^{-s\psi_i/\overline{\psi}}$$ (82)

where $\psi_i$ (m$^2$) is the area of a polygonal, and $\overline{\psi}$ (m$^2$) is the average polygonal area. From Eq. (82), we can deduce the probability of the packing fraction of a polygonal cell between $\phi_1$ and $\phi_2$:

$$P = \int_{\lambda_1}^{\lambda_2} s^s \frac{\lambda^s}{(s-1)!} e^{-s\lambda} d\lambda$$ (83)

where

$$\lambda_i = \frac{(1-\phi_i)\phi}{(1-\phi)\phi_i} \quad (i = 1, 2)$$ (84)

and $\phi$ is the packing fraction of the hollow fiber module. So the Voronoi tessellation technique can provide us with a way to determine the distribution of polygonal area and the local packing fraction.

(2) Flow distribution

Consider the flow field of longitudinal laminar flow between cylinders arranged in triangular array, as shown in Fig. 55. The momentum equation for this flow field in cylindrical coordinates is

$$\frac{\partial^2 u}{\partial r^2} + \frac{1}{r}\frac{\partial u}{\partial r} + \frac{1}{r^2}\frac{\partial^2 u}{\partial \theta^2} = \frac{1}{\mu}\frac{dp}{dz}$$ (85)

An important variable for describing the flow distribution in the overall fiber bundle is the product of the friction factor and Reynolds number, . $f \cdot Re$ When solving Eq. (85), Sparrow [37] developed an effective method to calculate the $f \cdot Re$:

$$f \cdot Re = \left[\frac{r_0^4}{V\mu}\left(-\frac{dP}{dz}\right)\right]\left[\frac{\theta_0}{(1-\varepsilon)^2} + \theta_0 - \frac{2\theta_0}{1-\varepsilon}\right]$$ (86)

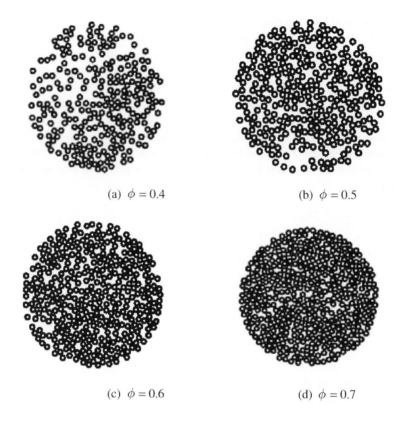

(a) $\phi = 0.4$             (b) $\phi = 0.5$

(c) $\phi = 0.6$             (d) $\phi = 0.7$

Fig. 53. Schematic representation of non-uniform distribution of hollow fibers in membrane module; adapted from the source [57].

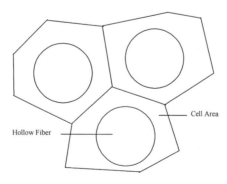

Fig. 54. Illustration of Voronoi tessellation of the module cross section.

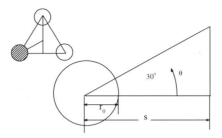

Fig. 55. Cylinders arranged in triangular array.

where $V$ (m$^3$ h$^{-1}$) is the volume flowrate of the fluid, $dP / dz$ (Pa m$^{-1}$) is the pressure gradient, and $\varepsilon$ is the porosity of the hollow fiber membrane module. Under the condition that (1) identical pressure drop in all cells; (2) the total mass flowrate is equal to the sum of the individual cell mass flowrate, the effective $f \cdot \mathrm{Re}$ in the overall fiber bundle can be determined by

$$\frac{1}{\left(f \cdot \mathrm{Re}\right)_e} = \sum_{i=1}^{n_f} \frac{1}{\left(f \cdot \mathrm{Re}\right)_i} \left(\frac{P_w}{P_{wi}}\right)^2 \left(\frac{A_{Ci}}{A_C}\right)^3 \tag{87}$$

where $P_{wi}$ (m) and $A_{Ci}$ (m$^2$) is the wetted perimeter and the flow area of the $i$th polygonal cell, respectively; $P_w$ (m) and $A_C$ (m$^2$) is the total wetted perimeter and total cross-sectional area for flow in the overall bundle, respectively; $n_f$ is the total number of fibers in the overall bundle [34].

Like the lumen side, a histogram consisting of $m$ categories to describe the distribution of the cells' area in the module is introduced. The effective $f \cdot \mathrm{Re}$ is then given by

$$\frac{1}{\left(f \cdot \mathrm{Re}\right)_e} = \sum_{j=1}^{m} \frac{1}{\left(f \cdot \mathrm{Re}\right)_j} n_{f,j} \left(\frac{P_w}{P_{w,j}}\right)^2 \left(\frac{A_{C,j}}{A_C}\right)^3 \tag{88}$$

where $n_{f,j}$ is the number of fibers or cells in the $j$th category. Except $n_{f,j}$, all the terms at the right of Eq. (88) can be replaced by the local packing fraction of a category $\phi_j$ and the module's packing fraction $\phi$. Introduce $P_j = n_{f,j} / n_f$, and it is obtained [38]:

$$\frac{1}{\left(f \cdot \mathrm{Re}\right)_e} = \sum_{j=1}^{m} \frac{1}{\left(f \cdot \mathrm{Re}\right)_j} \left[\frac{\left(1-\phi_j\right)\phi}{\left(1-\phi\right)\phi_j}\right]^3 P_j \tag{89}$$

where $P_j$ is the probability of the $j$th category occurring, and is determined from Eq. (83). Once $\left(f \cdot \mathrm{Re}\right)_j$ and $\left(f \cdot \mathrm{Re}\right)_e$ are obtained, the flow distribution at the shell side of the module can be determined by using the fractional flow $W_{s,j}$ $(W_{s,j} = F_{s,j} / F_s)$ at the shell side for the $j$th category [38]:

$$W_{s,j} = P_j \left[ \frac{(f \cdot Re)_e}{(f \cdot Re)_j} \right] \left[ \frac{(1 - \phi_j)\phi}{(1 - \phi)\phi_j} \right]^3 \tag{90}$$

Because in this case there is no polydispersity of fiber inner diameter, the lumen side flowrate in a category is proportional to its membrane area (or number of fibers). Therefore, the fractional flow at the lumen side of the $j$th category, $W_{t,j}$ $(W_{t,j} = F_{t,j} / F_t)$, is:

$$W_{t,j} = P_j \tag{91}$$

Once the flowrates at the lumen and shell sides of the module for each category are known, Eqs. (72) – (75) can be used to describe each category. The total flux of the module and the outlet temperatures of the feed and permeate can also be derived from Eqs. (80) and (81), respectively.

*5.2.6. Results and discussion*
(1) The effect of the polydispersity of fiber inner diameter
A specific hollow fiber module is used as an example to study the effect of polydisperisty of fiber inner diameter. The characteristics of the module studied are listed in Table 7.

Table7
Characteristics of the module studied

| Membrane area (m²) | Packing fraction | Fiber inner diameter (mm) | Number of fibers | Fiber length (m) |
|---|---|---|---|---|
| 1.0 | 0.6 | 0.30 | 3000 | 0.34 |

Fig. 56 shows the effect of the polydispersity of fiber inner diameter on the module flux at various flowrates of the feed and permeate, the standard deviation (SD) of fiber inner diameter $\sigma$ (m) as $x$-coordinate and the correction factor as $y$-coordinate. The correction factor is defined as the module flux with a positive value of $\sigma$ divided by that with $\sigma = 0$ (for the ideal hollow fiber module). Therefore, the correction factor is always less than 1.0, which means that the polydispersity of fiber inner diameter has a negative effect on the module flux. The reason is due to flow maldistribution at the lumen side. From Eqs. (78) and (79), it can be seen that flow distribution between the lumen and shell sides is different. The relationship of the cumulative fraction of flow at the lumen side, $\Sigma F_{t,j} / F_t$, with the cumulative fraction of membrane area (or of fibers), $\Sigma A_j / A$, is given in Fig. 57. If $\Sigma F_{t,j} / F_t$

vs $\Sigma A_j / A$ could be represented by the diagonal in Fig. 57, we can say that there is no

maldistribution at the lumen side. However, the curves' deflexion from diagonal in Fig. 57 indicates that flow maldistribution takes place at the lumen side. Because of the polydispersity of fiber inner diameter, the fiber bundle can be looked on as a combination of many regions with different fiber inner diameters. In some regions, the flowrate at the lumen side is larger than at the shell side; however, in other regions it may be reverse. In the regions where the flowrate at one side is higher than at the other side, the temperature in the latter should be close to that in the former, which means a reduction of the driving force in MD process.

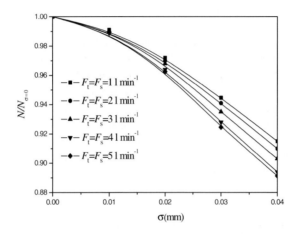

Fig. 56. Correction factor as a function of standard derivation of fiber inner diameter.

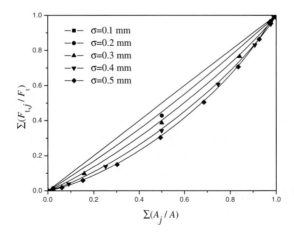

Fig. 57. Cumulative fraction flow at the lumen side as a function of cumulative membrane area.

(2) The effect of non-uniform packing

The fibers bundle of non-uniform packing is divided into a number of categories of different packing fraction. The calculation for each category is done individually, and the results are collected to get the overall performance of the module. Table 8 lists the regional results for a module with overall packing fraction of 0.5, and characteristics of the module comes from Table 1. For the purpose of comparison, the simulation results for the ideal hollow fiber module are also given at the bottom of Table 8. It can be seen that:

1. MD mostly occurs in the regions with the local packing fraction $\phi_i$ between 0.3 and 0.6. 93% of the production rate comes from these regions, but only occupying 75% of the overall membrane area. In the regions with $\phi_i$ larger than 0.6, the permeate flowrate is much less than the feed flowrate, so the temperature of the permeate is close to that of the feed. This means that more than 20% of the feed goes through the module almost without any driving force for MD. That is to say, more than 20% membrane area is ineffective.

2. Even in the regions of $\phi_i$ between 0.3 and 0.6, the permeate flux is less than that of the ideal hollow fiber module. In the two regions with $\phi_i$ equal to 0.526 and 0.588, the permeate flowrate is still less than the feed flowrate, and the temperature difference between the two sides alongside the module is very small. Moreover, the temperature in the permeate outlet is almost equal to that in the feed inlet. This corresponds to a very inactive membrane area, about 35% of the total. This means that some membrane area is ineffective.

3. The performance in the region with $\phi_i$ equal to 0.435 is the closest to ideal hollow fiber module. The reason is due to its $F_f / F_p$ close to 1.0. Unfortunately, this region occupies only 10.40% of the total membrane area, so its contribution is limited.

4. In the regions with $\phi_i$ between 0.25 and 0.4, about 70% of the total permeate contacts with only 19% of the total feed. In this case flow maldistribution brings out excess driving force in these regions, and the permeate flux is even higher than that in the ideal hollow fiber module. But this occurs in a very small portion of membrane area, so the contribution to the overall performance of the module is also limited.

On the other hand, at the shell side the flow imbalance (or flow maldistribution) may largely reduce the driving force of MD. This reduction will make only a small part of membrane area (30% in Table 8) provide the approximate flux of ideal hollow fiber module, but most of membrane area is less or almost not effective.

Fig. 58 shows the effect of packing fraction $\phi$ on the permeate flux of a non-uniformly packed module. The correction factor as $y$-coordinate is defined as mass flux of a non-uniformly packed module divided by mass flux of an ideal module. The reason why an inflexion appears in the curve is: in the case of $\phi$ below 0.6, the number of the nearest neighbors for a fiber is assumed to be $s = 4$; but in the case of $\phi$ above 0.6, the number is assumed to be $s = 6$. By this means, $s$ in Eq. (83) is determined. This has been proved to be effective by Chen's experiment [31].

Table 8

Summarized regional performance of a non-uniformly packed module with $\phi = 0.5$; the operation conditions are: $F_f = F_p = 2$ l min$^{-1}$, $t_f^{in} = 70$ ℃, $t_p^{in} = 25$ ℃, and the permeate is at the shell side.

| Local packing fraction, $\phi_j$ | Membrane area fraction (%) | $F_{f,j}$ (%) | $F_{p,j}$ (%) |
|---|---|---|---|
| 0.909 | 0.00 | 0.00 | 0.00 |
| 0.769 | 7.00 | 7.00 | 0.10 |
| 0.667 | 14.20 | 14.20 | 0.60 |
| 0.588 | 17.60 | 17.60 | 2.20 |
| 0.526 | 16.90 | 16.90 | 4.80 |
| 0.476 | 13.90 | 13.90 | 8.00 |
| 0.435 | 10.40 | 10.40 | 10.70 |
| 0.400 | 7.20 | 7.20 | 12.30 |
| 0.370 | 4.70 | 4.70 | 12.70 |
| 0.345 | 3.00 | 3.00 | 11.90 |
| 0.323 | 1.80 | 1.80 | 10.40 |
| 0.303 | 1.10 | 1.10 | 8.50 |
| 0.286 | 0.60 | 0.60 | 6.70 |
| 0.270 | 0.30 | 0.30 | 5.00 |
| 0.256 | 0.20 | 0.20 | 3.60 |
| 0.256 | 0.10 | 0.10 | 2.50 |
| Non-uniformly packed module | 100 | 100 | 100 |
| Ideal hollow fiber module | 100 | 100 | 100 |

| $t_f^{out}$ (℃) | $t_p^{out}$ (℃) | Flux (kg m$^{-2}$ h$^{-1}$) | Flux × Area (kg h$^{-1}$) |
|---|---|---|---|
| 70.00 | 70.00 | 0.00 | 0.0000 |
| 69.33 | 70.00 | 0.12 | 0.0084 |
| 68.01 | 70.00 | 0.35 | 0.0497 |
| 64.07 | 69.99 | 1.02 | 0.1795 |
| 56.35 | 69.99 | 2.27 | 0.3836 |
| 41.51 | 69.85 | 4.27 | 0.5935 |
| 26.06 | 48.73 | 6.15 | 0.4428 |
| 25.57 | 40.30 | 6.20 | 0.2914 |
| 25.42 | 35.51 | 6.20 | 0.1860 |
| 25.42 | 32.25 | 6.20 | 0.1116 |
| 25.32 | 30.43 | 6.21 | 0.0683 |
| 25.31 | 28.77 | 6.22 | 0.0373 |
| 25.30 | 27.53 | 6.24 | 0.0187 |
| 28.10 | 62.508 | 5.700 | 5.700 |
| 28.103 | 62.508 | 5.700 | 5.700 |
| 49.516 | 43.821 | 2.995 | 2.995 |

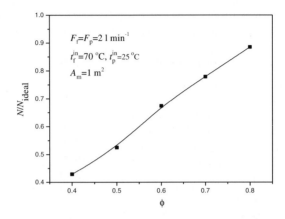

Fig. 58. The effect of packing fraction on mass flux of a non-uniformly packed module.

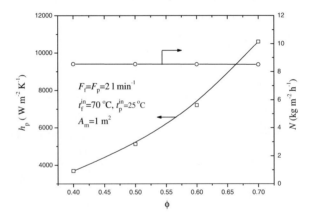

Fig. 59. The effect of packing fraction on heat transfer coefficient and mass flux of ideal hollow fiber module.

It can be seen from Fig. 58 that packing fraction has a significant effect on the performance of a non-uniformly packed module. There are two possible reasons for the mass flux enhancement by increasing packing fraction. It is known that heat transfer at the shell side can be improved by the increment of packing fraction in a tube and shell heat exchanger. This is also valid for hollow fiber module. Fig. 59 shows that packing fraction almost has no influence on the mass flux of an ideal module. This implies that the controlling step in hollow fiber is no longer the heat transfer in the thermal boundary layer. In fact, because of the small size of flow ducts, the heat transfer coefficients at both sides of the membrane are relatively high, so temperature polarization can almost be neglected. This is the first reason. Another reason is that the increment of packing fraction may mitigate flow maldistribution at the shell side, so the regional temperature difference between the two sides of the membrane increases. The shell side flow distribution at various packing fractions is shown in Fig. 60 by comparing cumulative flow fraction of the shell side, $\Sigma W_i$, with the cumulative fraction of membrane area, $\Sigma P_i$. Similar to the situation in the lumen side, the deflexion of these curves from diagonal is the evidence of flow maldistribution at the shell side of the module. With the increase of packing fraction, the curve is more and more close to the diagonal. This means that the shell side experiences a more uniform flow at a higher packing fraction, and a large temperature difference between the two sides is achieved.

In contrast with the above results, for some MD modules with low packing fraction, e.g. $\phi = 0.1$, mass flux of the module was significantly higher than that with high packing fraction, e.g. $\phi = 0.4$. However, it should be mentioned that the range of packing fraction involved in the above text is 0.4-0.8. The possible reason for the results obtained from the module with low packing fraction is that the transverse flow or radial mixing at the shell side of the module becomes serious with the decrease of packing fraction This has been pointed out by Wu and Chen [24], who obtained a minimum mass transfer coefficient at $\phi = 0.5$ when studying the effect of packing fraction on mass transfer within the range of $\phi = 0.1$-0.8.

They also analyzed the pressure drop as a function of axial flow velocity. It was found that velocity exponent $n$ versus the pressure drop at the shell side was higher than 1.4 at $\phi$ below 0.5, and $n$ dropped to 1.15 at $\phi = 0.5$. This indicates that turbulent flow plays an important role in the pressure drop of the module with low packing fraction. The transverse flow may exert a positive effect on the MD module performance by making the temperature profile on the cross section of the shell more uniform. But with the increase of packing fraction, transverse flow or radial mixing would become unimportant and flow maldistribution is then the crucial factor in MD.

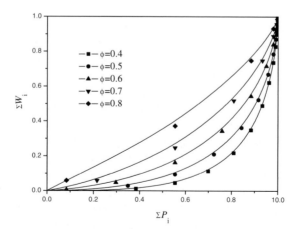

Fig. 60. Cumulative shell side flow fraction as a function of cumulative fraction of membrane area.

Herein, a conclusion is made for the performance of hollow fiber module used in MD. Since the hollow fiber module subjects to flow maldistribution, the influence on the module performance should be taken into account. To describe flow maldistribution mathematically, sub-division of the module is employed to divide the fiber bundle into a number of sub-regions with different characteristics. The regional performance is obtained in terms of ideal module, while the overall performance of a practical module by the combination of its sub-regions. At the lumen side, flow maldistribution is caused by the polydispersity of fiber inner diameter, and a Gauss distribution is adopted to divide the fiber bundle. At the shell side flow maldistribution is caused by the non-uniformity of fiber packing. The Voronoi tessellation technique is adopted to determine the distribution of polygonal cell area. The effect of flow maldistribution at two sides of the membrane is evaluated individually. The result shows that the polydispersity of fiber inner diameter has negative effect on the module permeate flux, but the negative influence of non-uniformity of fiber packing on the module performance seems to be more important. At the shell side, the module flux increases with the rising of packing fraction.

## 6. APPLICATIONS OF MD

MD can be applied in the areas such as desalination, water removal from blood, milk and juice concentration, separation of alcohol-water mixtures, waste water treatment and the cases where high temperature applications lead to degradation of the fluids processed.

### 6.1. Desalination

Many places around the world, such as the Middle East, the majority of Africa and the Northwest of China, are suffering from lack and scarcity of pure drinkable water, or the increase of ground water salinity. These districts are characterized by brackish water resource and infrequent rainfall, but fortunately an abundant source of solar energy. So the solar desalination is considered to be a feasible technique for solving this crisis. As a separation technique for water purification, the solar desalination has been known for a long time, and various types of solar stills for this purpose have been developed and applied worldwide [39]. The solar desalination system is of low operation and maintenance costs, but a major drawback is very low thermal efficiency. For a given productivity, usually a substantially large installation area is required, which in turn results in high initial investment costs.

The employment of membrane technology in desalination is another choice for water purification. In 1995, the production of pure water by means of membrane based technologies accounts for about 35% of the world's desalination capacity [40]. The main advantages of membrane based desalination process are: (1) membranes are of high selectivity for water molecules to pass through; (2) high mass transfer rates; (3) high packing density within membrane modules. In comparison with solar desalination, the membrane based desalination processes requires much smaller installation areas and thus a lower initial investment cost, but, on the other hand, the operation and maintenance costs are relatively higher.

Analyze the characteristics of these two desalination techniques, and it is straightforward to couple them to bring on a preferable technique. However, this hasn't been realized until the invention of the membrane distillation technique in the late 1960s. Since the 1980s MD has been intensively studied as an alternative technology for the separations conventionally finished by conventional distillation or reverse osmosis. In comparison with traditional desalination process, MD as a thermal driven process can successfully be carried out under substantially lower temperature levels, below 100 ℃. This provides the MD process with the merit to utilize alternative energy sources such as the free available solar energy.

Hogan et al. [41] were among the first to couple MD technique with a solar powered system. With the aid of a computer simulation program, the optimized results were used for the design of a solar power membrane distillation (SPMD) pilot plant. As shown in Fig. 61, this pilot plant mainly consists of a hollow fiber membrane module to perform MD process, and a heat exchanger to recover part of energy that is transferred from the feed steam to the permeate stream while they contact in membrane module. As the energy can't be 100% recovered, an auxiliary cooler is used to further decrease the temperature of the permeate coming out of the recovery heat exchanger before it returns to the membrane module.

This SPMD unit was tested in Sydney, and found to be technically feasible with the membrane process compatible with the transient nature of the energy source. With membrane area 1 m$^2$, collector area 4 m$^2$ and KA of recovery heat exchanger 800 W K$^{-1}$, the average daily product in this pilot plant is about 20 kg pure water, which is somewhat lower than expected. This may be caused by a poor fluid dynamics in the membrane module. Flow maldistribution at the shell side of the module reduced the effectiveness of membrane area.

Another SPMD unit was constructed in 1994 [42], as shown in Fig. 62. At that time an air gap membrane distillation (AGMD) module was used instead of DCMD module used by Hogan et al [41]. The latent heat recovery process in this pilot plant was integrated with AGMD in a spiral wound membrane modules. This module received the feed steam as the coolant to make the permeated vapor condense on the plate, and the feed was heated by the latent heat released. So the recovery heat exchanger adopted by Hogan et al. is omitted in this plant. The temperature of feed stream increases and then it is introduced to a heater, which is driven by the heat agent coming from the solar energy collector, to receive more energy. However, the additional mass transfer resistance caused by the air gap in the module resulted in a large reduction in the transmembrane water flux. With membrane area 10 m$^2$ and collector area 100 m$^2$, the capacity of this plant varied from 15 to 25 kg h$^{-1}$, which was also lower than expected.

Koschilkowski et al. [43] investigated a membrane module similar to that used by Bier et al. for the SPMD pilot plant. According to their calculation, with membrane area 7 m$^2$ and collector area only 6m$^2$, the plant can distill 150 liter of water per day in the summer of a southern country without any heat storage. Capacity of such a SPMD pilot plant is very attractive, and worthy of being accepted commercially. However, no experimental results were reported.

Banat et al [44] integrated a membrane distillation module with a solar still, as shown in Fig. 63, to produce potable water from simulated seawater. Here the solar still was used for both seawater heating and potable water production. The effects of temperature, brine flowrate, salt concentration and solar irradiation on the desalination process were tested in both indoor and outdoor experiments.

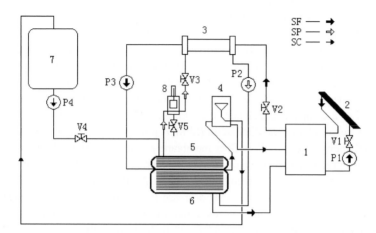

Fig. 61. Flow sheet of SPMD pilot plant developed by Hogan et al; 1 - feed tank, 2 - solar power collector, 3 - hollow fiber membrane module, 4 - overflow tank, 5 – cooler, 6 - recovery heat exchanger, 7 - raw tank, 8 - liquid level detector, P – pump, V – valve, SF - stream of the feed, SP - stream of the permeate; SC - cool water; adapted from the source [41].

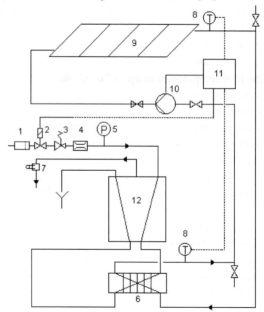

Fig. 62. Flow sheet of SPMD pilot plant developed by Bier et al; 1 – filter, 2 - electromagnetic valve, 3 - reducing valve, 4 - flow meter, 5 - pressure gauge, 6 - heat exchanger, 7 - conductivity meter, 8 - temperature gauge, 9 - solar energy collector, 10 – pump, 11 - control system, 12 - membrane module. adapted from the source [42].

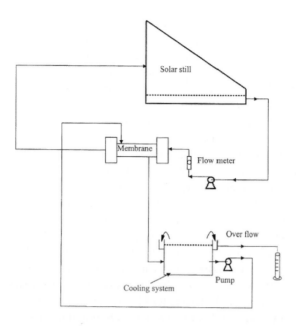

Fig. 63. Flowsheet of a SPMD pilot plant developed by Banat et al; adapted from the source [44].

It is interesting to find that the contribution of the solar still in the distillate process was no more than 20% of the total flux in the outdoor experiments and less than 10% in the indoor experiments. When the feed temperature is up to 70 ℃, the maximum mass flux obtained from this work was less than 2 kg m$^{-2}$ h$^{-1}$, which is much lower than that obtained from the experiment. It indicates that much work is needed to improve the MD module performance.

The energy resource adopted by MD desalination needn't be limited in solar power, some other choices, such as low-grade waste heat energy source in industry (the cooling water from engines, the condensed water of low-pressure vapor), geothermal energy and so forth, are also attractive energy resources for MD. Unfortunately, there is no report about the application of these energy resources in MD process.

## 6.2. Concentration of aqueous solution

Although membrane distillation was first developed as an alternative to traditional desalination technique, now it is receiving more attention in the field of the concentration of aqueous solution, especially containing heat sensitive substances. The concentration of fruit juice may be the most common one being tackled on by MD.

Usually, the freshly squeezed fruit juice contains 85% or more water by weight. To reduce the storage and shipping costs and achieve long time storage, in the former fruit juices were concentrated by multi-stage vacuum evaporation. This process results in a loss of fresh juice flavors, color degradation and a "cooked" taste because of the thermal effects. Since many consumers prefer the flavor, aroma, appearance and mouth feel of freshly squeezed juices, the juice industry has developed complex essence recovery, process control and blending techniques to produce a good quality concentrate that is acceptable to consumers, but still easily distinguishable from fresh juice. Many efforts have been devoted to develop an alternative method, such as freeze concentration, sublimation concentration and membranes concentration (ultrafiltration and reverse osmosis) for concentrated juice processing [45]. However, it is difficult to reach concentrations larger than 25–30 Brix, which is below the value of 45–65 Brix for standard products obtained by evaporation with a single-stage RO system due to the limitation of high osmotic pressure.

New membrane processes, including MD, osmotic distillation (OD), and the integrated membrane processes are still being developed in concentrated fruit juice processing to improve product quality and reduce energy consumption.

OD is a new membrane process, which is very similar to MD process. Both use the same membranes as the supporters of liquid-vapor interfaces for the evaporation of volatile components, and the mechanisms that regulate the mass transfer through the membrane are almost the same. The only difference between them lies in the way by which the driving force for mass transfer is exerted. OD has been developed since the late 1980s. In OD a membrane separates the feed and the salt water of high concentration at two sides by means of the force of osmosis pressure. The vapor from the feed will pass through the membrane and enter into the side of salt water. The result is that the feed is concentrated and the salt water is diluted. It seems that the driving force of this process is the pressure difference exerted by the vapor pressure drop of water containing salt. But the calculated result shows that the mass transfer coefficient in OD is much larger than in MD. So it is emphasized that the mechanism of OD isn't the same as that of MD, and more attention should be paid to studying its nature. OD can be performed at or under room temperature, so it is very suitable for treating heat sentitive substances, such as fruit juice, jam, enzyme, protein and so on.

In Australia, an OD pilot plant for the concentration of fruit juice and vegetable juice has been built up with the capacity of 50 L and the concentration of 65-70 wt%, and its maximum capacity is $100 \ 1 \ h^{-1}$ [46]. The benefit of this concentration process should be considerable. As fruit juice contains oily component, it should be cautious to guarantee the hydrophobic property of the membrane used.

Recently, Sanjay Nene et al. [47] applied DCMD for the concentration of raw cane-sugar syrup. In a conventional cane-sugar manufacturing process, the clarified juice (17-20 Brix) which comes as an overflow from the Dorr at a temperature of 100 ℃ or more proceeds to a film evaporator to remove some water. The consumption of steam for the removal of water from the juice is considerable. Sanjay Nene et al. pointed out that it's technically feasible to introduce a membrane distillation step to increase the solids in juice from 17 to 30 Brix. They designed an experiment, as shown in Fig. 65. The aim was to study the flux decay in membrane distillation when the run conditions were similar to those in the sugar industry. The obtained results indicate that it is possible to continuously remove water from the cane sugar solution at steady-state value with the capacity of about $10.0 \ kg \ m^{-2} \ h^{-1}$. And they claimed that MD saves energy in removing water from clarified cane juice by utilizing sensible heat.

In addition, due to the chemical stability of the membrane, MD can also be applied for the concentration of acid, even the strong inorganic acid. The work in this field has been done by many researchers. Among them, the work of Tomaszewska et al. [48] is noticeable. They studied the concentration of sulfur acid, hydrochloric acid and nitric acid by both DCMD and VMD, and the highest concentration of acid is up to 60%.

The advantages of MD are brought out not only in its energy saving, but also in its ability to tackle aqueous solution at the higher concentration than the maximum concentration acquired in RO or UF. This technique is even capable of getting crystal products from solution, so it is suitable for some high value-added heat sensitive substances. Wu et al. [49] studied the recovery of taurine from wastewater solution using anisomerous PVDF membrane, and got some useful results. Additionally, the treatment of other kinds of solutions was also involved. Sakai et al. [50] studied the dialysis of blood under the operation temperature below 50℃, and successfully concentrated blood plasma and cattle blood.

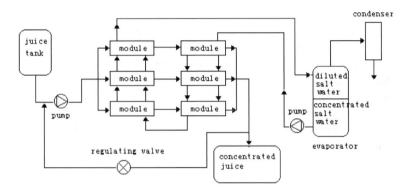

Fig. 64. The pilot plant for the concentration of fruit juice and vegetable juice using OD technique; adapted from the source [46].

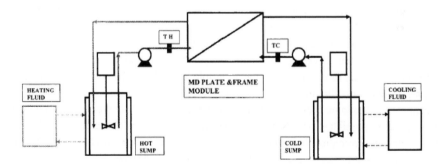

Fig. 65. Sanjay Nene's experimental equipment for the concentration of cane-sugar juice; adapted from the source [47].

## 6.3. Separation of volatile components

MD can also be used to treat aqueous solution containing other volatile components. Unlike desalination process where only water vapor passes through the membrane, two or more components vaporize and permeate through the membrane in this process. In this case, the selectivity of MD process is no longer theoretically 100%, and it largely depends on the relative volatility of the different components in the system. An example of the separation of volatile components from water is the concentration of ethanol-water solution. Gostoli et al. [51] investigated the separation of alcohol-water and found that when the system was kept under a relatively small flux, the concentration of alcohol in the permeate would become twofold of that in feed solution. The separation of alcohol-water by VMD was also studied in detail by Bandini et al. [52], and the concentration of alcohol in the permeate was tenfold of that in feed solution. A developed mechanical model can accurately describe the effect of various operation parameters on permeate flux. For removing alcohol (methanol, ethanol) from water, VMD is commercially competitive because of its large flux, high thermal efficiency and low energy cost. Moreover, the pore size of the membrane used in VMD is relatively small, so it improves the selectivity of alcohol to water. However, the separation of volatile component from water by MD isn't limited to alcohol-water system. The removal of trace amount of organic component from water by MD is being received more attention as an alternative to present popular techniques, i.e. ordinary distillation or adsorption.

Although MD is now being a hotspot because of its potential applications, many problems have to be solved before it is applied in industry as an effective technique for the separation of mixture. The problems left in MD are:

(1) Membranes characterized by high porosity, small thickness and low price (cheaper at least one order of magnitude than present MF membrane) are still to be fabricated;

(2) Membrane modules with good fluid dynamics are still to be designed to mitigate the influence of temperature polarization and/or intensify the mass and heat transfers in MD.

(3) The fouling phenomena need to be extensively studied so as to make the MD process more challenging in industry.

(4) Thermal efficiency of MD needs to be further improved.

(5) It is easy to break the azeotropic point for the mixture forming azeotrope by MD, but difficult to get high purity of products. So it is advisable to combine MD with other separation processes, i.e. ordinary distillation.

In attempts to overcome these challenges standing in front of MD process, investigators [56-80] launched all possible efforts to prove its validity as a successful separation technique. Benefiting from the developments in membrane fabrication, several approaches to make the MD as a viable separation technique were proposed. These approaches ranged from finding new areas of applicability that can benefit from the MD technique, to researches devoted in understanding its drawbacks and finding ways to prevent or eliminate them (by applying some enhancement techniques), in addition to the studies anticipated in finding out the effects of factors and variables affecting the MD process.

Even though the effects of the design and operating variables are well understood and documented in the MD literatures, scattered or inconsistency information exists. Some of the recognized MD configurations are often investigated, while others are rarely studied and far from being fully understood. Attempts to provide successful mathematical models that can describe the transport process in MD are widely available in the literatures. Despite their precisions in estimating effects of different parameters on permeate flux rate, those models based on special experimental facilities are rarely used. Although the effect of some factors and variables are generally agreed upon, new researches still appear with almost the same results. For example, effects of operating variables such as feed and permeate temperatures and flow velocities as well as feed concentration on the flux of MD process have been widely investigated (with each study providing a relatively different model). On the other hand, the factors such as pore size distribution, tortuosity and fouling are rarely studied or even ignored despite the important role that they play in the MD process.

## REFERENCES

[1] J.D. Seader and E.J. Henley (eds.), Separation Process Principles, Wiley, New York, 1998.

[2] K.W. Lawson and D.R. Lloyd, J. Membr. Sci., 124 (1997) 1-25.

[3] C.A. Smolder and A.C.M. Franken, Desalination, 72 (1989) 249-261.

[4] M. Khayet and T. Matsuura, Desalination, 158 (2003) 51-56.

[5] L. Martinez-Diez, M.I. Vaquez-Gonzaez and F.J. Florido-Diaz, J. Membr. Sci., 144 (1998) 45-56.

[6] T. Y. Cath, V. D. Adams and A.E. Childress, J. Membr. Sci., 228 (2004) 5-16.

[7] P. K. Weyl, Recovery of Demineralinzed Water from Saline Waters, US Patent No. 3 340 186 (1967).

[8] K.W. Lawson and D.R. Lloyd, J. Membr. Sci., 120 (1996) 123-133.

[9] M.C. Garcia-Payo, M.A. Izquierdo-Gil and C. Fernandez-Pineda, J. Membr. Sci., 169 (2000) 61-80.

[10] Z. Ding, W. Liu, G. Zhang and R. Ma, Modern Chemical Industry (China) 22 (2002) 26-29.

[11] K. Mohamed, G. Paz and I. M. Juan, J. Membr. Sci., 165 (2000) 261-272.

[12] J. Phattaranawik, R. Jiraratananon and A.G. Fane, J. Membr. Sci., 212 (2003) 177-193.

[13] W. Kast and C.R. Hohenthanner, Int. J. Heat Mass Transf., 43 (2000) 807-823.

[14] W.L. McCabe, J.C. Smith and P. Harriott (eds.), Unit Operations of Chemical Engineering (6th edition), McGraw-Hill, New York, 2001.

[15] A. F. Mills (ed.), Mass Transfer (2nd edition), Prentice-Hall, New Jersey, 2001.

[16] W. Fred Ramirez, Computational Methods for Process Simulation (2nd edition), Butterworth-Heinemann, London, 1997.

[17] M.A. Izquierdo-Gil, M.C. Garcia-Payo and C. Femandez-Pineda, J. Membr. Sci., 155 (1999) 291-307.

[18] R.W. Schofield, A.G. Fane and C.J.D. Fell, J. Membr. Sci., 33 (1987) 299-313.

[19] K.W. Lawson and D.R. Lloyd, J. Membr. Sci., 120 (1996) 111-121.

[20] T.K. Sherwood, R.L. Pigford and C.R. Wilke, Mass Transfer, McGraw-Hill, New York, 1976.

[21] R. Ratnam and V.S. Patwardhan, Chem. Eng. Prog., 89 (1993), 85-91.

[22] J.D. Rogers and R.L.Jr. Long, J. Membr. Sci., 134 (1997), 1-17.

[23] A. Gebelman A and S.K. Hwang, J. Membr. Sci., 159 (1999) 61-106.

[24] J. Wu and V. Chen, J. Membr. Sci., 172 (2000) 59-74.

[25] I. Noda, D.G. Brown-West, C.C. Gryte, J. Membr. Sci, 5 (1979) 209-225.

[26] C.J. Tompkins, A.S. Michaels and S.W. Peretti, J. Membr. Sci., 75 (1992) 277-292.

[27] J. Lemanski and G.G. Lipscomb, AIChE J., 41 (1995) 2322-2326.

[28] P. Prasad and K.K. Sirkar, AIChE J., 33 (1987) 1057-1066.

[29] M.J. Costello, A.G. Fane, P.A. Hogan and R.W. Schofield, J. Membr. Sci., 80 (1993) 1-11.

[30] S.R. Wickramasinghe, M.J. Semmens and E.L. Cussler, J. Membr. Sci., 84(1993) 1-14.

[31] V. Chen and M. Hlavacek, AIChE J., 40 (1994) 606-612.

[32] J.D. Rogers and R. Long, J. Membr. Sci., 134 (1997) 1-17.

[33] J.K. Park and H.N. Chang, AIChE J., 32 (1985) 1937-1947.

[34] S.R. Wickramasinghe, M.J. Semmens and E.L. Cussler, J. Membr. Sci., 69 (1992) 235-250.

[35] R.O. Crowder and E.L. Cussler, J. Membr. Sci., 134 (1997) 235-244.

[36] D.Y.C. Chan, B.D. Hughes and L. Paterson, Transp. Porous Media, 3 (1988) 81-94.

[37] E.M. Sparrow and A.J.Jr. Loeffler, AIChE J., 5 (1959) 325-330.

[38] Z.W. Ding，Study on Module Performance and Dynamic Process of Membrane Distillation (Ph.D. Thesis), Beijing University of Chemical Technology, Beijing, 2001.

[39] M.T. Chaibi, Desalination, 127 (2000) 119-133.

[40] T. Yoshio, Desalination, 110 (1997) 21-35.

[41] P.A. Hogan, Sudjito, A.G. Fane and G.L. Morrison, Desalination, 81 (1991) 81-90.

[42] C. Bier and U. Plantikow, Solar Powered Desalination by Membrane Distillation (in IDA World Congress on "Desalination and Water Science"), ABU Dhabi, 1995.

[43] J. Koschikowski, M. Wieghaus and M. Rommel, Desalination, 156 (2003) 295-304.

[44] M. Banat, R. Jumah and M. Garaibeh, Renew. Energy, 25 (2002) 293-305.

[45] B. Jiao, A. Cassano and E. Drioli, J. Food Eng., 63 (2004) 303-324.

[46] P.A. Hogan, R.P. Canning and R.A. Johnson, Chem. Eng. Prog., 34 (1998) 49-68.

[47] S. Nene, S. Kaur, K. Sumod, B. Joshi and K.S.M.S. Raghavarao, Desalination, 147 (2002) 157-160.

[48] M. Tomaszewska, M. Gryta and A.W. Morawski, J. Membr. Sci., 102 (1995) 113-122.

[49] Y. Wu, Y. Kong, J. Liu, J. Zhang and J. Xu, Desalination, 80 (1991) 235-242.

[50] K. Sakai, T. Koyano, T. Muroi and M. Tamura, Chem. Eng. J., 38 (1988) B33-B39.

[51] C. Gostoli and G.C. Sarti, J. Membr. Sci., 33 (1989) 211-224.

[52] B. Bandini, C. Gostoli and G.C. Sarti, J. Membr. Sci., 73 (1992) 217-229.

[53] R.C. Reid, J.M. Prausnitz and B.E. Poling (eds.), The Properties of Gases and Liquids, McGraw-Hill, New York, 1987.

[54] B.E. Poling, J.M. Prausnitz and J.P. Oconnell (eds.), The Properties of Gases and Liquids (fifth edition), McGraw-Hill, New York, 2000.

[55] J.S. Tong (ed.), The Fluid Thermodynamics Properties, Petroleum Technology Press, Beijing, 1996.

[56] D. Wang, K.Li and W.K.Teo, J. Membr. Sci., 163 (1999) 211-220.

[57] P.A. Hogan, Doctor of Philosophy Thesis, The University of New South Wales, Sydney, 1994.

[58] M. Khayet and T. Matsuura, Desalination, 158 (2003) 51-56.

[59] M.A. Izquierdo-Gil and G.Jonsson, J. Membr. Sci., 214 (2003) 113-130.

[60] L. Martínez, F.J. Florido-Díaz, A. Hernández and P. Prádanos, Sep. Purif. Technol., 33 (2003) 45-55.

[61] S. Bouguecha, R. Chouikh and M. Dhahbi, Desalination, 152 (2003) 245-252.

[62] S. Bouguecha and M. Dhahbi, Desalination, 152 (2003) 237-244.

[63] M.Gryta, K. Karakulski and A.W. Morawski, Water Res., 35 (2001) 3665-3669.

[64] L. Martínez, F.J. Florido-Díaz, A. Hernández and P. Prádanos, J. Membr. Sci., 202 (2002) 15-27.

[65] M. Gryta, Desalination, 142 (2002) 79-88.

[66] R.A. Johnson, J.C. Sun and J. Sun, J. Membr. Sci., 209 (2002) 221-232.

[67] M. Khayet, K.C. Khulbe and T. Matsuura, J. Membr. Sci., 238 (2004) 199-211.

[68] Z. Wang, F. Zhen and S. Wang, J. Membr. Sci., 183 (2001) 171-179.

[69] A.O. Imdakm and T. Matsuura, J. Membr. Sci., 237 (2004) 51-59.

[70] J.I. Mengual, M. Khayet and M.P. Godino, Int. J. Heat Mass Transf., 47 (2004) 865-875.

[71] R. Bagger-Jørgensen, A.S. Meyer, C. Varming and G. Jonsson, J. Food Eng., 64 (2004) 23-31.

[72] P.J Foster, A.Burgoyne and M.M. Vahdati, Sep. Purif. Technol., 21 (2001) 205–217.

[73] M. Gryta, Sep. Purif. Technol., 24 (2001) 283-296.

[74] M. Gryta, M. Tomaszewska, J. Grzechulska, and A.W. Morawski, J. Membr. Sci., 181 (2001) 279-287.

[75] J. Phattaranawik and R. Jiraratananon, J. Membr. Sci., 188 (2001) 137-143.

[76] K.B. Petrotos and H.N. Lazarides, J. Food Eng., 49 (2001) 201-206.

[77] M.Courel, E. Tronel-Peyroz, G.M. Rios, M. Dornier and M. Reynes, Desalination, 140 (2001) 15-25.

[78] M.A. Izquierdo-Gil, J. Abildskov and G. Jonsson, J. Membr. Sci., 239 (2004) 227-241.

[79] A. Cassano, B. Jiao and E. Drioli, Food Res. Int., 37 (2004) 139-148.

[80] C.H. Lee and W.H. Hong, J. Membr. Sci., 188 (2001) 79-86.

# Chapter 7. Pressure-swing distillation

In the field of special distillation processes, pressure-swing distillation (PSD) is the one able to be utilized to separate the mixture with close boiling point or forming azeotrope, but no new additive added. So PSD is an environment-friendly process. However, the application of PSD is very limited, the well-known system almost only involving tetrahydrofuran (THF) / water. The reason may be that for most of the systems the change of $x$-$y$ curve at different pressures isn't so distinct. This chapter tries to present a wide aspect of PSD in operation principle, operation mode and conceptual design.

## 1. INTRODUCTION

### 1.1. Separation principle

The separation principle of pressure-swing distillation (PSD) is based on the fact that a simple change in pressure can alter relative volatility of the mixture with close boiling point or forming azeotrope. In some cases, this results in a significant change in the azeotropic composition or enlarging the relative volatility of close boiling point components, which allows the recovery of feed mixture without adding a separating agent. Therefore, an outstanding advantage of PSD is that it belongs to environment-friendly process. However, PSD is particularly challenging for the separation of homogeneous azeotropes. Table 1 lists some azeotropes that have the potential to be separated by PSD technique.

### 1.2. Operation modes

There are three types of operation modes for PSD, i.e. continuous operation, batch operation and semi-continuous operation [4]. In continuous operation, the separation is performed using two columns maintained at different pressures. The distillates, which approach the azeotropic compositions at high and low pressure, are cycled between the two columns.

In batch operation, only a single column is used for the separation. It is supposed that component A is the ultimate product in the mixture consisting of components A and B. The column is initially charged with the feed mixture (F) and operates at pressure $P_1$, as shown in Fig. 1. Component B is removed from the bottom of the tower, while at the top the mixture $D_1$ approaching the azeotropic composition. Then, the column is recharged with the mixture $D_1$ and operates at pressure $P_2$. Thus, component A is obtained from the bottom, while at the top the mixture $D_2$ approaching the azeotropic composition. Only can this process be carried out on the condition that the composition of the feed mixture $x_F$ is in the range of $x_{azeo2}$ and 1. To provide high recovery, this cycle may be repeated many times.

Table 1
Examples of PSD binary azeotropes, adapted from the references [1-3]

| No. | Components |
|---|---|
| 1 | carbon dioxide - ethylene |
| 2 | hydrochloric acid - water |
| 3 | water - acetonitrile |
| 4 | water – ethanol |
| 5 | water - acrylic acid |
| 6 | water – acetone |
| 7 | water - propylene oxide |
| 8 | water - methyl acetate |
| 9 | water - propionic acid |
| 10 | water - 2-methoxyethanol |
| 11 | water - 2-butanone (methyl ethyl ketone (MEK)) |
| 12 | water – tetrahydrofuran (THF) |
| 13 | Carbon tetrachloride - ethanol |
| 14 | carbon tetrachloride - ethyl acetate |
| 15 | carbon tetrachloride - ethyl acetate |
| 16 | carbon tetrachloride - benzene |
| 17 | methanol-acetone |
| 18 | methanol - 2-butanone (MEK) |
| 19 | methanol - methyl propyl ketone |
| 20 | methanol - methyl acetate |
| 21 | methanol - ethyl acetate |
| 22 | methanol - benzene |
| 23 | methanol - dichloromethane |
| 24 | methylamine - trimethylamine |
| 25 | ethanol - dioxane |
| 26 | ethanol - benzene |
| 27 | ethanol - heptane |
| 28 | dimethylamine - trimethylamine |
| 29 | 2-propanol - benzene |
| 30 | propanol - benzene |
| 31 | Propanol - cyclohexane |
| 32 | 2-butanone (MEK) - benzene |
| 33 | 2-butanone (MEK) - cyclohexane |
| 34 | isobutyl alcohol - benzene |
| 35 | benzene - cyclohexane |
| 36 | Benzene - hexane |
| 37 | phenol - butyl acetate |
| 38 | aniline – octane |

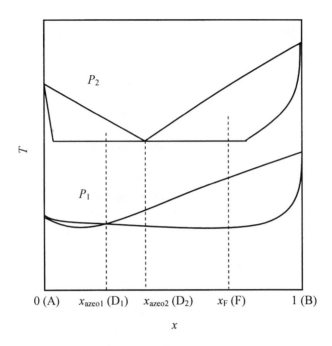

Fig. 1. *T-x-y* diagram at different pressures $P_1$ and $P_2$.

In semi-continuous operation, only a single distillation column is involved. But the column operates continuously and periodically. It isn't emptied or recharged. Liquid levers are maintained on the trays or packing, stream is fed continuously to the reboiler, and cooling water is fed continuously to the condenser. The advantage is lower investment costs and shorter downtimes when the mixture to be separated is changed. An optimal-control algorithm has been programmed by Phimister and Seider [4]. Simulation results for the dehydration of tetrahydrofuran indicate that the column achieves production rates near 89% of the maximum throughput of a single column in the continuous process, which shows superior performance when compared with batch operation. So, undoubtedly, among the three types of operation modes, the mode of continuous operation is more desirable in most cases.

## 2. DESIGN OF PSD

### 2.1. Column sequence

A main problem in PSD is how to arrange the column sequence at pressures $P_1$ and $P_2$. It can be seen from Fig. 1 that if first at pressure $P_1$ and then at pressure $P_2$, the composition of the feed mixture $x_F$ is in the range of $x_{azeo2}$ and 1; inversely, if first at pressure $P_2$ and then at pressure $P_1$, the composition of the feed mixture $x_F$ is in the range of 0 and $x_{azeo1}$. But if the composition of the feed mixture $x_F$ doesn't fall into this desirable range, extra stream should be replenished to make into a new mixture and change the original composition.

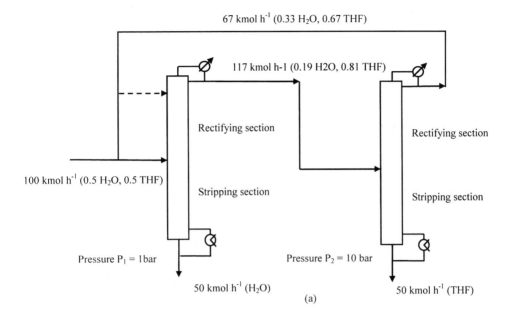

67 kmol h$^{-1}$ (0.33 H$_2$O, 0.67 THF)

117 kmol h-1 (0.19 H2O, 0.81 THF)

Rectifying section

Rectifying section

100 kmol h$^{-1}$ (0.5 H$_2$O, 0.5 THF)

Stripping section

Stripping section

Pressure P$_1$ = 1bar

Pressure P$_2$ = 10 bar

50 kmol h$^{-1}$ (H$_2$O)

50 kmol h$^{-1}$ (THF)

(a)

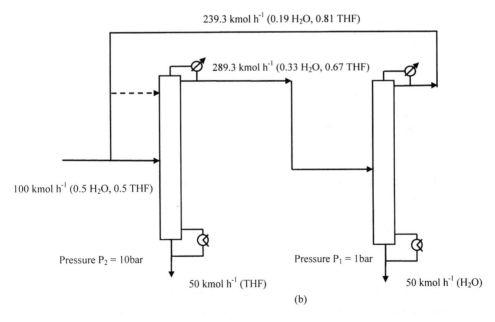

239.3 kmol h$^{-1}$ (0.19 H$_2$O, 0.81 THF)

289.3 kmol h$^{-1}$ (0.33 H$_2$O, 0.67 THF)

100 kmol h$^{-1}$ (0.5 H$_2$O, 0.5 THF)

Pressure P$_2$ = 10bar

Pressure P$_1$ = 1bar

50 kmol h$^{-1}$ (THF)

50 kmol h$^{-1}$ (H$_2$O)

(b)

Fig. 2. Column sequence for the dehydration of THF (a) first at pressure P$_1$ and then at pressure P$_2$, adapted from the reference [4]; (b) first at pressure P$_2$ and then at pressure P$_1$.

For example, a typical PSD process, involving the dehydration of tetrahydrofuran (THF), is illustrated in Fig. 2, where the material balance is marked [4, 5]. It is shown that the cycled flowrates when first at pressure $P_2$ and then at pressure $P_1$ are much larger than those when first at pressure $P_1$ and then at pressure $P_2$. This means that the column sequence of first at pressure $P_1$ and then at pressure $P_2$ is more economical.

Yet PSD isn't restricted to binary mixtures only. It can be extended to separate multi-component mixtures containing distillation boundaries. The presence of a boundary in a ternary mixture means that the three pure components can't be separated without the addition of an entrainer. However, if the boundary (boundaries) can be shifted by adjusting the pressure, a pressure-swing sequence can be used, as shown in Fig. 3.

## 2.2. Column number

The minimum number of columns required for a given separation (but not necessarily the optimum, especially for dilute feeds) can be calculated from [1, 6]

$$N_{col} = N_P + N_B - 1 \tag{1}$$

where $N_{col}$ is the minimum number of columns required, $N_P$ is the number of pure component products; $N_B$ is equal to the number of boundaries crossed for PSD, but this doesn't include boundaries that disappear as the pressure changes. For extractive distillation, $N_B$ is equal to unity. For non-azeotropic system, this reduces to the familiar equality

$$N_{col} = N_P - 1.$$

For example, in Fig. 3 $N_P = 3$ and $N_B = 1$ because the distillation boundary ② is crossed once by material balance line $D_2D_1B_2$ as the pressure is $P_1$ in column 2. Therefore, $N_{col} = 3 + 1 - 1 = 3$ and at least three columns are needed. The process of extractive distillation for separating ternary mixture into its three constituent pure components is also illustrated in Fig. 4, in which also three columns are needed. That means $N_B = 1$ for extractive distillation. Consequently, it is unlikely that PSD will be advantageous when more than one boundary is crossed because the minimum number of columns required increases.

Noted that the azeotropic point for homogeneous azeotropes may vanish as the pressure decreases to some extent, and thus the column number is reduced ( $N_B = 0$ ). In principle, two pure components can be obtained at this pressure by ordinary distillation. However, this phenomenon isn't exploited commercially because in most cases the relative volatility remains close to 1.0, which results in a big reflux ratio and a large number of equilibrium stages.

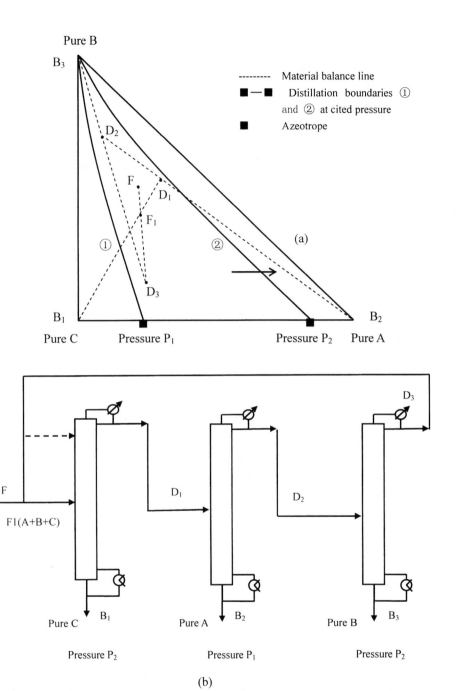

Fig. 3. PSD for separating ternary mixture into its three constituent pure components: (a) material balance lines; (b) column sequence; adapted from the reference [1].

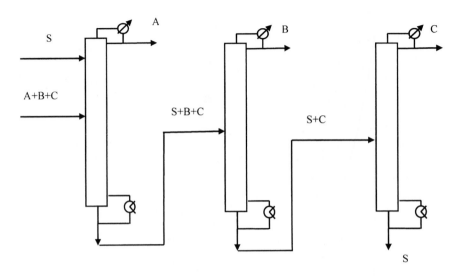

Fig. 4. Extractive distillation for separating ternary mixture into its three constituent pure components.

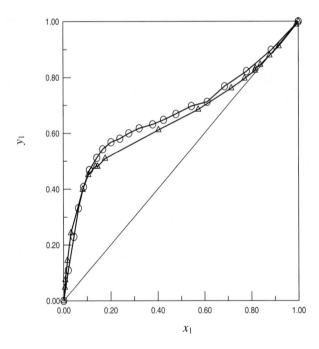

Fig. 5. x-y phase diagram of the system of ethanol (1) / water (2); x and y represents the mole fraction in the liquid and vapor phases, respectively; ○ – at pressure 50 mmHg; △- at pressure 760mmHg; the data come from the reference [8].

As can be seen from Fig. 5, the VLE curve at pressure 50 mmHg is very close to diagonal at high ethanol concentration, although in this case the binary azeotrope is eliminated. Black [7] has indicated that a vacuum distillation carried out at 65 mmHg to break the ethanol / water azeotrope and produce anhydrous ethanol overhead isn't economically competitive with other azeotrope-breaking schemes. For this reason, it is thought that PSD takes advantage of a change in azeotropic composition with pressure, but maybe isn't perfect in getting high-purity of product.

The topic on energy optimization of PSD system has been studied by Hamad and Dunn [9]. It is claimed that using global energy optimization strategies can significantly reduce the energy consumption as opposed to local energy optimization strategies for the minimum-boiling homogeneous azeotropic system of THF / water.

# REFERENCES

[1] J.P. Knapp and M.F. Doherty, Ind. Eng. Chem. Res., 31 (1992) 346-357.
[2] T.C. Frank, Chem. Eng. Prog., No. 4 (1997) 52-63.
[3] L.H. Horsley and R.F. Gould (eds.), Advances in Chemistry 116, American Chemical Society, Washington, 1973.
[4] J.R. Phimister and W.D. Seider, Ind. Eng. Chem. Res., 39 (2000) 122-130.
[5] S.I. Abu-Eishah and W.L. Luyben, Ind. Eng. Chem. Res., 24 (1985) 132-140.
[6] J.R. Knight, Synthesis and Design of Homogeneous Azeotropic Distillation Process, Ph.D. Dissertation, The University of Massachusetts at Amherst, 1991.
[7] C. Black, Chem. Eng. Prog., No. 9 (1980) 78-85.
[8] J. Gmehling and U. Onken (eds.), Vapor-liquid Equilibrium Data Collection (Aqueous-organic Systems), Dechema, Frankfurt, 1977.
[9] A. Hamad and R.F. Dunn, Ind. Eng. Chem. Res., 41 (2002) 6082-6093.

# Chapter 8. Other distillation techniques

Some distillation processes, other than special distillation processes, are carried out in special circumstances relative to ordinary distillation. In this chapter they are emphasized and can be divided into three types:

(1) those which have been applied in special distillation processes, such as high viscosity material distillation and thermally coupled distillation.

(2) those which have the potential to use in special distillation processes, but not yet, such as heat pump and multi-effect distillation.

(3) those which themselves, in principle, can be taken on as special distillation processes for separating mixtures with close boiling point or forming azeotrope, but actually aren't economic, such as molecular distillation.

In this chapter, only the first type is clarified in detail since it is related to our topic, and the others are briefly mentioned.

## 1. HIGH VISCOSITY MATERIAL DISTILLATION

### 1.1. Introduction

The system of high viscosity materials is often met in the distillation plants, such as the separation of C5 mixture by extractive distillation in which polymerization of isoprene occurs and dimer or even heavier polymers with high viscosity will form. Here the PVA (polyvinyl alcohol) production is exemplified. In the polymerization stage of PVA production, VAC (vinyl acetate) would react with PVA to form PVAC (polyvinyl acetate). But first of all, we should separate the polymer (PVAC) and its monomer (VAC) in the NO. 1 polymer distillation column. However, the viscosity of the polymer coming from the reactor is generally very large because of high polymer content, which gives rise to the difficulty of subsequent distillation separation. Formerly, the bubble tray column was adopted, which often led to flooding and thus shutting down of the distillation column. So the NO. 1 polymer distillation column has to be improved to overcome this problem. A strategy, that is, using a new type of tray, high-efficiency flow-guided tray, is put forwarded for this purpose.

A typical feeding composition of the NO. 1 polymer distillation column is PVAC 30%, VAC 47.28%, MeOH (methanol) 22.6%, $H_2O$ 0.1% and the residual 0.02% respectively. The flowrate is 6857 kg $h^{-1}$. The requirement for the concentration of the product is: VAC 60%, MeOH 39%, $H_2O$ 0.1% at the top of the column, PVAC 30-33%, VAC < 0.08%, MeOH 69.92% at the bottom of the column (to ensure that the yield of VAC is 99.7%).

While designing the high-efficiency flow-guided sieve tray [1-3], an advanced chemical engineering simulation software (e.g. PROII, ASPEN Plus, HYSYS, etc.) was used to simulate the separation of PVAC and VAC. In the simulation, the relationship of equipment parameter (i.e. tray number $N$) and operation parameter (i.e. reflux ratio $R$) was investigated.

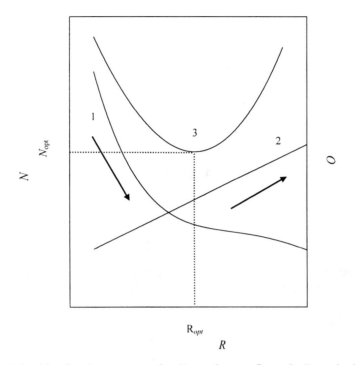

Fig. 1. The relationship of optimum tray number $N_{opt}$, optimum reflux ratio $R_{opt}$ and minimum total expense $Q_{min}$: curves 1, 2 and 3 represent the relationship of reflux ratio $R$ and tray number $N$, reflux ratio $R$ and operation expense $Q$, and reflux ratio $R$ and total operation expense $Q$, respectively.

It was found that tray number $N$ is reverse to reflux ratio $R$ in the case of same separation efficiency. It is known that tray number $N$ reflects equipment expense, and reflux ratio $R$ reflects operation expense [4]. Therefore, there exists a minimum total expense that is a trade-off of these conflicting effects. As shown in Fig. 1, at the point of minimum total expense, tray number $N$ and reflux ratio $R$ are optimum. When those are determined, the operation condition in the distillation column can be obtained by mathematical simulation. On this basis, a high-efficiency flowed-guided sieve tray can be properly designed.

## 1.2. Design of high-efficiency flow-guided sieve tray
### 1.2.1. Characteristics of high efficiency flow-guided sieve tray

Since the concentration of PVAC in the feeding material was up to 30%, the viscosity of 10000 - 50000 Pa s was so high that the material flowed very slowly in the No. 1 polymer distillation column. The former adopted bubble tray column was densely covered by the bubbles. Thus, the flow on the tray was greatly hindered, which was prone to lead to a high pressure drop and abnormal operation such as flooding. Moreover, the gas space over the bubble was decreased due to high bubble height, which brought on excessive liquid entrainment.

bubble-promoting devices

Fig. 2. The structure of high-efficiency flow-guided sieve tray.

In order to overcome these problems, the former bubble tray was replaced by high-efficiency flow-guided sieve tray. This type of tray is different from ordinary sieve tray. It involves two modification: one is to open a number of proper flow-guided sieves, which ensures that the gas and liquid flow is reasonable, and the other is to install the bubble-promoting devices near the entrance of the liquid, which ensures that there is a less fall of liquid layer from the inlet to outlet on the tray. The structure of flow-guided sieve tray is portrayed in Fig. 2.

### 1.2.2. Pressure drop of high-efficiency flow-guided sieve tray

The holes on the tray are composed of the flow-guided holes and the common sieve holes. In order to compare pressure drop of high-efficiency flow-guided sieve tray with that of the common sieve tray, the following equation is used:

$$h_{DF} = h_{DS} K \tag{1}$$

where $h_{DF}$ and $h_{DS}$ are pressure drops of high-efficiency flow–guided sieve tray and the common sieve tray (Pa), respectively; $K$ is the correlation factor.

It is obvious that the key to calculate $h_{DF}$ is to solve $K$, since $h_{DS}$ is given by

$$h_{DS} = 0.051(\frac{u_0}{c_0})^2 \frac{\rho_v}{\rho_l}(1-\phi)^2 \tag{2}$$

where $u_0$ (m s$^{-1}$) is the gas velocity passing through the sieve hole, $c_0$ is the tray coefficient, $\rho_V$ and $\rho_L$ (kg m$^{-3}$) are the gas and liquid density, respectively, and $\phi$ is the hole area ratio.

The correlation factor $K$ can be correlated by experimental data. A set of experimental apparatus has been established to measure the tray hydrodynamics including pressure drop, leakage and entrainment. Water and air were used as the medium. The measuring column had three trays, i.e. entrainment tray, measuring tray and gas distribution tray. It was made up of organic glass, having 500 mm I.D., 15 mm thickness and tray distance 300 mm. The common sieve tray was firstly selected as the measuring tray so that the reliability of experimental apparatus can be verified. Then high-efficiency flow–guided sieve tray with the same parameters as the common sieve tray was used. By regressing experimental data, we obtained the following equation describing correlation factor $K$:

$$K = (1+0.45\phi)^{-2} \tag{3}$$

where the hole area ratio $\phi$ is used as the independent variation because the difference between these two trays is arrangement and type of the holes.

It can be seen from Eq. (3) that $K$ is always smaller than unity. So it is evident that $h_{pF} < h_{pS}$. That is to say, pressure drop of high-efficiency flow-guided sieve tray is lower than that of the common sieve tray under the same operation condition and tray parameters.

*1.2.3. Momentum transfer of high-efficiency flow–guided sieve tray*

Fig. 3. illustrates the controlling volume on the tray when deducing the momentum conversation equation. The momentum conservation equation [5-14] is given by

$$\frac{Q_L^2 \rho_L}{h_{L1} \cdot W} + a(\frac{Q_G^2 \rho_G}{A_t^2} \cdot A_g + P_G \cdot A_g) + \frac{1}{2} h_{L1}^2 W \rho_L g = \frac{Q_L^2 \rho_L}{h_{L2} W} + \int_0^L (W + 2h_L) \tau_0 dz + \frac{1}{2} h_{L2}^2 W \rho_L g \tag{4}$$

where $Q_L$ and $Q_G$ (m$^3$ s$^{-1}$) are respectively the liquid and gas flowrate, $h_{L1}$ and $h_{L2}$ (m) are respectively the height of surfaces 1 and 2, $A_t$ and $A_g$ (m$^2$) are respectively the area of all the holes and the flow-guided holes, $W$ (m) is the width of liquid flowing path, $g$ (m$^2$ s$^{-1}$) is acceleration of gravity, $a$ (m$^2$ m$^{-3}$) is the interfacial area per unit volume, $P_G$ (Pa) is the gas pressure drop passing through the holes, and $\tau_0$ (s) is residence time.

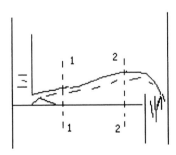

Fig. 3. Controlling volume on high-efficiency flow–guided sieve tray.

For high efficiency flow-guided sieve hole, Eq. (4), after integral, can be rewritten as

$$\int_0^t (W + 2h_L)\tau_0 dz = \frac{Q_L^2 \rho_L}{W}(\frac{1}{h_{L1}} - \frac{1}{h_{L2}}) + \frac{1}{2}W\rho_L g(h_{L1}^2 - h_{L2}^2)$$ (5)

When solving Eq. (4), it is required to determine the gas pressure drop $P_G$. The curve of the

gas pressure drop $P_G$ versus gas velocity passing through the holes $u_0$ is obtained by

experiments and is shown in Fig. 4. So it is now easy to deduce the velocity distribution on the whole tray according to Eq. (4).

### 1.3. Industrial application of high-efficiency flow-guided sieve tray

High efficiency flow-guided sieve tray has been applied in the No.1 polymer distillation column in one certain plant used for producing PVA. The former bubble tray was replaced. After the modification, the separation effect is shown in Table 1.

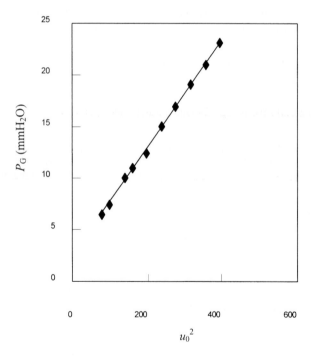

Fig. 4. The relationship of $P_G$ and $u_0^2$ used in Eq. (4).

Table 1
The separation effect of high-efficiency flow-guided sieve tray

| Items | Concentration at the top of the column (mass fraction) | | Concentration at the bottom of the column (mass fraction) | | | VAC yield |
|---|---|---|---|---|---|---|
| | VAC | MeOH | PVAC | VAC | H$_2$O | |
| Former target | 50 | 49 | 30 | 0.26 | 0.15 | 99.3 |
| Desired target | 60 | 39 | 30 - 33 | ≤ 0.08 | ≤ 0.10 | 99.8 |
| Practical target | 60 | 39 | 30 - 33 | 0.02 - 0.03 | 0.05 | 99.8 - 99.9 |

It can be seen from Table 1 that it is advisable to use high-efficiency flow-guided sieve tray in the PVA production. The economic analysis is made as follows:

(1) Investment
The structure of high-efficiency flow-guided sieve tray is simple, and can be manufactured easily and cheaply. So the technology investment is small, about RMB 0.5 million.

(2) Energy saving
Because the return ratio decreases from 1.3 to 0.85, the technology can save energy 2164 MJ h$^{-1}$ (i.e. 8000 t water vapor per year), amounting to RMB 0.48 million annually.

(3) Water saving
This modification can save energy 240000 t water vapor per year, amounting to RMB 96000 annually.

(4) Increasing the VAC yield
After modification, the VAC concentration at the bottom of the column decreases to 0.02-0.03%, amounting to RMB 0.638 million annually.

(5) Improving the product quality
As mentioned above, due to the decreasing VAC concentration at the bottom of column, the product quality is improved, amounting to RMB0.8 million annually.

(6) The return period on investment
The return period on investment is about 2.7 months.


## 2. THERMALLY COUPLED DISTILLATION

### 2.1. Introduction

Thermally coupled distillation (TCD) is only related with multi-component separation process. Among all possible new schemes for multi-component distillation process, the thermally coupled distillation (TCD) schemes are very promising for both energy and capital cost saving [15-18]. Before introducing different TCD processes, we provide the following definitions to help our understanding.

(1) Single source:

Most often the problem to be considered has only a single source which is to be split into all the desired products.

(2) Sharp separation:

If each column separates the feed into products with no overlap in the components between them, it is performing a "sharp" separation. An example is to split a mixture containing components A, B, C and D into the pure-component product A and the mixture of B, C and D. It is a sharp separation solution only if one column is used for this separation.

(3) Simple columns:

Distillation columns having one feed and producing two products are denoted as "simple" columns. However, if the top and/or bottom of the column are connected by liquid and vapor streams, this column can also be taken on as simple column. As shown in Fig. 5, suppose that at the top a total condenser is added afterwards, and then

$$D = V_D - L_D \tag{6}$$

$$R = L_D / D \tag{7}$$

where $D$ (kmol s$^{-1}$) is the flowrate of the top product, and $R$ is reflux ratio; at the bottom, suppose that a reboiler is added afterwards, and then

$$B = L_B - V_B \tag{8}$$

where $B$ (kmol s$^{-1}$) is the flowrate of the bottom product; at the feed,

$$F = L_c - V_c \tag{9}$$

where $F$ (kmol s$^{-1}$) is the flowrate of the feeding mixture. In the shortcut design of simple distillation, the thermal quality of the feeding mixture, $q$, is needed and defined by

$$q = \frac{\text{the heat required for transferring the present state into saturated vapor state per kmol}}{\text{the latent heat of vaporization of the feeding mixture per kmol}}.$$

Apparently, $q = 1$ for saturated liquid, $q = 0$ for saturated vapor, $0 < q < 1$ for the mixture of liquid and vapor, $q > 1$ for subcooled liquid and $q < 1$ for superheated liquid. Sometimes $q$ can be regarded as the liquid fraction of the feeding mixture, provided that the molar latent heat of vaporization is equivalent among the components to be separated, of course. Under this condition,

$$q = \frac{L_c}{L_c + V_c} \tag{10}$$

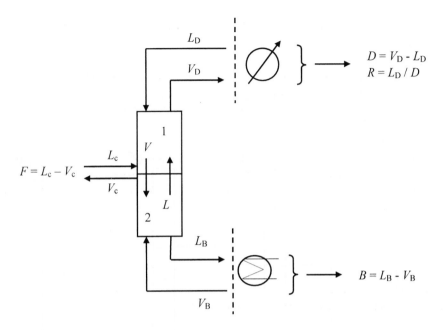

Fig. 5. One form of simple distillation column.

(4) Column section:

For instance, for a simple distillation column as shown in Fig.5, there are two column sections, i.e. rectifying section and stripping section, indicated by the number "1" and number "2" respectively.

(5) Separation sequence

Thompson and King [19] provide the following formula to determine the number of simple distillation sequences for separating an $N$-component mixture into $N$ pure-component products:

$$\text{Number of simple distillation sequences} = \frac{(2(N-1))!}{N!(N-1)!} \tag{11}$$

For instance, for a ternary separation, the tree structure and flowsheet of separation sequences are illustrated in Fig. 6, where A is the lightest component, B the middle and C the heaviest component.

*Example:* Please draw the tree structure of a quaternary separation.
*Solution:* In terms of Eq. (11), in total there are six kinds of separation sequences, the tree structure of which is drawn in Fig. 7.

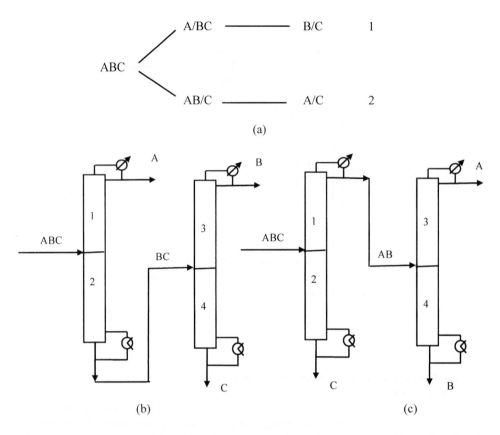

Fig. 6. The tree structure and flowsheet of ternary separation; (a) tree structure; (b) one flowsheet in which the lightest component A is obtained in the first column; (b) another flowsheet in which the heaviest component C is obtained in the first column.

In general, for convenience, it is better to substitute tree structure for flowsheet in the synthesis of separation sequence.

Note that the following assumptions are implied in this text if without particular specification:

(1) Each column section has enough theoretical stages to implement a sharp separation.

(2) The feeding mixture belongs to the categories: ideal mixture, nearly ideal mixture and nonideal mixture but without azeotropes formed.

(3) The relative volatilities between the feeding mixture are higher than the general limitation in which special distillation processes have to be used, i.e. $\alpha > 1.2$.

(4) Components are ranked according to their volatilities. For example, for the mixture A, B and C, A is the most volatile, while C is the least volatile in the decreasing order.

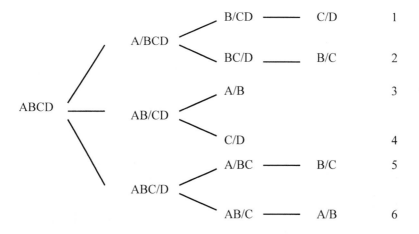

Fig. 7. The tree structure of quaternary separation.

## 2.2. Design and synthesis of TCD

For a $N$-component mixture, the minimum number of columns in a separation sequence is:

$$N_{col} = N - 1 \tag{12}$$

which coincides with the result discussed in chapter 7.

The corresponding number of column sections is:

$$N_{sec} = 2(N - 1) \tag{13}$$

Consequently, we can understand the source of Eq. (11), in which the numerator reflects the total distinguishable sites that are occupied by both pure components ($N$ sites) and the mixtures ($N$ - 1 sites), and the denominator reflects the $N$ undistinguishable sites occupied by pure components and $N$ - 1 undistinguishable sites occupied by the mixture. Only until this combination is finished, we can distinguish the $N$ pure components by the series of A, B, C, etc.

Actually, Eq. (11) is only suitable for simple distillation sequences, excluding complex distillation sequence, such as TCD. Unfortunately, up to date, there isn't an uniform formulation to account for the number of TCD separation sequences.

Based on the above background knowledge, we can understand TCD more easily. Then what is TCD? The most significant characteristics in TCD configuration is that at least one end of a distillation column has both liquid and vapor exchange with another column, and thus either the reboiler or the condenser is eliminated from this end of the distillation column. For ternary component mixtures, there are only three feasible thermally coupled configurations called the side stripping (SS), the side rectifying (SR) and the fully coupled (FC) (or Petlyuk configuration), respectively [20, 21]. However, in the whole FC configuration there is only one condenser and one reboiler.

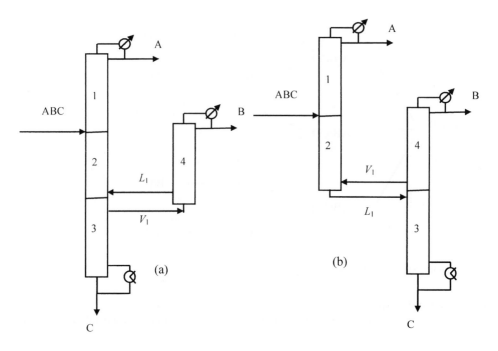

Fig. 8. The side rectifying (SR) TCD column (a) and its equivalent configuration (b).

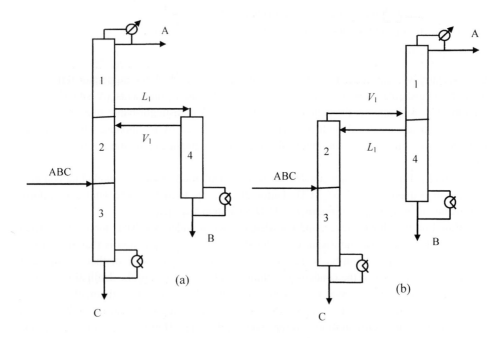

Fig. 9. The side stripping (SS) TCD column (a) and its equivalent configuration (b).

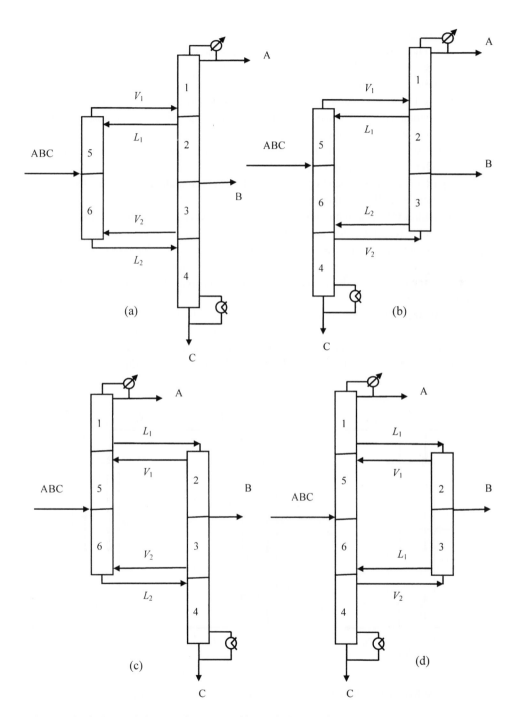

Fig. 10. The fully coupled (FC) column (a) and its equivalent configurations (b), (c) and (d).

Three types of TCD column configurations are illustrated in Figs. 8, 9 and 10, respectively. Over ternary component mixture, the feasible configurations increases rapidly.

It can be seen from Figs. 8 and 9 that the SR and SS TCD columns formally have thermodynamic equivalent configurations, which are transferred from the three basic units shown in Fig. 11 [22, 23]. As in Fig. 5, these equivalent configurations can be looked on as simple distillation column, which, however, simplifies the design of complex separation sequences. But for FC TCD column (see Fig. 10), it isn't easy to do such a simplification, and by far the solution to this problem is only limited to ternary separation system. The interested reader can refer to the references [15, 24, 25].

A separation sequence made up of TCD column configurations is constructed with the following units, excluding FC configuration:

(1) Main column (MC): a main column in a complex scheme is a column with an overall condenser and a reboiler while connecting with side columns. For a complex scheme, to separate five-component mixtures, there may be two such main columns, of which the one with feedstock is called main column. In Figs. 8a, 9a and 10d each has a MC.

(2) Side rectifying column (SRC): a column with only one overall condenser. In Fig. 8a, there is a SRC.

(3) Side stripping column (SSC): a column with only one reboiler. In Fig. 9a, there is a SSC.

(4) Side column (SC): a column without either one overall condenser or one reboiler. It is used as a unit between the columns in a complex TCD column configuration.

But at the same time, for a feasible TCD configuration, the following constraints should be abided:

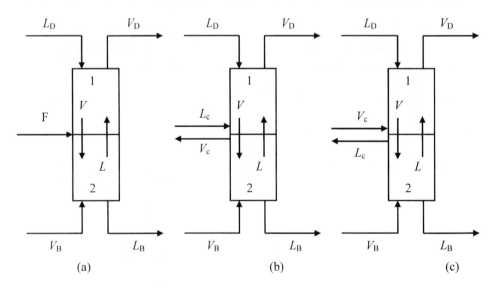

Fig. 11. Three basic units of thermodynamic equivalent configurations; adapted from the reference [23].

(1) The number of column sections in a TCD configuration is the same as of a simple column sequence, i.e. 2 ($n$-1). $n$ is the number of products in a simple column sequence with all sharp separations. For example, in Figs. 8 and 9, the number of column sections is: 2× (3-1) = 4. Note that FC configuration is exceptional, having 6 column sections as shown in Fig. 10.

(2) Side stripping columns are connected with the rectifying section of a column, while side rectifying columns are connected with the stripping section of a column.

(3) A side column can be connected with the main column, as well as with the other side columns in a complex TCD configuration.

(4) There aren't the same products from the different column sections in a flowsheet.

(5) Only one product exists in a SRC or SSC, i.e. the distillate in a side rectifying column and bottoms product in a side stripping column.

(6) No product exists in a SC. If having one product, SC becomes SRC or SSC; if having two products, SC becomes MC.

In summary, for a given multi-component separation task, a lot of possible separation sequences can be sought, which can be divided into three types:

(1) Simple distillation sequences. The number is counted by using Eq. (11).

(2) TCD sequences, including SS, SR, FC and their hybrids.

(3) Combination of simple distillation sequences and TCD sequences.

However, there is some certain relationship between simple distillation sequences and TCD sequences. As an example of separation of five-component mixture, a simple distillation sequence is illustrated in Fig. 12. Now we can modify this process in the following steps:

(1) In the first step, feedback vapor streams are added in Fig. 13 and denoted by dashed lines.

(2) In the second step (see Fig. 14), the reboilers in the front three columns can be eliminated since the vapor streams have been supplied at the bottom.

(3) In the third step, in terms of the principle of thermodynamic equivalent configuration, Fig. 14 can be re-drawn in another form (see Fig. 15).

Therefore, a simple distillation sequence is finally transferred into FC configuration sequence. Note that column sections are equal in both cases.

In industry, an important multi-component separation process is so-called crude oil atmospheric and vacuum distillation, where products A, B, C, D and E are not pure component, but a mixture of narrow boiling-point hydrocarbons. Thus, the Fig. 15 process can be further modified, and three SS columns are incorporated into one column (see Fig. 16). This process has been claimed to maintain a higher distill-off ratio than the traditional one by about 0.9% [59]. Additionally, the boiling point range of products becomes narrower, which means that the quality of extracted oil is improved.

342

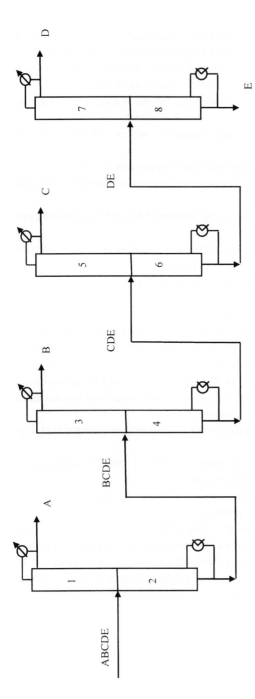

Fig. 12. A simple distillation sequence for separation of five-component mixture.

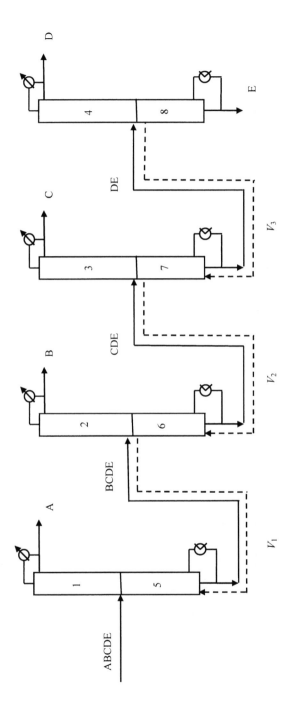

Fig. 13. Modified process in the first step.

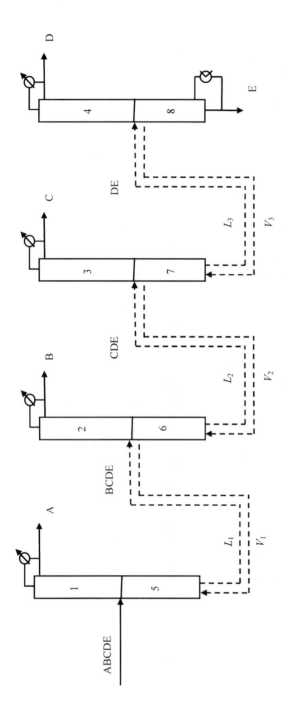

Fig. 14. Modified process in the second step.

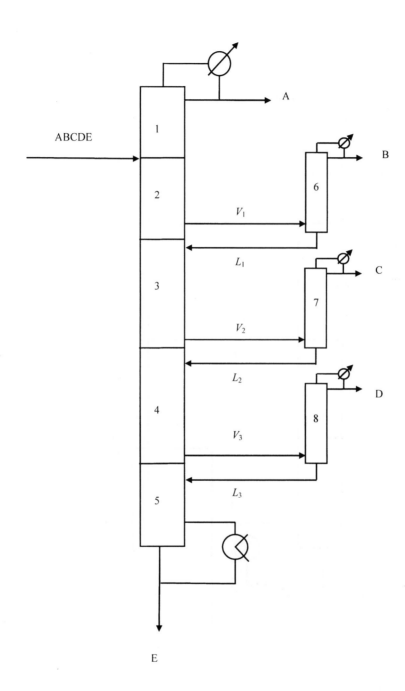

Fig. 15. Modified process in the third step.

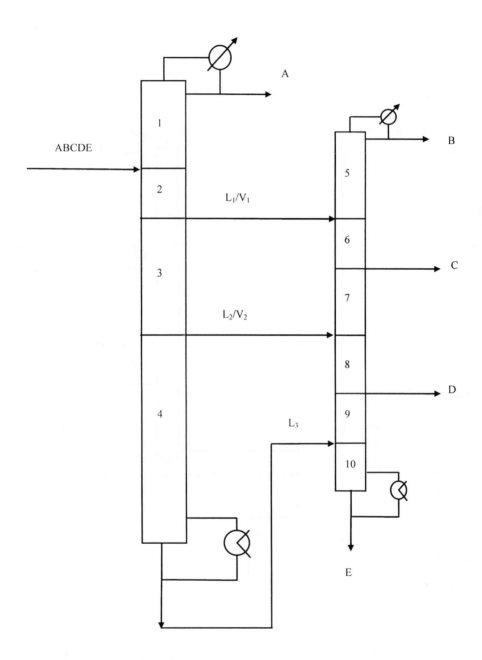

Fig. 16. Modified process in the fourth step.

## 2.3. Application of TCD in special distillation processes

For multi-component separation in special distillation processes, the problem on selection of suitable separation sequences is still urgent. This is slightly different from simple distillation sequences, TCD sequences, and/or their combination in that a separating agent is added into the separation system and $N + 1$ or even more components are involved.

On the basis of Fig. 15, Fig. 17 illustrates a FC configuration of extractive distillation for separating ternary system. Applying Fig. 17 to the separation of C4 mixture by extractive distillation as discusses in Chapter 2, it is interesting to find that this process is suitable for C4 separation. In this case, products, A, B and C, are replaced by butane (butene), 1,3-butadiene and VAC (vinylacetylene), respectively.

*Example:* Please enumerate other flowsheets for the separation of C4 mixture by extractive distillation.

*Solution:* In chapter 2, some flowsheets have been put forward and discussed deeply. Herein, one new flowsheet of the combination of simple distillation and TCD sequences is drawn, as shown in Fig. 18. The interested readers can draw other flowsheets and the corresponding equivalent configurations.

Then, one question arises, i.e. which possible flowsheet is the best? When selecting a feasible separation sequence, many factors, such as energy consumption, the number of columns, the number of condensers and reboilers, the number of column sections, operation complexity and so on, should be taken into account. These factors may be divided into three kinds:

(1) Directly obtained from the flowsheet, i.e. the number of columns, column sections, condensers and reboilers.

(2). Obtained by short cut and rigorous calculations, i.e. energy consumption. Since many complex distillation columns can be simplified as simple distillation column, the minimum vapor flow in each column is calculated using Underwood's method. The outcome from shortcut calculation can be used as the initial value of rigorous calculation.

(3). Qualitative property, i.e. operation complexity. It is believed that comparing to simple distillation sequences, TCD sequences are more difficult to control because a main column and a prefractionator are interlinked, more degrees of freedom are involved.

It is thought that TCD sequences, especially FC configurations, are of high thermodynamic efficiency and low capital cost. FC configurations have been effectively implemented for the separation of a ternary system in naphtha reforming plant [26]. However, Agrawal and Fidkowski [27] showed that the thermodynamic efficiency of FC configurations isn't as high as suggested in earlier studies for some cases. The thermodynamic efficiencies computed from minimum work of separation and energy loss of a conventional system and the fully thermally coupled distillation are compared. A striking result is that, for FC configurations, the range of values of feed composition and relative volatilities is quite restrained to become the most thermodynamically efficient configuration. When the compositions of feed and liquid in feed tray are different, the introduced feed is mixed with the tray liquid and the mixing causes irreversibility to decrease the thermodynamic efficiency.

So now it is the tendency to combine simple distillation sequences and TCD sequences in the process synthesis in order to take advantage of their individual merits.

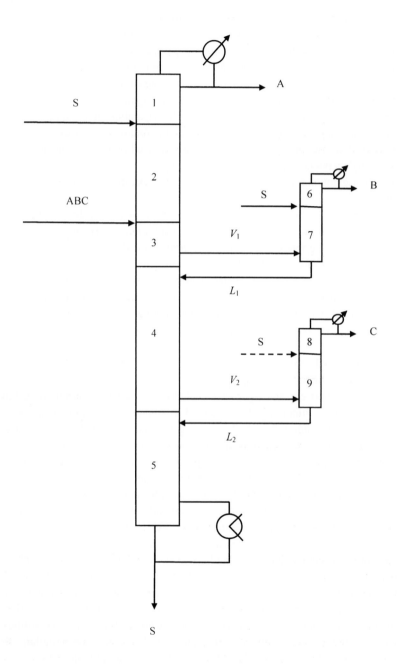

Fig. 17. A FC configuration of separating ternary system by extractive distillation; the dashed line represents that separating agent may be added or not.

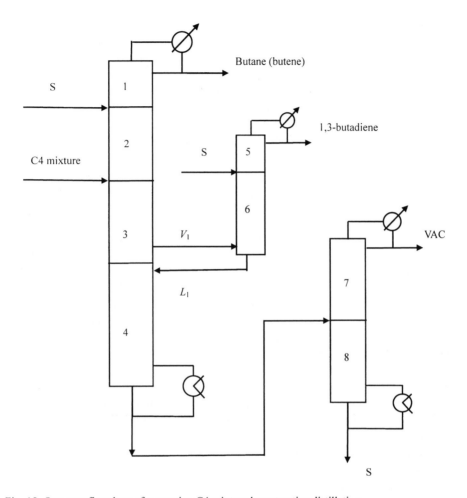

Fig. 18. One new flowsheet of separating C4 mixture by extractive distillation.

## 3. HEAT PUMP AND MULTI-EFFECT DISTILLATIONS

There are few reports about heat pump distillation in the international journals, only one [28] found by title search in ISI Web of Science (http: //wos5.isiknowledge.com). It implies that this technique of heat pump distillation isn't attractive in theoretical and practical aspects, and gains no wide attention in application, especially in special distillation processes.

The thermodynamic efficiency of a distillation column is usually quite low while at the same time the energy consumption is high. This is due to the fact that energy separating agent in the distillation process depends directly on the amount of energy used and the recovery of this energy is often low. A brief survey has been carried out by Bjorn et al. [28] of the possibilities of improving the energy conservation quotient and increasing the thermodynamic efficiency for a single distillation column system utilizing various heat pump arrangements, paying attention to the various aspects of both open and closed types of systems. The use of more elaborate systems involving intermediate heat exchanger is, theoretically, necessary in order to achieve a higher thermodynamic efficiency; in practice, however, these systems don't always turn out to be economically viable when compared with simpler arrangements.

Heat pump distillation is promising in the case where the recovering energy is very large and the temperature difference between the top and the bottom is not distinct, often below 20 ℃. For example, in the propane / propylene distillation column, the amount of feeding mixture to be dealt with is often so high, up to 10-100 t h$^{-1}$, the temperature difference being about 10 ℃. Hence, the absolute value of energy heated at the bottom is very large. In order to relieve the energy, heat pump is employed by using a part of bottom product as refrigerating agent to connect top and bottom of the distillation column. It is reported that if a heat pump of open type with adiabatic flash and compression can improve thermodynamic efficiency from 20% to 70% [29, 30].

Unlike heat pump distillation, multi-effect distillation is widely used in practice, especially in seawater desalination [31-37]. Strictly speaking, multi-effect distillation is exactly multi-effect evaporation with only one theoretical stage, i.e. the reboiler. One theoretical stage is enough because salt doesn't appear in the vapor phase. The design and synthesis of various multi-effect distillations are present convincingly by Richardson and Harker [38]. It is beyond our scope to extend this content. The interested reader can refer to it.

## 4. MOLECULAR DISTILLATION

Molecular distillation is generally accepted as the safest method to separate and purify thermally unstable compounds and substances having low volatility with high boiling point [39]. The process is distinguished by the following features: short residence time in the zone of the molecular evaporator exposed to heat; low operating temperature due to vacuum in the space of distillation; a characteristic mechanism of mass transfer in the gap between the evaporating and condensing surfaces.

The separation principle of molecular distillation is based on the difference of molecular mean free path. The passage of the molecules through the distillation space should be collision free. Their mean free path, $\langle \lambda \rangle$, is defined by the following relation, derived from the theory of ideal gases:

$$\langle \lambda \rangle = \frac{1}{\sqrt{2}\pi d^3 n} = \frac{kT}{\sqrt{2}\pi d^3 P} = \frac{RT}{\sqrt{2}\pi d^2 N_A P} \tag{14}$$

where $d$ (m) is the molecular diameter, $N_A$ ($6.023 \times 10^{23}$ mol$^{-1}$) is Avogadro constant, $P$ (Pa) is pressure and $T$ (K) is temperature.

In theory, molecular distillation can also be used for separating mixtures with close boiling point or forming azeotrope because the constituents' molecular diameters are frequently not identical. The constituent with small molecular diameter should be more volatile than that with big molecular diameter according to Eq. (14). However, the motivation of molecular distillation arises not for this purpose, but for separating heat-sensitive compounds. By far, this technique is mostly applied in the fields of medicine and biology. Table 2 lists some applications of molecular distillation. The reason why molecular distillation isn't used as a special distillation process for separating the mixture with close boiling point or forming azeotrope may be attributed to:

(1) A very limited theoretical stages, when compared with common distillation column.

(2) Low production scale.

(3) Complicated equipment and high investment cost in order to achieve high vacuum degree.

Table 2
Application of molecular distillation

| No. | Cases |
|-----|-------|
| 1 | High concentration of Monoglycerides [40] |
| 2 | Fractionation of dimers of fatty-acids [41] |
| 3 | Separation of radioactive nuclides from melts of irradiated media [42] |
| 4 | Get the oils of vacuum pumps [43] |
| 5 | Compartmentalization of secretory proteins in pancreatic zymogen [44] |
| 6 | Inhibition of tumor-cell growth by low boiling point structured fatty-acids [45] |
| 7 | Cell culture supports for slam-frozen [46] |
| 8 | Preparation of dunaliella-parva for ion localization studies by X-Ray-Microanalysis [47] |
| 9 | Sample preparation for electron-microscopy using cryofixation [48] |
| 10 | Separation of lanolinic alcohols [49] |
| 11 | Molecular distillation drying of mammalian-cells [50] |
| 12 | Post embedding immunolabeling [51] |
| 13 | Lanolin purification [52] |
| 14 | Recovering biodiesel and carotenoids from palm oil [53] |
| 15 | Recovering vitamin E from vegetal oils [54] |
| 16 | Concentration of monoglycerides [55] |
| 17 | Recovery of carotenoids from palm oil [56] |
| 18 | Synthesis of pure diglicyde ether of bisphenol-A [57] |
| 19 | Fructose monocaprylate and polycaprylate purification [58] |

**REFERENCES**

[1] Q.S. Li, Chemical Industry and Engineering Progress (China), 14 (1995) 38-41.

[2] Q.S. Li, Chemical Industry and Engineering Progress (China), 15 (1996) 61-64.

[3] Compiling committee of chemical engineering (eds.), Handbook of Chemical Engineering, Chemical Industry Press, Beijing, 1996.

[4] T.E. Tan, B.X. Mai and H.H. Ding (eds.), Principle of Chemical Engineering, Chemical Industry Press, Beijing, 1999.

[5] C.J. Liu, H. Chen, X.G. Yuan and G.C. Yu, Journal of Tianjin University (China), 34 (2001)

741-744.

[6] X.L. Wang, C.J. Liu and G.C. Yu, Chem. Ind. Eng. (China), 18 (2001) 390-394.

[7] C.J. Liu, X.G. Yuan, S.Y. Wang, Y. Gao and G.C. Yu, Chem. Eng. (China), 30 (2002) 7-11.

[8] B.T. Liu and C.J. Liu, Chinese J. Chem. Eng., 10 (2002) 517-521.

[9] C.J. Liu and X.G. Yuan, Chinese J. Chem. Eng., 10 (2002) 522-528.

[10] C.J. Liu, X.G. Yuan, K.T. Yu and X.J. Zhu, Chem. Eng. Sci., 55 (2000) 2287-2294.

[11] H.R. Mortaheb, H. Kosuge and K. Asano, Chem. Eng. J., 88 (2002) 59-69.

[12] J.M.V. Baten, J. Ellenberger and R. Krishna, Catal. today, 66 (2001) 233-240.

[13] S.V. Makarytchev, T.A.G. Langrish and D.F. Fletcher, Chem. Eng. J., 87 (2002) 301-311.

[14] C.V. Gulijk, Comput. Chem. Eng., 22 (1998) S767-S770.

[15] Z. Fidkowski and L. Krolikowski, AIChE. J., 32 (1986) 537-546.

[16] A.W. Westerberg, Comput. Chem. Eng., 9 (1985) 421-429.

[17] M. Emtir, E. Rev and Z. Fonyo, Appl. Therm. Eng., 21 (2001) 1299-1317.

[18] R. Agrawal, Trans. IchemE., 77 Part A (1999) 543-553.

[19] R.W. Thompson and C.J. King, AIChE J., 18 (1972) 941.

[20] F.B. Petlyuk, V.M. Platonov and D.M. Slavinskii, International Chemical Engineering, 5 (1965) 555-561.

[21] R. Agrawal, Ind. Eng. Chem. Res., 35 (1996) 1059-1071.

[22] B.G. Rong, A. Kraslawski and L. Nystrom, Comput. Chem. Eng., 24 (2000) 247-252.

[23] B.G. Rong, A. Kraslawski and L. Nystrom, Comput. Chem. Eng., 25 (2001) 807-820.

[24] Y.H. Kim, Chem. Eng. J., 85 (2002) 289-301.

[25] Y.H. Kim, Chem. Eng. J., 89 (2002) 89-99.

[26] R. Agrawal and Z.T. Fidkowski, Ind. Eng. Chem. Res., 37 (1998) 3444-3454.

[27] J.Y. Lee, Y.H. Kim and K.S. Hwang, Chem. Eng. Proc., 43 (2004) 495-501.

[28] I. Bjorn, U. Gren and K. Strom, Chem. Eng. Proc., 29 (1991) 185-191.

[29] F.C. Ouyang, C.G. Sun, M.J. Guo and C.J. Yue, Journal of Northeast China Institute of Electric Power Engineering (China), 14 (1994) 28-33.

[30] W.P. Gao, Q.Y. Wei, Y.L. Zhang and Y. Xue, Journal of JiLin Institute of Chemical Technology (China), 11 (1994) 1-14.

[31] M.A. Shammiri and M. Safar, Desalination, 126 (1999) 45-59.

[32] L. Tian, Y.Q. Wang and J.L. Guo, Desalination, 152 (2002) 223-228.

[33] L. Tian and Y.Q. Wang, Desalination, 133 (2001) 285-290.

[34] V. Dvornikov, Desalination, 127 (2000) 261-269.

[35] L.G. Rodriguez and C.G. Camacho, Desalination, 122 (1999) 205-214.

[36] F.A. Juwayhel, H.E. Dessouky and H. Ettouney, Desalination, 114 (1997) 253-275.

[37] A. Ophir and A. Gendel, Desalination, 98 (1994) 383-390.

[38] J.F. Richardson, J.H. Harker and J.R. Backhurst (eds.), Chemical Engineering Particle Technology and Separation Process, Butterworth-Heinemann, Oxford, 2002.

[39] J. Lutisan and J. Cvengros, Chem. Eng. J., 56 (1995) 39-50.

[40] H. Szelag and W. Zwierzykowski, Fett. Wiss. Technol., 85 (1983) 443-446.

[41] H. Szelag and W. Zwierzykowski, Przem. Chem., 62 (1983) 337-338.

[42] Z.V. Ershova, B.V. Petrov and Y.G. Klabukov, Sov. Radiochem., 24 (1982) 487-489.

[43] Z. Kawala, R. Kramkowski and J. Kaplon, Przem. Chem., 61 (1982) 13-15.

[44] D. Gingras and M. Bendayan, Biol. Cell, 81 (1994) 153-163.

[45] S. Nakamura, Y. Nishimura and K. Inagaki, Cancer Biochem. Bioph., 14 (1994) 113-121.

[46] S.L. White, D.A. Laska and P.S. Foxworthy, Microsc. Res. Techniq., 26 (1993) 184-185.

[47] M.A. Hajibagheri and T.J. Flowers, Microsc. Res. Techniq., 24 (1993) 395-399.

[48] J.G. Linner, S.A. Liversey and D. Harrison, Inst. Phys. Conf. Ser., 93 (1988) 331-332.

[49] Z. Kawala, R. Kramkowski and J. Kaplon, Przem. Chem., 68 (1989) 555-557.

[50] S.R. May, J. Linner and S. Livesey, Cryobiology, 25 (1988) 575-576.

[51] J.G. Linner and S.A. Livesey, J. Histochem. Cytochem., 35 (1987) 1021-1021.

[52] H. Szelag and W. Zwierzykowski, Przem. Chem., 63 (1984) 255-257.

[53] C.B. Batistella, E.B. Moraes and R. Maciel, Appl. Biochem. Biotech., 98 (2002) 1149-1159.

[54] C.B. Batistella, E.B. Moraes and R. Maciel, Appl. Biochem. Biotech., 98 (2002) 1187-1206.

[55] J. Kaplon, K. Minkowski and E. Kaplon, Inz. Chem. Procesowa., 22 (2001) 627-632.

[56] C.B. Batistella and M.R.W. Maciel, Comput. Chem. Eng., 22 (1998) S53-S60.

[57]R. Kramkowski, J. Kaplon and M. Berdzik, Przem. Chem., 76 (1997) 483.

[58] I. Redmann, D. Montet and F. Bonnot, Biotechnol. Tech., 9 (1995) 123-126.

[59] L. Wang, Z. Duan and R. Zhou, J. Chem. Ind. Eng. (China), 51 (2000) 383-389.

# Subject Index